AF813680

Autorenkollektiv

Biomimetic and Bioorganic Chemistry

AKADEMIE-VERLAG · BERLIN

Autorenkollektiv

Biomimetic and Bioorganic Chemistry

Autorenkollektiv

Biomimetic and Bioorganic Chemistry

With 111 Figures and 45 Tables

AKADEMIE-VERLAG · BERLIN
1985

Die Originalausgabe erscheint im Springer-Verlag
Berlin · Heidelberg · New York · Tokyo
in der Schriftenreihe „Topics in Current Chemistry",
Volume 128

Vertrieb ausschließlich für die DDR und die sozialistischen Länder

Alle Rechte vorbehalten
© Springer-Verlag Berlin · Heidelberg 1985

Erschienen im Akademie-Verlag Berlin, DDR-1086 Berlin, Leipziger Straße 3–4
Lizenznummer: 202 · 100/541/85
Printed in the German Democratic Republic
Gesamtherstellung: VEB Druckerei „Thomas Müntzer", 5820 Bad Langensalza
Umschlaggestaltung: Karl Salzbrunn
LSV 1275
Bestellnummer: 763 491 6 (3072/128)
13800

Managing Editor:

Dr. *Friedrich L. Boschke*
Springer-Verlag, Postfach 105280, D-6900 Heidelberg 1

Editorial Board:

Prof. Dr. *Michael J. S. Dewar* Department of Chemistry, The University of Texas Austin, TX 78712, USA

Prof. Dr. *Jack D. Dunitz* Laboratorium für Organische Chemie der Eidgenössischen Hochschule Universitätsstraße 6/8, CH-8006 Zürich

Prof. Dr. *Klaus Hafner* Institut für Organische Chemie der TH Petersenstraße 15. D-6100 Darmstadt

Prof. Dr. *Edgar Heilbronner* Physikalisch-Chemisches Institut der Universität Klingelbergstraße 80, CH-4000 Basel

Prof. Dr. *Shô Itô* Department of Chemistry, Tohoku University, Sendai, Japan 980

Prof. Dr. *Jean-Marie Lehn* Institut de Chimie, Université de Strasbourg, 1, rue Blaise Pascal, B. P. Z 296/R8, F-67008 Strasbourg-Cedex

Prof. Dr. *Kurt Niedenzu* University of Kentucky, College of Arts and Sciences Department of Chemistry, Lexington, KY 40506, USA

Prof. Dr. *Kenneth N. Raymond* Department of Chemistry, University of California, Berkeley, California 94720, USA

Prof. Dr. *Charles W. Rees* Hofmann Professor of Organic Chemistry, Department of Chemistry, Imperial College of Science and Technology, South Kensington, London SW7 2AY, England

Prof. Dr. *Fritz Vögtle* Institut für Organische Chemie und Biochemie der Universität, Gerhard-Domagk-Str. 1, D-5300 Bonn 1

Prof. Dr. *Georg Wittig* Institut für Organische Chemie der Universität Im Neuenheimer Feld 270, D-6900 Heidelberg 1

Table of Contents

Solvation and Ordered Structure in Colloidal Systems

Gotthold Ebert

Fachbereich Physikalische Chemie — Polymere. Universität D-3550 Marburg, FRG

Table of Contents

1 Introduction

As a contribution to the topic "Biomimetic and Bioorganic Chemistry" this review deals mainly with colloidal systems consisting of biomolecules forming association colloids or existing as macromolecular colloids (biopolymers). The associated monomolecular units are in a more or less ordered state in micelles, vesicles or in the membrane itself. However the molecular order of solubilized molecules is also strongly influenced. Biopolymers are able to attain highly ordered, periodic structures depending on various parameters e.g. temperature, pH, electrolyte concentration etc. Usually the solvation effects or changes in the solvation of these molecules are involved in the formation of ordered structures.

Because biomolecules normally exist in liquid water, this article will be largely concerned with their ordered structures in aqueous media and therefore with hydration effects. In order to understand better the influence of solute-solvent interactions on molecular order, also solvation in organic liquids will be considered to some extent.

"Solvation" and especially "hydration" are rather complex phenomena and little is known about them. Depending on the kind of molecular groups, atoms or ions interacting with the solvent, one can differ between lyo- or hydrophilic and lyo- or hydrophobic solvation or hydration. Due to these interactions the so-called "liquid structure" is changed. Therefore it seems to be unavoidable to consider, at least very briefly, the intermolecular interactions and the main features of liquids, especially "water structure" before dealing with solvation/hydration and their effects on the formation of ordered structures in the colloidal systems mentioned above.

2 Intermolecular Forces and Structures in Liquids

2.1 Apolar Liquids

This type of liquid is characterized by direction independent, relatively weak dispersion forces decreasing with r^{-6}, when r is the distance between neighbouring molecules. A simple model for this type of liquid, which accounts for many properties, was given by Luck [1,2]: it is represented by a slightly blurred lattice-like structure, containing hole defects which increase with temperature and a concentration equal to the vapor concentration. Solute molecules are trapped within the holes of the liquid thus reducing their vapor pressure when the latter is negligible.

2.2 Interactions in Polar Liquids

In this case the interaction between neighbouring molecules is not only stronger but also orientation dependent. The mean statistical energy of dipole-dipole interactions $\bar{U}_{dd}$ also decreases with r^{-6}, but depends on μ^2 (μ: dipole moment) according to [3]

$$\bar{U}_{dd} = \frac{2\mu_1^2\mu_2^2}{3r^6kT} \quad (1)$$

As a consequence of such interaction there are some preferential mutual orientation disturbed by the thermic motion of the molecules.

Ion-dipole interactions interact over longer distances. Their mean interaction energy $\bar{U}_{id}$ decreases with r^{-4} [3]:

$$\bar{U}_{id} = \frac{e_0^2\mu^2}{3r^4kT} \tag{2}$$

Ion-ion interactions are long-range interactions decreasing only with r^{-1} and their mean interaction energy $\bar{U}_{ii}$ is given by [3]

$$\bar{U}_{ii} = \frac{1}{\varepsilon}\frac{z_i z_j e_0^2}{r} \tag{3}$$

Hydrogen bonds play a very important role in the formation of structures in liquids. Their interaction energy is relatively high (3–10 kcal/mole) and is strongly dependent on the direction between donor and acceptor [1,3-7]. The maximum of hydrogen bond energy is attained if proton donor and acceptor are arranged linearly and the angle β between both is zero. Deviations from the linear arrangement up to $\beta = 30°$ have only a negligible influence on the bonding energy [1,4-7] and in fact discrete values of β characteristic for ring shaped associates have been observed [1]. Therefore, in order to enable dipole molecules to participate in hydrogen bonding, the maximum of intermolecular interaction must occur in the O—H ... O-direction and not in that of the dipole moment which lies in the direction of the H—O—H angle bisector, as in the case of water.

The typical properties of water arise from the ability of the water molecule to participate in four hydrogen bonds due to its two protons and its two lone electron pairs $(2s)^2(2p_z)^2$ which act as proton acceptors. In the condensed state, the angle between the $2p_x$ and the $2p_y$ orbital of oxygen is enlarged by hydridisation to a mixture of s- and p-state to 109°. Because both of the free electron pairs are situated in a plane

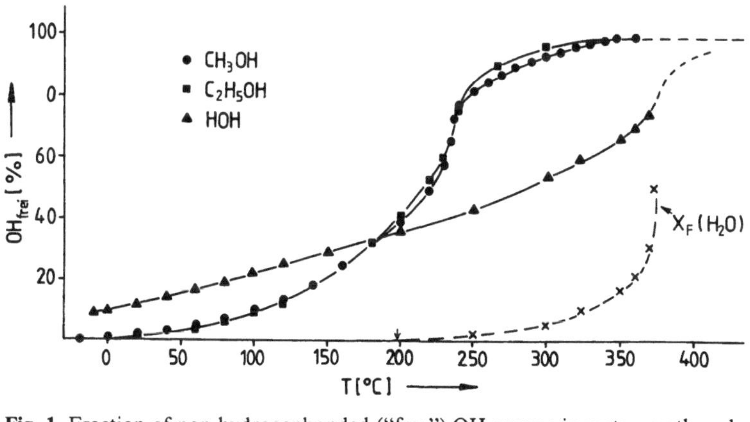

Fig. 1. Fraction of non-hydrogenbonded ("free") OH-groups in water, methanol and ethanol determined by IR-spectroscopy. x_F is the fraction of hole defects for water [7]

perpendicular to that of the H—O—H, each water molecule in the ice modification ice I is surrounded tetrahedrically by four others, so forming a hexagonal lattice similar to the SiO_2-tridymite containing large holes. When the ice melts, it is only the long-range order which is lost, wheras a remarkable degree of short-range order is maintained over a wide temperature range. This was shown by X-ray analysis [8,9], IR-, Raman-, NMR-spectroscopy [11-13], dielectric measurements [14], thermodynamical considerations [1,6,15] etc. reviewed by several authors.

For this reason the number of "free", non-hydrogen bonded molecules is on the average of time at near room temperature rather small, not only in water but also in alcohols. This is shown in Fig. 1, in which the fraction of "free" OH-groups is plotted against temperature [7]. There are many theories on the structure of liquid water [5,6,16], but no one is able to offer an interpretation of all the properties of liquid water. One of the most advanced theories on liquid water is offered by Nemethy and Scheraga [17] and allows for the evaluation of many of its thermodynamic properties. The authors assume "flickering clusters" with life-times of $\approx 10^{-11}$ sec and a size of ≈ 60 water molecules at 20 °C in equilibrium with "monomeric" water. Unsatisfactorily the number of broken hydrogen bonds in this model is too high in comparison with experimental results. According to a model developed recently by Luck [7] and valid up to 180 °C, there are structures of "liquidified ice" and hydrogen bond defects which have the tendency to attain a packing as in apolar liquids by means of a cooperative mechanism.

It should be mentioned that in the last few years super-cooled water has attracted the interest of many scientists because of its exceeding properties and life at temperatures below 0 °C [18,19]. Speedy recently published a model which allows for the interpretation of the thermodynamic anomalies of supercooled water [20]. According to this model there are hydrogen bonded pentagonal rings of water molecules which have the quality of self-replication and association with cavities.

Pentagonal rings and other cyclic structures have been suggested not only for liquid water [22a-c] but also for clathrate hydrates [21].

2.3 Water Structure in Aqueous Solutions

More complicated and less known than the structure of pure water is the structure of aqueous solutions. In all cases, the structure of water is changed, more or less, by dissolved substances. A quantitative measure for the influence of solutes on the structure of water was given in 1933 by Bernal and Fowler [23], introducing the terminus "structure temperature, T_{st}". This is the temperature at which any property of pure water has the same value as the solution at 20 °C. If a solute increases T_{st}, the number of hydrogen bonded water molecules is decreased and therefore it is called a "water structure breaker". Vice versa, a T_{st} decreasing solute is called a "water structure maker". Concomitantly the mobility of water molecules becomes higher or lower, respectively.

2.3.1 Hydrophilic Hydration

Hydrophilic hydration signifies that a strong energetically favored direct interaction exists between dissolved polar or ionic particle and the surrounding water molecules by ion-dipole-, dipole-dipole-interactions and/or hydrogen bonds.

It is possible to indicate by thermodynamic considerations [24,25,27], by spectroscopic methods (IR [28], Raman [29], NMR [30,31]), by dielectric [32] and viscosimetric measurements [26], that the mobility of water molecules in the hydration shell differs from the mobility in pure water, so justifying the classification of solutes in the water structure breaker and maker, as mentioned above.

The order of cations and anions regarding these structure breaking and making properties is related to their position in the lyotropic or Hofmeister series, such as

$$Al^{3+} > Mg^{2+} > Ca^{2+} > Sr^{2+} > Ba^{2+} > Li^+ > Na^+ > K^+ \approx NH_4^+ > Rb^+ > Cs^+$$

and

$$PO_4^{3-} > SO_4^{2-} > CO_3^{2-} > F^- > Cl^- > Br^- > I^- > SCN^- > ClO_4^-$$

On the left-hand side there are strong hydrated multivalent ions with a high charge density on the ion surface responsible for the strong ion-dipole interaction. On the right hand side one finds large monovalent ions with a lower charge density. This is shown in Table 1 for the cations [33].

Table 1. Ionic radii and charge densities of alkali and alkali earth cations (Hilgenfeld, R., Saenger, W. [33])

Ion	r [Å]	Charge/density × 10^{20} [Coulomb/Å^2]
Li^+	0,68	12,1
Na^+	0,95	4,46
K^+	1,33	1,62
Rb^+	1,48	1,18
Cs^+	1,69	0,79
Be^{2+}	0,35	178
Mg^{2+}	0,66	26,8
Ca^{2+}	0,99	7,90
Sr^{2+}	1,12	5,46
Ba^{2+}	1,34	3,17

It should be emphasized that the position of some ions in the series mentioned above depends somewhat on the method used for its determination [5,34]. Recently it was shown by Lang that in supercooled water and under low pressure all kinds of ions, e.g. Mg^{2+} and Li^+ act as water structure breaker [35,36]. Besides ions, non-ionic but polar solutes also influence the mobility of H_2O molecules by mutual interactions which change the water structure [24,31]. Solutes containing proton acceptors or proton donators are especially able to form hydrogen bonds with water molecules and act as water structure makers or breakers [37–40].

2.3.2 Hydrophobic and Lyophobic Solvation

As shown by Frank and Evans [41], solutions of apolar substances in water are characterized by a large entropy of mixing, leading to a high positive free energy of dissolving.

Table 2. Partial excess quantities for mixing pure solvent and pure solute to infinitely diluted solution according to Chan et al.[42]

Quantity	Hydrophobic	Normal
$\Delta\mu_2^E$ (cal · mol^{-1})	$\geqq 5000$	$\approx \pm 600$
ΔH_2^E (cal · mol^{-1})	$\approx \pm 1000$	$\approx \pm 1000$
ΔS_2^E	$T\lvert\Delta S_2^E\rvert \gg \lvert\Delta H_2\rvert$	$T\lvert\Delta S_2^E\rvert < \lvert\Delta H_2\rvert$
$\Delta C_{p_2}^E$ (cal · mol^{-1} deg^{-1})	$\geqq 50$	± 3
ΔV_2^E (cm^3 · mol^{-1})	$\leqq 15$	± 1

This effect is explained by a structuring of the solvent surrounding the apolar solute. Table 2 shows a comparison of the thermodynamical excess quantities for mixing the pure solvent with the pure solute to an infinitely diluted solution for hydrophobic and non-hydrophobic solutes, according to Chan et al.[42].

Besides the large partial excess free energy $\Delta\mu_2^E$ due to the highly negative excess entropy ΔS_2^E, the large excess heat capacity $\Delta C_{p_2}^E$ indicates that there is a remarkable energy necessary for "melting" the ordered solvent structure by raising the temperature. Therefore, a clustering of apolar solutes is favored by gaining entropy due to a decrease in the number of ordered water molecules. Theoretical treatment of the hydrophobic effects have been given by several authors[20,24,42–46].

The question is raised as to whether lyophobic effects also occur in nonaqueous, high polar solvents. As a result of energy and interfacial energy measurements on mixtures of polar, aprotic solvents and n-decane, Greco[47] comes to the conclusion that at a colloidal level hydrogen bonds are unnecessary for entropy losses as in the case of hydrophobic effect. Indeed there seems to be a higher degree of order in the solvation shell of polar molecules surrounding an apolar one because, for energetical reasons, the self-interaction of these solvent molecules is favored compared with the polar-apolar interaction. For this reason the extent of the freedom of orientation in the solvation shell may be decreased according to a loss in entropy.

3 Solvation of Amphiphilic Molecules and Ordered Structures

3.1 Solvation and Ordered Structures in Reversed Micelles

Molecules consisting of polar and apolar moieties behave as amphiphiles: they are simultaneously hydro- and lipophilic or hydrophilic and hydrophobic. Due to this fact, such molecules are able to associate equally well in both aqueous and hydrocarbon solutions above a certain concentration — the critical micelle concentration (c.m.c.) — forming thermodynamic stable micelles in water or reversed micelles in hydrocarbons. Detailed reviews of these phenomena were published a few years ago in this series[48,49].

Micelles and reversed micelles are able to solubilize substances which are insoluble in the bulk phase of the system considered. This solubilization is due to a solvation by the amphiphile and concomitantly a change in the order of the solubilized molecules may occur as a consequence of its modified solvation shell. In this sense reversed micelles of detergents in hydrophobic solvents with solubilized water in the core are of special interest. A detailed molecular model of reversed micelles has been published by Kitahara and Kon-No [50].

Inside the small water pool there are strong interactions of the water molecules with the ionic head groups of the surfactant and between the water molecules themselves. Eicke and coworkers [51,52] have studied the role of water for the formation of reversed micelles experimentally and theoretically in detail. They have shown for AOT (bis-(2-ethyl-hexyl)-2-sulfonatosuccinate)/isooctane/water-systems, that hydrogen bonding is very important for micellisation of surfactants in hydrocarbons [51]. From photon correlation spectroscopy measurements between —85 and +95 °C, an unexpected high thermal stability of AOT reversed micelles was found [52]. This is probably due to the reinforced formation of extended networks by hydrogen-bonded water molecules in the water pool.

Senō et al. [53–56] have studied the properties of water solubilized in the core of reversed micelles of various surfactants such as docecylammoniumpropionate (DAP), hexadecyltrimethylammoniumbromide (HTAB) and sodium bis-(2-ethylhexyl)-2-sulfonatosuccinate (AOT or SBSS) in different organic solvents (CH_2Cl_2, $CHCl_3$, C_6H_6, n-C_6H_{14}) by IR and NMR-spectroscopy. It was found that there are two kinds of water molecules within the water-pool: one of them (type A) interacts quite strongly with the ionic head groups of the surfactant, the other kind (type B) is less bound and interacts with the Type A water molecules of the primary hydration layer. According to Senō et al. [54,55] the properties of these solubilized water molecules are closely related to the enzyme-like behaviour of DAP/hexane/water systems. This becomes obvious by the hydrolysis of adenosine-5′-triphosphate (ATP) greatly enhanced by Mg^{2+} in reversed micelles [53]. FT-IR studies [56] have shown that the maximum of the bands of water associated with Na-sulfonate and tetrabutylammoniumbromide (TBAB) are close to that of AOT and HTAB. This seems to indicate that the interactions of water with the ionic head groups in reversed micelles are very similar to those in the hydration shell of ions in aqueous solution. From difference-spectra obtained by subtracting the spectra of the NaBr solution from that of the TBAB-solution it is possible to estimate the approximate frequencies of water associated with Na^+ and TBA ions. A band at $\approx 3400\ cm^{-1}$ is most likely caused by water molecules interacting with ammoniumgroups. Probably due to the structure-making effect of quaternary alkylammonium ions the O—H stretching band of the solubilized water is shifted to lower frequencies.

The strong interactions between the water molecules also become obvious from NMR measurements by Tsujii et al.. [57] ^{13}C-NMR experiments were used for determining the microviscosity of water in reversed micelles of dodecylammoniumpropionate with ^{13}C glycine cosolubilized. It was found that the apparent viscosity of the water-pool corresponds to the viscosity of a 78 % aqueous glycerol solution, obviously as a consequence of the extended network formation by strong hydrogen bonding.

Similar conclusions were obtained from ^1H and ^{31}P NMR and also from IR studies of egg phospholecithin reversed micelles in benzene by Boicelli et al. [58–61]. According to the results of these experiments the water structure within the reversed phospholecithin micelles alters considerably compared with water in bulk. This becomes evident from the shortening of the relaxation time T_1 of the water protons split into two relaxation times T_{1A} and T_{1B}, indicating that there are at least two populations, A and B, of water molecules [60]. This is in agreement with Senō et al. [54, 55]. If amino acids or small peptides are cosolubilized, the relative amounts of A and B are changed. One can assume that due to the hydration of the solute the number of water molecules available for one polar head group will be decreased. As can be easily understood, the solute molecules prefer to interact with water molecules belonging to population B, rather than with those of population A firmly bound to the phospholipid polar head-groups.

With increasing water content the reversed micelles change via "swollen micelles" [62] into a lamellar crystalline phase, because only a limited number of water molecules may be entrapped in a reversed micelle at a distinct surfactant concentration. Tamamushi and Watanabe [62] have studied the formation of reversed micelles and the transition into liquid crystalline structures under thermodynamic and kinetic aspects for AOT/isooctane/water at 25 °C. According to the phase-diagram, liquid crystalline phases occur above 50—60% H_2O. The temperature dependence of these phase transitions have been studied by Kunieda and Shinoda [63].

Sodium octanoate (NaO) forms reversed micelles not only in hydrocarbons but also in 1-hexanol/water. The hydration of the ionogenic NaO headgroups plays an important role in this case too. For this reason Fujii et al. [64] studied the dynamic behaviour of these headgroups and the influence of hydration water with ^{13}C and ^{23}Na NMR measurements. Below $w_0 = [H_2O]/[NaO] \approx 6$ the ^{23}Na line-width

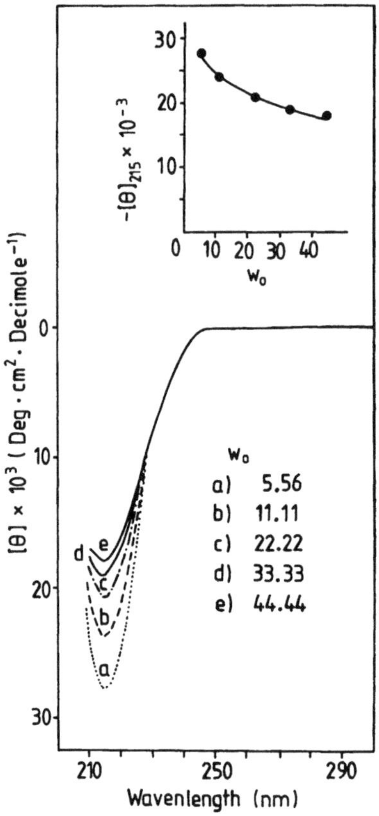

Fig. 2. CD-spectra of (L-Lys · HBr)$_n$ in reversed micellar systems of AOT/octane/H_2O at different w_0-values [65]. T: 20 °C

increases very steeply due to a restricted mobility of the Na^+ in the water pool. Furthermore it was found that the OH-group of 1-hexanol is probably in contact with water molecules entered among the polar NaO headgroups and that the mobility of methylen carbons 5 and 6 of NaO is affected by these water molecules.

As mentioned above, water structure in reversed micelles deviates considerably from the structure in the bulk-phase. Therefore, the hydration shell of macromolecules entrapped in reversed micellar systems should be changed and thus also their conformation. According to the results of several authors this is indeed the case.

Senō et al. [65] have used basic poly-α-aminoacids (BPAA) and copolymers of basic aminoacids with L-leucin as model substances cosolubilized in a reversed micellar system AOT/octane/water. As a result the basic homopolyaminoacids $(Orn)_n$, $(Lys)_n$ and $(Arg)_n$ form Iβ conformation in reversed AOT micelles (Fig. 2) whereas in AOT solutions at low concentrations $\approx 10^{-4}$ to $\approx 10^{-3}$ mol/l CD spectra similar to a superposition of IIβ and α-helix are observed (Fig. 3). With increasing w_0 value ($[H_2O]/[AOT]$ ratio), *i.e.* increasing water content of the reversed micelles, the specific ellipticity $[\theta]_{218}$ decreases. This may be explained by a weakening of the interactions between the $-SO_3^-$ and the basic side group of the BPAA reponsible for β-structure formation by electrostatic shielding. A copolymer (L-Lys $HBr^{0,50}$, L-$Leu^{0,50})_n$ shows CD spectra similar to that of α-helix in 5×10^{-5} — $9{,}4 \times 10^{-4}$ m AOT solutions,

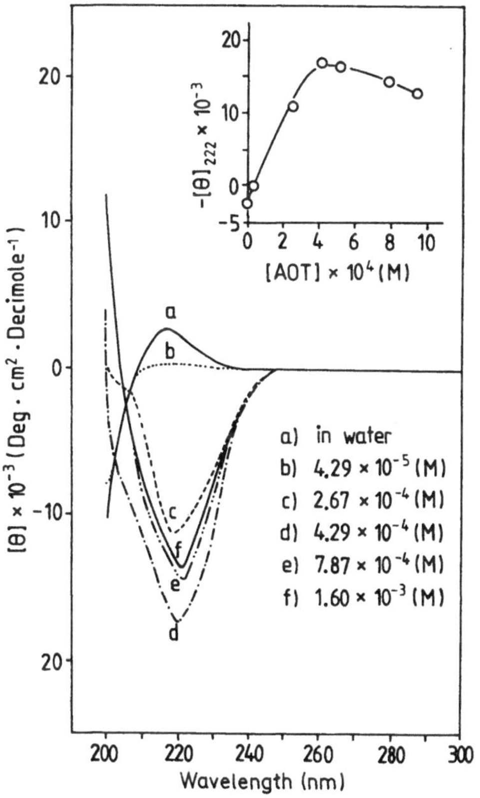

Fig. 3. CD-spectra of (L-Arg · HBr)$_n$ in water and in AOT-solutions at pH 6,2, T: 20 °C

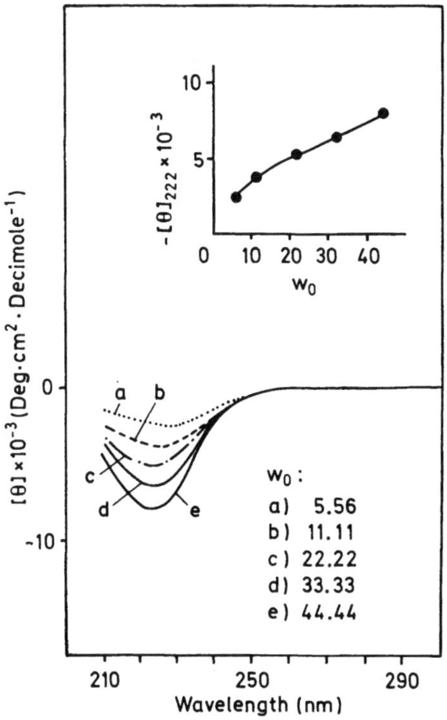

Fig. 4. CD-spectra of (L-Lys · $HBr^{0,5}$,L-$Leu^{0,5}$)$_n$ in reversed micellar systems of AOT/octane/H_2O at different w_0-values [65]

however, only a shallow minimum near 225 nm similar to a IIβ-structure increasing with w_0 (Fig. 4) in reversed micelles. Especially L-leucine rich copolymer (L-Lys · $HBr^{0,37}$-L-$Leu^{0,63}$)$_n$ with a very stable α-helical conformation in water up to at least 90 °C and also in concentrated electrolyte solutions [66] shows a dramatic change of the CD spectra as is shown in Fig. 5 and is most pronounced at low w_0-values (5,56). It seems that in this case all water molecules present are located within the ionic groups whereas the apolar isobutyl side-chains of the L-leucyl residues are not exposed to water but to the apolar medium. As a consequence, hydrophobic interactions stabilizing the α-helix in aqueous phases would be lacking and a conformation change would take place. As one can see from Fig. 6, this copolymer attains an α-helical conformation at low AOT-concentrations in aqueous solution.

The conformation of bovine myelin basic protein (MBP) in AOT/isooctane/water reversed micellar systems was studied by Waks *et al.* [67]. This MBP is an extrinsic water soluble protein which attains an extended conformation in aqueous solution [68] but is more densily packed at the membrane surface. The solubilization of MBP in the AOT reversed micelles depends on the water/AOT-ratio w_0 [68]. The maximum of solubilization was observed at a w_0-value as low as 5.56. The same value was obtained for another major protein component of myelin, the Folch-Pi proteolipid [69]. According to fluorescence emission spectra of MBP, accessibility of the single tryptophane residue seems to be decreased in AOT reversed micelles. From CD-spectra one can conclude that there is a higher conformational rigidity in reversed micelles and a more ordered aqueous environment.

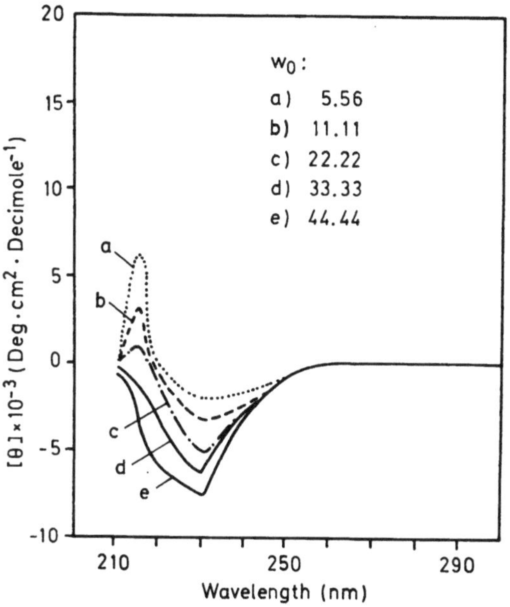

Fig. 5. CD-spectra of (L-Lys · $HBr^{0,37}$, L-$Leu^{0,63}$)$_n$ in reversed micellar systems of AOT/octane/water at different w_0 values. T: 20 °C [65]

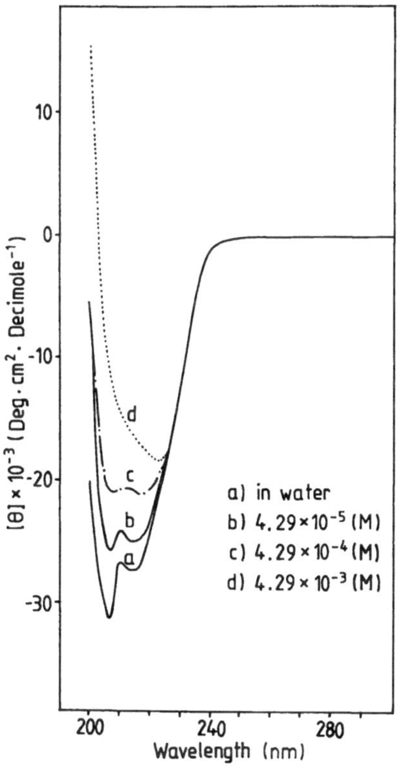

Fig. 6. CD-spectra of (L-Lys · $HBr^{0,37}$, L-$Leu^{0,63}$)$_n$ in water and in AOT-solutions at pH 5,9. T: 20 °C [65]

The behaviour of enzymes in reversed micelles was recently studied by the Luisi group and clear solutions of enzymes in AOT/hydrocarbon-systems with a water content of only 1 % were obtained. The enzymes studied are ribonuclease, γ-chymotrypsin, trypsin, horse liver aldehyde dehydrogenase, lipoxygenase, peroxidase, lysozyme, alkaline phosphatase [70, 71a–c]. Steinmann et al. reported on the behaviour of lysozyme in these systems by means of various spectroscopic methods, by activity measurements and by sedimentation behaviour in an analytical ultracentrifuge [72]. Denaturation of lysozyme was observed in the reversed micelles as became obvious from the CD and fluorescence spectra. By the addition of saccharide to the native enzyme a stabilization against denaturation in the reversed micelles is observed. The lower the water content and the larger the saccharide, the higher the stabilization effect will be. According to the results of activity measurements with a distinct substrate, the activity of the enzyme at a low water content is higher than in bulk water.

3.2 Formation of Liposomes and Solvation

Phospholipids e.g. form spontaneously multilamellar concentric bilayer vesicles [73] if they are suspended e.g. by a mixer in an excess of aqueous solution. In the multilamellar vesicles lipid bilayers are separated by layers of the aqueous medium [74–78] which are involved in stabilizing the liposomes. By sonification they are dispersed to unilamellar liposomes with an outer diameter of 250–300 Å and an internal one of 150–200 Å. Therefore the aqueous phase within the liposome is separated by a bimolecular lipid layer with a thickness of 50 Å. Liposomes are used as models for biological membranes and as drug carriers.

The stability of these vesicles is changed by the addition of salts depending on their position in the lyotropic series. This was recently investigated by Nakagaki et al. [77]. In the case of positively charged liposomes the destabilization effect decreased in the order $SCN^- > I^- > Br^- > Cl^-$, whereas for negatively charged liposomes $Li^+ > Na^+ > K^+ > Cs^+$ was found. If the net charge is zero, the destabilizing effect of anions is interestingly inversed according to $Cl^- > Br^- > SCN^- > I^-$, indicating the strength of dehydration from hydrophilic colloid. In the case of D,L-α-Dipalmitoylphosphatidylcholine (DPPC) at pH 6 the stability of the liposomes is lowered in the order

$$Cl^- > Br^- \approx NO_3^- > SO_4^{2-} > I^- > SCN^- .$$

It is surprising that in contrast to the usual order sulfate is situated between Br^-/NO_3^- and I^-.

Cationic quaternary ammonium compounds such as distearyldimethylammonium-chloride (DSDMAC) used as a softener and as an antistatic, form hydrated particles in a dispersed phase having a similar structure to that of the multilayered liposomes or vesicles of phospholipids [77, 79]. This liposome-like structure could be made visible by electron microscopy using the freeze-fracture replica technique as shown by Okumura et al. [79]. The concentric circles observed should be bimolecular lamellar layers with the sandwiched parts being the entrapped water. In addition, the longest spacings of the small angle X-ray diffraction pattern can be attributed to the inter-lamellar distances. These liposome structures are formed by the hydrated detergent not only in the gel state but also at relatively low concentrations.

A new class of compounds able to form lamellar structures consisting of bilayers separated by water-layers are dialkylammonium-salts of 2-[((trifluormethyl)-phenyl)-amino]-benzoic acid (Flufenamic acid) as was found by Eckert and Fischer [80].

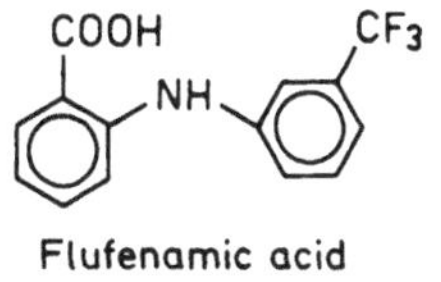

Flufenamic acid

The ability of flufenamic acid — especially of the diethylammonium salt (DEAF) — to form lyotropic mesophases is mainly due to the presence of the trifluoromethyl-group and to the organic cation. Cations favoring the formation of mesophases form ion pairs with the flufenamic-anion decreasing the solubility in water. As a result of ^{19}F-NMR studies the authors suggest an interaction of the CF_3-groups with an interlamellar water layer responsible for the stabilization of the lamella [80,81]. At high DEAF concentrations ($>54\%$) reversed micelles are formed [80].

4 Stabilization of Ordered Structures of Biopolymers by Solvation Effects

4.1 Conformation and Solvation of Polypeptides

The biological function of biopolymers such as polypeptides, proteins, nucleic acids etc. depends strongly on their ordered structure which is determined by the pattern of inter- and intramolecular interactions given by the primary structure.

The factors accounting for the stabilization of protein conformation are shown in

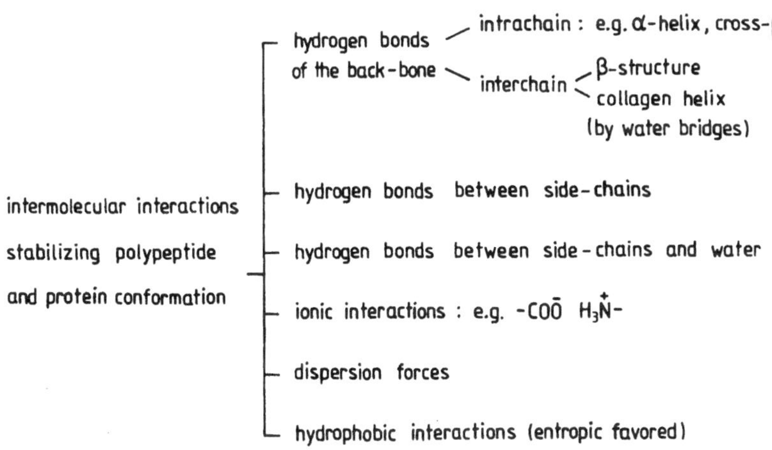

Fig. 7. Scheme of the intermolecular forces stabilizing ordered structures in polypeptides and proteins [82]

Fig. 7. Besides direct interactions between functional groups of the biopolymer molecule itself there are also various kinds of interactions with water molecules. These hydrophilic and hydrophobic interactions are essential for stabilizing the native conformation of biopolymers. In the last few years some progress was made in elucidating the hydration of these molecules.

Furthermore it can be shown that besides the direct influence of hydrophilic and hydrophobic hydration on the conformation, the interaction of charged groups with ions is also strongly influenced by the hydration of the groups involved. Such studies were made largely by using relatively simple poly-α-aminoacids with ionogenic side chains as model substances.

The interaction of ions with ionic side groups of aminoacids depends on various parameters, e.g. on the hydration of these charged particles which was studied using several kinds of acidic and basic poly-α-aminoacids (BPAA). The latter are especially useful for studies regarding the influence of ion hydration on the cation-anion interaction and the conformation of BPAA. This is because there are various cationic groups such as ammonium, guanidinium, trialkylammonium etc. different in size, basicity and ability to form hydrogen bonds with solvents and solutes etc. Poly-α-aminoacids with charged side chains exist in a more or less extended conformation due to the electrostatic repulsion of the ionic groups [83,84]. The role of hydration in this case was discussed by Barteri and Pispisa [85].

Several years ago Makino et al. [86] studied the influence of anions on the conformation of poly-[L-methionine-S-methylsulfonium] salts in solution. They found that especially perchlorate will induce α-helix formation whereas Cl^- and Br^- do not. Since then several authors [87–92] have found a similar α-helix inducing effect in the case of poly-L-lysine $(Lys)_n$ and other BPAA at low pH-values where the polymer molecules usually attain an extended conformation due to the electrostatic repulsion of the ammonium groups. Therefore, the α-helix inducing effect is obviously an

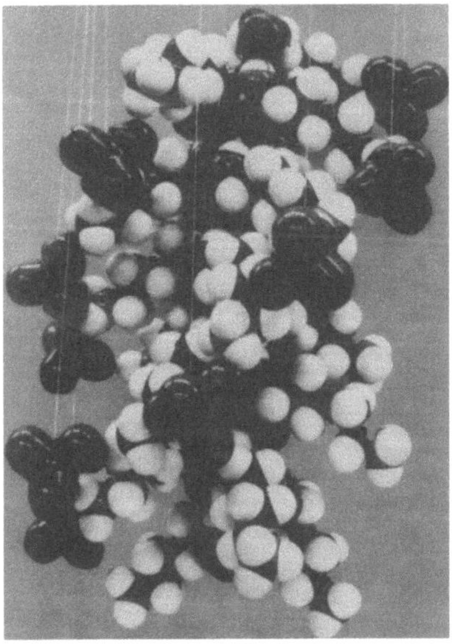

Fig. 8. Space filling model of a poly-L-lysine α-helix with ClO_4^- anions inserted between the NH_3^+-side groups forming a left-handed superhelix [91]

electrostatic shielding phenomenon cancelling the electrostatic repulsion of the cationic side groups due to a specific interaction with the perchlorate anions. It was suggested that by an insertion of the ClO_4' anions a left-handed superhelix around the right-handed α-helix is formed, as shown in Fig. 8 [91,93]. Such a specific interaction occurs when the ions involved have a water structure breaking character. This assumption gained further credibility through the studies of Sugai and coworkers on the anion-induced conformational transitions of poly[L-homo-arginine] [92] $(Har)_n$. According to Murai et al. [92] perchlorate and thiocyanate which are strong water structure breakers are bound more easily by the guanidinium side-groups of $(Har)_n$ compared with $(Lys)_n$ because this side-group is a stronger water structure breaker than the NH_3^+-group. This was confirmed by Miyazaki et al. [94] who also studied this effect in the case of poly-[L-arginine] $(Arg)_n$ and poly-[L-α-amino-γ-guanidinobutyric acid] $(Agb)_n$ at neutral pH caused by I^-, ClO_4^- and SCN^-.

The authors determined the enthalpy of transferring the BPAA from water into the aqueous $NaClO_4$ solution (ΔH_z) calorimetrically as a function of the electrolyte concentration. The curves obtained by plotting ΔH_z against $NaClO_4$ concentration obviously reflect the conformation change. Because the dilution enthalpy was experimentally compensated, the ΔH_z values contain the transition enthalpy change ΔH_0 from the charged extended conformation to the uncharged helix as well as the enthalpy of binding the anion to the BPAA-side chain ΔH_0^b:

$$\Delta H_z = f_H \Delta H_0 + \gamma \Delta H_0^b \tag{4}$$

(f_H is the helix fraction and γ the degree of association of the anion with the BPAA).

For the apparent binding constant K_{app}, it yields [95],

$$K_{app} = \frac{\gamma}{(1-\gamma)\, C_s f_\pm} = K_0^b \exp 2w(1-\gamma) \tag{5}$$

when K_0^b is the intrinsic binding constant, C_s the salt concentration added, $f_\pm$ the average activity coefficient of the salt and w is the electrostatic interaction parameter for coil or helical conformation. Conio et al. [95] assumed proportionality of w to $\ln C_s$. Miyazaki et al. [94] used separate values of w for coil (w_c) and helix (w_h), an improvement, when taking into consideration that the ion-ion-interaction is different in the coil- and helical state. In the case of low salt concentrations and $f_H = 0$ the relation

$$\frac{1}{\Delta H_t} = \frac{1}{K_0^b\, \Delta H_0^b \exp(2w_c)\, C_s f_\pm} - \frac{1}{\Delta H_0^b} \tag{6}$$

was obtained as an approximation from equation (4) and (5).

From plotting $1/\Delta H_t$ against $1/[\exp(2w_c)\, C_s f_\pm]$ for $(Har)_n$ and $(Agb)_n$ in concentration ranges of $NaClO_4$ from 0,001—0,005 and 0,01—0,04 mol/l respectively, $\Delta H_0^b = -850 \pm 150$ cal/mol and $K_0^b = 1,2 \pm 0,2\ M^{-1}$ were obtained. For the binding of SCN^- to $(Har)_n$ K_0^b is also found as $1,2\ M^{-1}$, however with a higher degree of uncertainty. These values of K_0^b are four times as large as those obtained for the binding of SCN^- with $(Orn)_n$ and $(Lys)_n$ ($K_0^b = 0,23 \cdot M^{-1}$) [91]. This clearly indicates that, as mentioned above, the guanidinium groups bind the two anions ClO_4^- and SCN^- in a much stronger way than the ammonium group. It was concluded that this stronger binding of the anions is due to the higher water structure-breaking effect of the guanidinium group, studied by Miyajima et al.[96,97] and Ebert[98]. This assumption also gains some evidence from the fact that the strong water structure making SO_4^{2-} does not usually induce α-helix formation[91,93,99-101] of BPAA. These findings are of some interest because ClO_4^- and SCN^- denaturate natural occuring proteins by "salting-in" accompanied by a helix-coil transition, whereas sulfate and other water structure makers increase the denaturation temperature[102,103] ("salting-out" effect) as a consequences of stabilizing the native conformation of proteins.

Paudjojo was able to show by ultracentrifuge measurements that there is a strong specific interaction between $(Lys)_n$ and ClO_4^-, however no measurable one between $(Lys)_n$ and SO_4^{2-} [99-101]. According to her results, the sedimentation coefficient shows a concentration and temperature dependence similar to the example shown in Fig. 9. The results of the ultracentrifuge measurements obtained by Friehmelt[104,105] for $(Arg)_n$ and $(Har)_n$ in ClO_4^-, SCN^-, $H_2PO_4^-$ and Cl^- containing solutions have given further evidence toward proving that the topology of the arrangement of anions along the polycation is important for α-helix-inducing effect. No α-helix-inducing effect of SO_4^{2-} could be detected for $(Arg)_n$ in contrast to Ichimura et al.[102]. Methylsulfate induces at very high concentrations (>3.5 mol/l) an α-helix formation[93,107,108]. However, in this concentration range the primary hydration shell of $CH_3SO_4^-$ is incomplete, obviously favoring the interaction with the $(Lys)_n$ side chains. It is interesting to note that this effect could not be observed for Poly-[N,ε-trimethyl-L-lysine][107,108] which on the other hand interacts strongly with SCN^- and ClO_4^- as was found by Y. H. Kim[107,108]. In Fig. 10 the b_0 parameters of

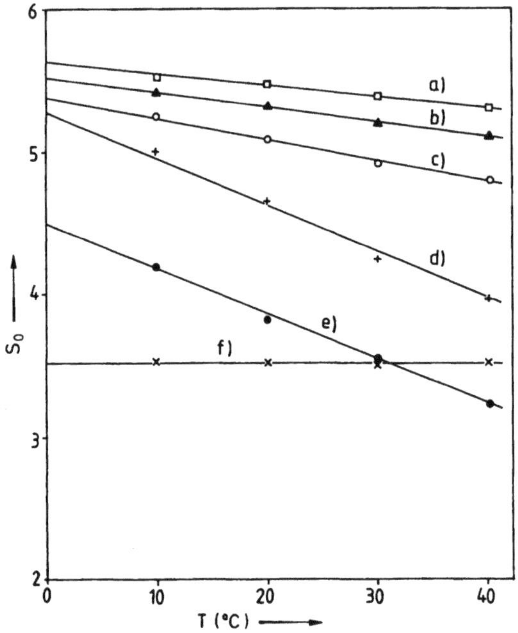

Fig. 9. Sedimentation coefficient of $(Lys)_n$ in different electrolyte solutions as a function of temperature a) 1,5 m $LiClO_4$ solution[100,101]; b) 1,0 m $LiClO_4$ solution; c) 0,5 m $LiClO_4$ solution; d) 0,2 m $LiClO_4$ solution; e) 0,1 m $LiClO_4$ solution; f) 0,1 and 1,5 m Li_2SO_4 solution

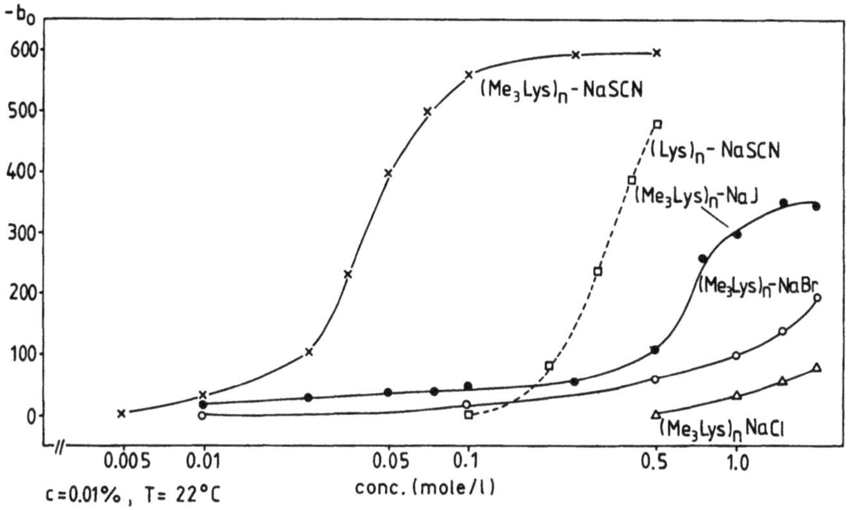

Fig. 10. b_0-values of poly-[N,ε-trimethyl-L-lysine] in various electrolyte solutions as a function of concentration [107, 108]

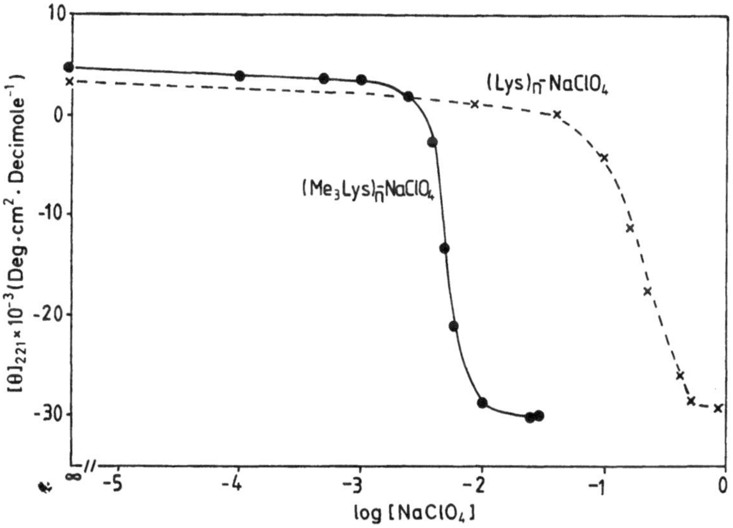

Fig. 11. Specific ellipticity $[\theta]_{221}$ of $(Lys)_n$ and $(Me_3Lys)_n$ as a function of $NaClO_4$-concentration [107, 108]

the Moffit-Yang equation in Cl^-, Br^-, I^-, SCN^--solutions are plotted against ion concentration. b_0 is assumed to be -630 for 100% α-helix content. Fig. 11 shows a comparison of $[\theta]_{221}$ for $(Lys)_n$ and $(Me_3Lys)_n$ in $NaClO_4$ solution. In spite of the fact that the quaternary ammonium group behaves as a water structure maker due to its hydrophobic substituents, an α-helix-inducing effect by a strong specific interaction is observed. This seems to be a contradiction to the observation of Sugais group [92, 94]. However, one has to consider that in contrast to the hydrophilic hydrated

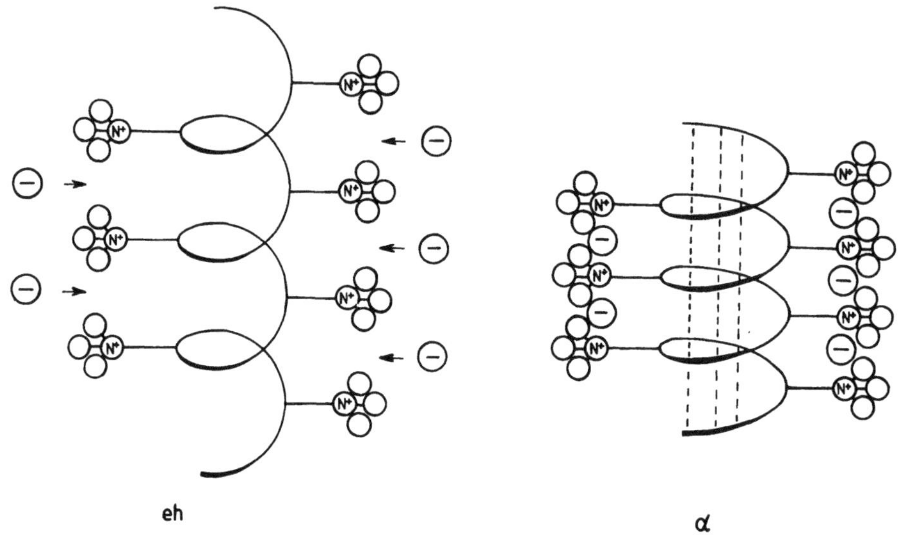

Fig. 12. Schematic representation of the α-helix-inducing effect of specific bound anions like ClO_4^- or SCN^- in the case of N-ω-trialkylated BPAA: the extended helical conformation contracts by the insertion of the anions. The water molecules of the hydration shells are omitted [109]

structure makers such as SO_4^{2-} the hydrophobic hydration of the quaternary ammonium group supports the entrapping of the anions as was recently pointed out and shown schematically in Fig. 12 [109]. The electrostatic shielding effect of the inserted naked anions and the hydrophobic interactions of the methyl groups outside are clearly moving in the same direction. These considerations are supported by the results of Zembala and Czarniecki [110] obtained on films of dodecyltriethyl-ammonium (DTEA)-films at mercury surfaces. Investigating the influence of anions on the stability of absorbed layers of DTEA, these authors found that ClO_4^- increases the life-time of the films remarkably compared with Cl^- and they similarly proposed an insertion of the ClO_4^- between the $—N^+\equiv$ as was suggested for the BPAA.

Nakagaki [111] has given a theoretical treatment of the electrostatic interactions by using the Gouy-Chapman equation for the relation between the surface charge density σ_e and surface potential ψ. The experimental data for $(Lys)_n$ agrees very well with the theoretical curve obtained.

Kuroyanagi has studied the influence of apolar L-leucine residues on the conformation of random and alternating copolymers with basic amino acids for eluciditation of ionic and solvation effects [112–115]. In the case of random copolymers with high L-leucine content e.g. (L-$Lys^{0,37}$, L-$Leu^{0,63}$)-α-helix formation occurs in pure aqueous solution at pH 7 probably due to the strong hydrophobic solvation of the leucyl side-groups. At lower Leu contents such as (L-$Lys^{0,5}$, L-$Leu^{0,5})_n$ at pH 7 CD-spectra characteristics for extended conformation are observed. However, at much lower ClO_4^--concentrations ($\geqq 3 \cdot 10^{-3}$ mol/l) than in the case of $(Lys)_n$ ($2 \cdot 10^{-1}$ mol/l) α-helix formation occurs. In contrast to $(Lys)_n$ sulfat anions act as α-helix inducing anions at rather low concentration $\geqq 5 \cdot 10^{-3}$ mol/l [112, 113]. These results give further evidence for the α-helix inducing effect of SO_4^{2-} by enhancing hydrophobic interactions of the apolar Leu side-groups.

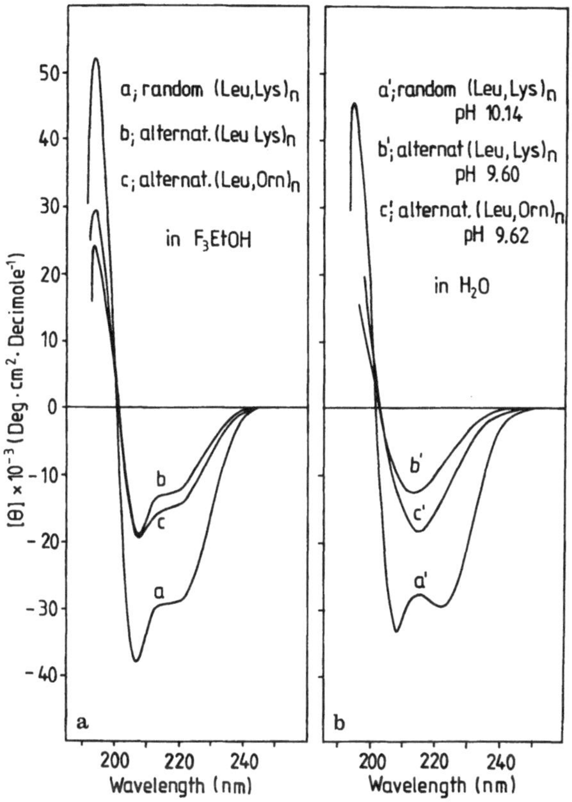

Fig. 13. CD-spectra of random and alternating (L-Leu, L-Lys)$_n$ [112, 119]. a) in water; b) in trifluoroethanol

Alternating copolymers of L-leucine and basic amino acids, in contrast to random copolymers of the same molar ratio, attain a β-conformation in perchlorate solutions. Also, at pH > 8,5 in the absence of other conformation influencing anion, an β-structure formation is observed [112, 114–118, 120] (Fig. 13b). It may be of great importance for the conformation of native proteins and polypeptides that the β-conformation of these alternating copolymers is also induced by sulfate concentrations as low as $2 \cdot 10^{-3}$ mol/l. The question: "why is the conformation dependent on the sequence of two aminoacids" can be easily answered by considering the fact that in an all-trans conformation of the polypeptide-chain all apolar side-groups are arranged on one side of the molecule and all ionogenic groups on the other. Therefore, by aggregation of two neighboured molecules the maximum number of hydrophobic interactions can be formed, whereas all hydrophilic groups are exposed to the aqueous medium. In this way the minimum of free energy of the conformation will be realised. If these alternating copolymers are dissolved in 2,2,2-trifluoroethanol, however, they adopt an α-helical conformation (Fig. 13a). In water-alcohol mixtures with increasing alcohol content depending on pH, a coil-α-helix or a β-structure-α-helix transition was reported [112, 119] (Fig. 14). Because α-helix formation also occurs at pH 7,0 of the starting solution, i.e. when the basic side groups are fully protonated, these results seem to indicate that the α-helix formation at a higher alcohol content

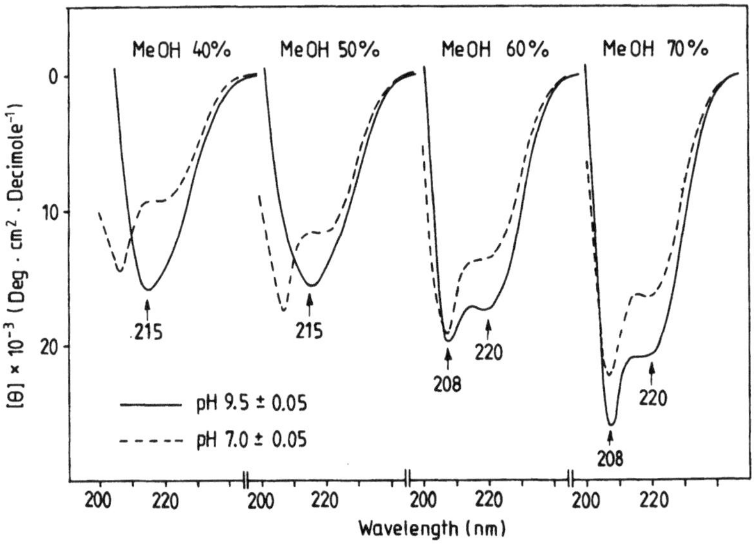

Fig. 14. CD-spectra of alternating (L-Leu-L-Lys)$_n$ in water-methanol mixtures at different pH of the starting solution [112,119]

is due to a change in the hydrophilic hydration of the ionic groups. Concomitantly their dissociation may be lowered so far that the electrostatic repulsion of these groups is cancelled. In this way, by modifying the solvation of the polycation a conformation change is obtained.

According to Epand and Scheraga the homopolymeric (Lys)$_n$ shows α-helix formation at >90% methanol content [120], whereas the (L-Leu^{0,48}, L-Lys^{0,52})-copolymer becomes 50% α-helical at 30% methanol and completely α-helical at 50% methanol content (Fig. 15). Similar results were obtained for [L-Leu, L-Lys] copolymers with low leucine content by Arfmann et al. [121]. This may be interpreted by an enrichment of the alcohol in the solvation shell of the polymer molecule due to the interaction with the apolar leucyl side groups, increasing the influence of the alcohol on solvation and dissociation of the ionogenic group as mentioned above. This assumption is supported by the increasing α-helix inducing effect with the chain length of the alcohol [112,119] or — when also considering fluorinated alcohols — with the hydrophobicity of the apolar part:

$$CH_3OH < C_2H_5OH < i\text{-}C_3H_7OH < 2,2,2\text{-trifluoroethanol}$$

For an α-helical fraction $f_H = 0{,}5 \cdot 30\%$ methanol, 20% ethanol, 15% i-propanol or 10% trifluoroethanol are necessary. Trifluoroethanol like perfluorinated alcohols, e.g. hexafluoroisopropanol is characterised on the hand by a strong acidic proton at the OG-group due to the —1-effect of the fluor atoms. On the other hand fluorocarbons are more hydrophobic than the hydrocarbons which is mainly due to the larger surface of the F compared with H. For this reason the critical micelle concentration of perfluorinated detergents is much lower than that of the corresponding hydrocarbon compounds. It was found that C_4F_7-derivatives act as detergents

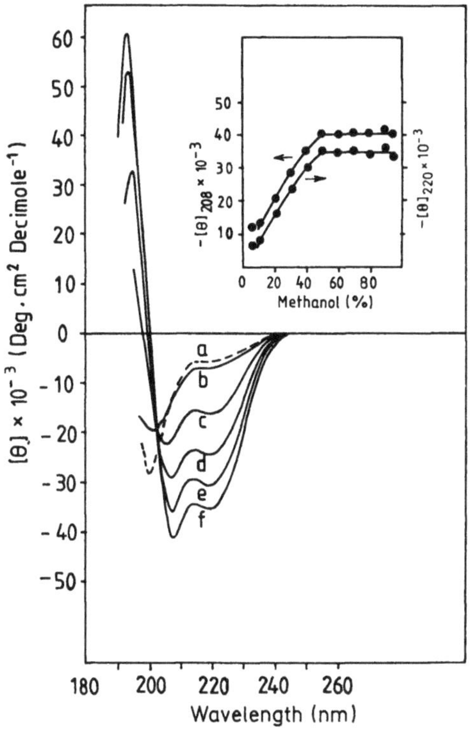

Fig. 15. CD-spectra of random (L-Leu^{0,48}, L-Lys^{0,52})$_n$ in water-methanol mixtures as a function of methanol concentration at 20 °C and pH 7,0. The insert shows $-[\theta]_{208}$ and $-[\theta]_{220}$ dependent on the methanol content [112,113]

whereas in the hydrocarbon-series a chain length of C_8 is required for behaving in this manner. An interaction of the OH-group with the carbonamide groups of the polymer molecule by hydrogen-bonds may hardly be responsible for the α-helix inducing effect of the fluorinated alcohol because such a solvation usually prevents the formation of ordered structures. Probably an interaction of the fluoratoms with the apolar groups lowering the water content of the solvation shell may cause the α-helix formation.

In this sense it should be mentioned that $(Ala)_n$ is dissolved by hexafluorisopropanol (HFIP) without conformation change and that stretched fibers of $(Leu)_n$ shrink in HFIP (20 °C) at about 60% only by disorientation of $(Leu)_n$ molecules, because no conformation change occurs according to X-ray measurements [122,123]. An interaction of the OH-group of the alcohol with the back-bone-CO-NH-groups is not very likely because, in this case, conformation changes should be observed.

As was reported by Ebina et al. [139] TFE has a greater effect on the hydrophobic interactions of proteins than isopropanol, which is described below. Timasheff et al. have investigated the relation between solvation and conformation of some biopolymers, especially in alcohols, by means of their preferential binding to proteins [128–130]. Very interesting studies on the solvation of poly-[L-lysine hydrobromide] $(Lys \cdot HBr)_n$ and poly-[sodium-L-glutamate] $(Glu \cdot Na)_n$ by high polar organic solvents like dimethylsulfoxide (DMSO), N-methylformamide (NMF), N,N-dimethylformamide (DMF), 1-methyl-2-pyrrolidone (NMP) 1,2-ethanediol (EG), 1- and 2-propanole (1-, 2-PrOH), t-butanole (t-BuOH) have been made by Iijima and coworkers [125–127].

Komiyama et al.[125] reported on measurements regarding the preferential binding of these compounds to $(Lys \cdot HBr)_n$ from polar solvent mixtures using the relation[129]

$$(\partial m_3/\partial m_2)^0_{\mu_1,\mu_2} = \frac{(M_2/M_3)}{1 - c_3\bar{v}_3}\{(\partial n/\partial C_2)_{\mu_1,\mu_3} - (\partial n/\partial C_2)_{m_3}\}/(\partial n/\partial C_3)_{m_2} \quad (7)$$

The subscripts 1,2,3 refer to the main solvent, the polymer, and the solvent added, respectively. The meanings of the other symbols are; n: refractive index; m: molarity of respective component in solvent 1; C: the concentration in g cm^{-3} of the solution; $\bar{V}$: the partial specific volume; μ: the chemical potential; M: molecular weight (for the polymer: per residue). The surscript ° indicates infinite dilution of the polymer.

As Fig. 16 shows, the preferential binding of DMSO, DMF and NMF from aqueous solution to $(Lys \cdot HBr)_n$ at low contents of the organic solvent x increases with its concentration. However, at approximately $x_3 = 0{,}2$ a maximum is reached and then preferential hydration between $x_3 = 0{,}3$ and 0,5 occurs. No preferential binding was observed for NMP, EG or 2 PrOH, however increasing hydration occured with x_3. Only in 2 PrOH at $x_3 > 0{,}3$ α-helix formation occured. Furthermore binding parameters for the systems NMP + DMSO, EG + DMSO and DMF + DMSO have been determined. An initial preferential binding of DMSO by $(Lys \cdot HBr)_n$, a maximum and a subsequently inversion of the binding parameter was also observed in these mixtures. The order of relative affinity is DMSO > DMF > EG > NMP. In DMF/DMSO-mixtures $(Lys \cdot HBr)_n$ attains an α-helical conformation above 20 vol.-% DMF and in 2-PrOH/water above 70 vol.-% 2 Pr-OH.

In order to determine the solvation numbers and the exchange constants it is assumed that one $(Lys \cdot HBr)_n$ residue consists of two parts, namely, the ionic group (numbered 2) and the less polar remainder of the molecule (numbered 1). Plotting

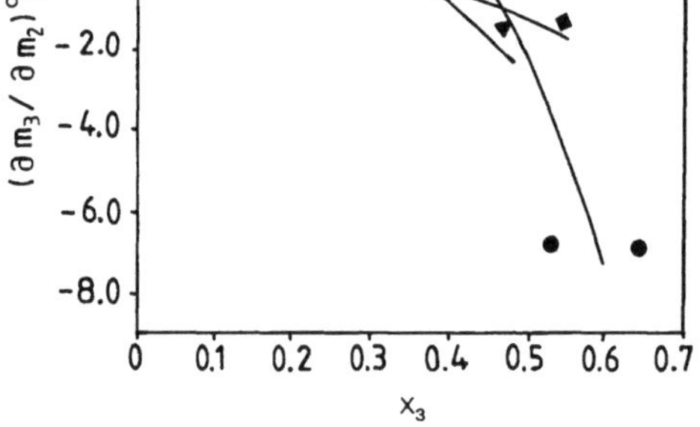

Fig. 16. Preferential binding of organic solvent to $(Lys \cdot HBr)_n$ dependent on solvent composition[125]. (●) in H_2O + DMSO; (▲) in H_2O + DMF; (◆) in H_2O + NMF

the preferential binding parameter against x_3/x_1 for water +2-PrOH systems a linear curve is obtained in the concentration range where no α-helix formation is observed. This indicates that the exchange constants between the solvation shell and the bulk is zero and that from the slope the total solvation number can be calculated [126]. In this case a hydration number 27 was obtained as the sum of 12 for the less polar part 1 and 15 for the ionic moiety of one residue. It should be mentioned that according to NMR-studies of Kuntz [131] the hydration number of L-lysine in the protonated as well as in the deprotonated state amounts to four and seven for $(Glu)_n$ in the dissociated, but only two in the undissociated one. However, these results may not be unambigous for primarily technical reasons [132]. Furthermore it may be that by these measurements the second loosely bound hydration shell can not be detected. Ifft and coworkers [134] have found 6,5 for $(GluCs)_n$ and 16,5 for $(Lys \cdot HBr)_n$ in aqueous CsCl solutions from buoyant density measurements. From water sorption hydration numbers of 3 and 4 were calculated [132].

The formation of α-helix is clearly accompanied by a decreasing affinity of the polymer molecule to the solvent and the release of some solvent molecules from the less polar part of the Lys-residue as became evident for the DMF + DMSO system. Mori et al. reported [127] that the hydration number 27 was also found in aqueous t-BuOH. Above $x_3 = 0{,}3$ an increase of the binding parameter was observed together with α-helix formation. This may be attributed to the decreasing dielectric constant ε of the solvent with increasing alcohol concentration favoring a stronger cation-anion interaction below $\varepsilon \approx 40$. Indeed, the preferential hydration of the two polymers studied decreased at a higher alcohol content. This tendency is most pronounced for $(Glu \cdot Na)_n$ in 2 PrOH/water compared with the other alcohols (Fig. 17) and for $(Lys \cdot HBr)_n$ in t-BuOH compared with 2-PrOH-water mixtures. As it is shown in Fig. 18 and 19 the α-helix formation starts at considerably higher x_3-values than at values where dehydration occurs. From conductivity measurements the authors came to the conclusion that for maximum helix content the hydration number is 3 instead of 20 in the case of $(Glu \cdot Na)_n$ and that helix formation starts at a hydration number

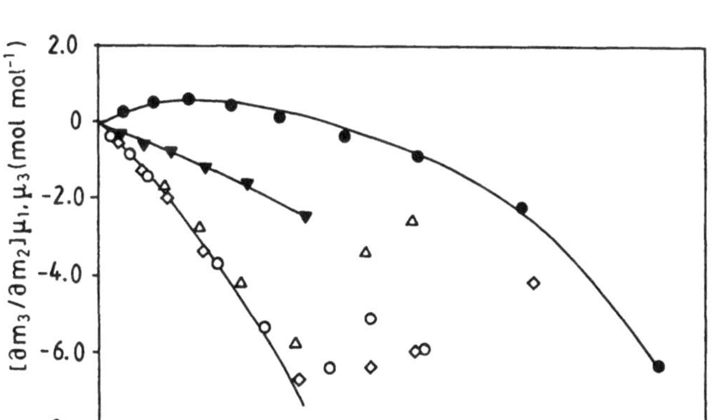

Fig. 17. Preferential binding of organic solvents to $(Glu \cdot Na)_n$ depending on solvent composition in mixtures of water with 1-PrOH (◇), 2-PrOH (△), t-BuOH(○), EG (●), DMSO (▼) [126]

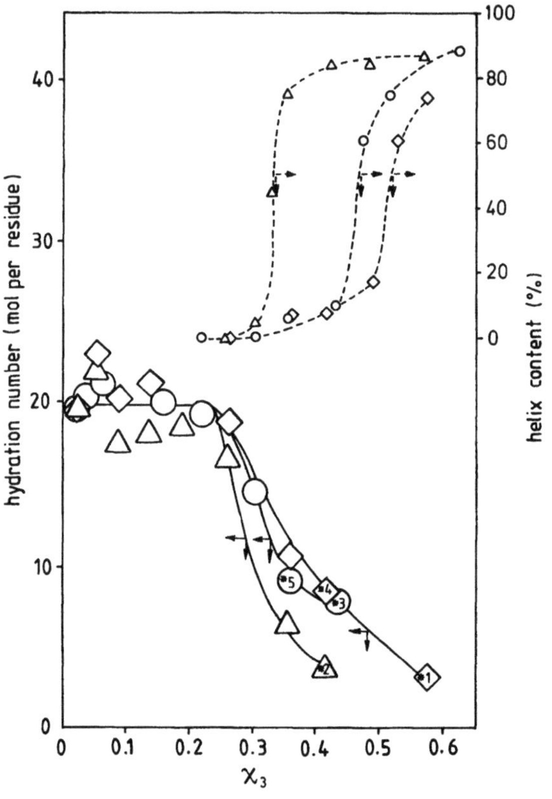

Fig. 18. Hydration number and α-helix content of $(Glu \cdot Na)_n$ as a function of solvent composition in mixtures of water with 1-PrOH (◇), 2-PrOH (△) and t-BuOH (○)[126)]

eight. From the model used it follows that in the random state 12 water molecules are bound to the ionic part and 8 to the less polar part of one residue from which 5 H_2O are released during helix formation. For $(Lys \cdot HBr)_n$ it was found that α-helix formation is accompanied by the loss of 20 water molecules: 15 of them are released from the ionic group by counter-ion binding and 5 from the less polar part. Compared with $(Glu \cdot Na)_n$ the α-helix formation of $(Lys \cdot HBr)_n$ starts in a given alcohol-water-system at a higher charge density than the former. This result seems quite reasonable considering the fact that there are stronger hydrophobic interactions between the four CH_2-groups in the lysine side chain than in the case of $(Glu)_n$ with two side chain CH_2-groups.

These results again confirm that a shielding of the electrostatic repulsing charged side-groups is responsible for the α-helix-formation, as postulated by several authors, and that the dehydration is necessary for a binding strong enough for this effect. The hydration of the charged groups is of course a consequence of the electrostatic ion-dipole interactions with water or alcohol molecules and decreases the electrostatic repulsion between neighboured charged groups proportional to $1/\varepsilon$ (ε is the microscopic dielectric constant in the solvation shell). There are probably changes in the hydration shell of the polar side-group accompanying the α-helix → coil transition, but they are scarcely responsible for this phenomenen, like it was postulated [81)]. On the other hand, the desolvation of the ionic side groups may be a prerequisite for the binding

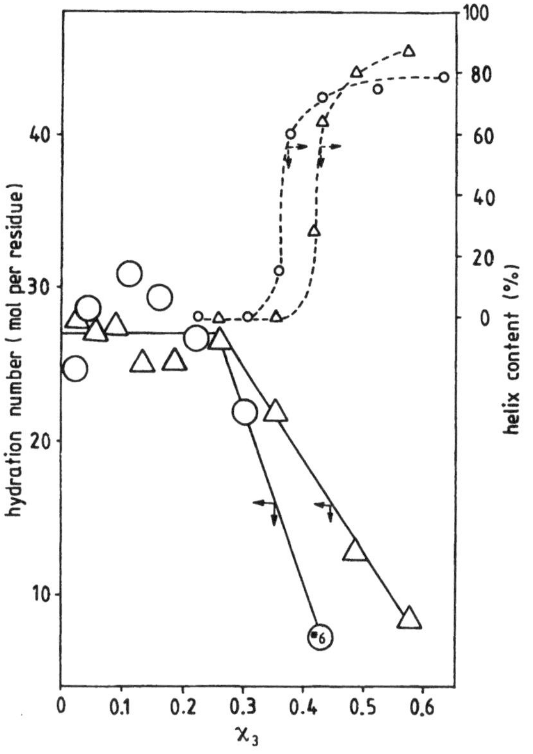

Fig. 19. Hydration number and α-helix content of (L-Lys · HBr)$_n$ as a function of solvent composition in mixtures of water with 2-PrOH (△) and t-BuOH (○) [126]

of counter ions close enough for an electrostatic shielding allowing α-helix formation.

It is known that the native conformation of proteins is affected more or less by water miscible cosolvents. The effect of alcohols in particular was studied by several authors. At a very early stage Zahn [136] pointed out that hydrophobic interactions in keratin fibers are lowered by alcohols. As a consequence, the ordered colloid structure of the fibers breaks down at lower temperatures than in pure water. Thereafter Scheraga and coworkers [137] have shown that the denaturation temperature T_m of ribonuclease decreases remarkably with alkylchain length of the alcohol. This is obviously due to the increasing hydrophobicity of the alcohols responsible for a lowering of the hydrophobic interactions stabilizing the native conformation of the protein. The decrease of T_m could be determined by means of the theory of hydrophobic interactions calculating a binding constant for the alcohol from the free enthalpy $\Delta F_{H\varphi}$. Because the denaturation in alcoholic solutions is caused by an exposition of the hydrophobic side groups — due to an almost unfolding — to the medium, the beginning of denaturation can be described in terms of changes in activity coefficients of the hydrophobic side groups, and the peptide unit. For this reason Ebina et al. [138] used ethyl esters of N-acetyl-amino acids with hydrophobic side groups (L-Pe, L-Tyr and L-Trp) as model substances for determining the activity coefficients in water and in mixtures of water with 2-halogenoethanols, 1-propanol and ethanol. If the activity coefficients obtained are plotted semilogarithmically against the alcohol concentration a similar behaviour was observed for all derivatives and alcohols studied. For determining an interaction which influences the activity coefficients, Setchenow constants and molar free energies of transfer from water to the alcoholic solutions were determined at low alcohol concentrations. A negative sign for all systems was investigated and according to their absolute values the following order was found:

$-H_2Br$ (bromoethanol) > $-H_2CH_3$ (propanol) > H_2Cl (chloroethanol) > F_3 (tri-fluoroethanol) > $-H_2F$ (fluoroethanol) > $-H_3$ (ethanol)

The authors concluded that the alcohols interact with any of the derivatives primarily via hydrophobic interactions, according to this order, and it was accepted as an order of ability to penetrate hydrophobic clusters in the protein interior. At higher alcohol concentrations however, some complications occur. The same authors studied the effect of 2-halogen-ethanols and 1-propanol as denaturants of proteins in the case of the unfolding reactions of chymotrypsinogen [139]. They used u.v. difference spectra of chymotrypsinogen, the viscosity of chymotrypsinogen solutions and the charge-transfer spectrum derived from nitro-acetophenyl-chymotrypsin for studying the distortion of the hydrophobic pocket-structure, which was found to be structurally changed. Also in this case there exists a similar order of effectiveness to that found for the model substances [139] mentioned above:

$$-H_2Br > H_2Cl > -F_3 > -H_2CH_3 > -H_2F > -H_3$$

However, as one can see, there is an inversed order of propanol and trifluoroethanol.

4.2 Solvation and Conformation of Proteins

The role of solvation in regard to conformation and properties of proteins was studied by Cordone et al. [140–144, 146] by changing the structure of the solvent due to the addition of organic cosolvents especially in the case of hemoglobin. This protein was chosen because it changes its conformation concomitantly with the uptake or release of oxygen. After binding oxygen, the so-called R conformation has a larger hydrophobic surface exposed to the solvent than the deoxygenated T conformation stabilized mainly by electrostatic interactions at the protein surface. Therefore it is possible to investigate the influence of the dielectric constant of the medium on the conformation and functional properties of the protein in different states. On the other hand, it allows to study the relation between the hydrophobicity of the solvent and the protein conformation.

The results obtained by measuring the affinity to oxygen in the presence of various monohydric alcohols (methanol, ethanol, 2-propanol, 1-propanol) [140–144] were interpreted in terms of the Monod-Wyman-Changeux model [145], by which the change of the standard free-energy difference between R and T state in the absence of oxygen, due to the addition of alcohol, can be determined, i.e.

$$\Delta\Delta G^\circ(C) = \Delta G^\circ(C) - \Delta G^\circ(O) \qquad (8)$$

C is the alcohol concentration. The $\Delta\Delta G^\circ(C)$ plots against the inverse dielectric constant of the solvent showed that at low alcohol concentration all points measured are on a single straight line for all the alcohols used. At higher alcohol concentrations, however, the $\Delta\Delta G^\circ$ values are situated on different curves [145]. One can conclude that

1) at low alcohol concentrations contribution to $\Delta\Delta G^\circ(C)$ linearly depending on the inverse dielectric constant of the solvent (bulk electrostatic contributions) are predominant:

$$\Delta\Delta G^\circ(C) = A\frac{1}{\varepsilon(C)} \tag{9}$$

(A is a parameter accounting for the effective number and positions of surface charge changed by the T → R transition.)

2) With increasing alcohol concentration non-bulk electrostatic contributions become relevant. Because these non-bulk electrostatic contributions depend on the concentration of the cosolvent as well on the size of the alkyl-group, one can conclude that there is a relation to the smaller free energy necessary for exposing hydrophobic surfaces to the medium. It yields:

$$\Delta\Delta G^\circ_{nes}(C) = \Delta\Delta G^\circ(C) - \Delta\Delta G^\circ_{es}(C) \tag{10}$$

A thermodynamic analysis regarding the relation between enthalpy and entropy values on the one hand and non-bulk electrostatic contributions on the other leads to the same result [144].

Investigations on the oxygen affinity of hemoglobin after the addition of some amides increasing the dielectric constant of the medium, have given further evidence for the importance of electrostatic and non-bulk electrostatic interactions for the T → R transition of hemoglobin [140]. As might be expected, the bulk electrostatic contributions had an inversed sign in comparison to the alcoholic solutions, whereas the non-bulk electrostatic ones kept not only the same sign but also the same dependence on cosolvent concentration and size of the apolar parts of cosolvent molecules. It was found [147] that the oxygen dissociation constants K_R and K_T for the R and T state respectively, are not changed by the alcohols.

In an analogous way the influence of alcohol on the kinetics of thermal denaturation of met-hemoglobin was studied successfully [144].

Further, for studying the role of pH and salt concentrations on bulk-electrostatic and non-bulk electrostatic contributions the same approach was made to experiments on the influence of the alcohols mentioned above on the oxygen affinity at various KCl concentrations and pH-values [144,146]. The results obtained indicate that at a low alcohol concentration the bulk-electrostatic contributions are dominant and that with increasing size of the alkyl group, alcohol and KCl concentration, the non-bulk electrostatic, hydrophobic contributions increase. Recent results of kinetic measurements of O_2 release show that cosolvents such as alcohols and formamide influence mainly the allosteric parameter L, i.e. the equilibrium between T and R conformation and that the separation of the alcohol effects into bulk-electrostatic and hydrophobic (non-bulk electrostatic) contributions is justified.

Interestingly, at 10 °C the protein-alcohol hydrophobic interactions are higher for 2-propanol than for 1-propanol in contrast to the results obtained at 21,8 [147] and 25 °C [140] and also to Schrier et al. [137]. However, according to Roux et al. [148] in aqueous solutions of 2-propanol and other branched alcohols microheterogenities occur with decreasing temperature in the concentration range used. This effect is ascribed to the formation of aggregates as a consequence of the reinforcement of the hydrophobic hydration due to the hydrogen bonding interaction of the hydroxy-group with water molecules [146, 148].

4.3 Hydration of Protein Crystals

Protein crystals contain between 25 and 65 vol% water, which is essential for the crystallisation of these biopolymers. A typical value for the water content of protein crystals is 45% according to Matthews et al. [149, 150]. For this reason it is possible to study the arrangement of water molecules in the hydration-shell by protein-water and water-water interactions near the protein surface, if one can solve the structure of the crystal by X-ray or neutron diffraction to a sufficiently high resolution [151–153].

A review on this topic was given by Finney in 1979 [154]. As was pointed out by the author, at this time it was likely that at least some deviations of the hydration shell from that in solution would occur because there would probably be some remarkable perturbations of the hydration shell due to the interactions with neighboured molecules. Furthermore, the pH values and the salt concentrations necessary for preparing protein crystals are not identical with those under the usual conditions for native proteins in solution which could give rise to deviations in the hydration shell for crystallized and dissolved proteins.

The information obtained from X-ray measurements on the arrangement of the water molecules naturally depends very much on the resolution and state of refinement of the crystal structure investigated. For detailed information on the organization of water molecules in the protein hydration shell at the surface and on the bulk water in the crystals a 1,2 to 1,8 Å resolution range is necessary [153].

Recently Blake et al. [153] made such studies in the case of human (HL) and tortoise egg-white (TEWL) lysozyme based on crystallographic refinements at 1,5 and 1,6 Å resolution, respectively. By these investigations they attempted to obtain information on the perturbations of water structure in the hydration shell by neighboured protein molecules and by high salt concentrations as well as on the degree of order of the bound water. The authors came to the conclusion that the number of ordered water molecules are 128 in TEWL and 140 in HL, whereas the overall content is made up of 650 and 350 water molecules per lysozyme molecule.

The results suggest further that the observated water structure is not affected significantly, neither by neighbour-effects nor by high salt concentrations. All the ordered water molecules in HL and TEWL are arranged in a single layer within 4.5 Å of the protein surface. They are situated at hydrogen bonding distances from polar protein groups or from other water molecules and almost all polar groups of the protein exposed are involved in hydrogen bond interaction.

4.4 Hydration and Conformation of Nucleic Acids

One of the most thoroughly investigated examples of polymeric biomolecules in regard to the stabilization of ordered structures by hydration are the DNAs. Only shortly after establishing the double-helix model by Watson and Crick 1953 it became clear, that the hydration shell of DNA plays an important role in stabilizing the native conformation. The data obtained by the authors working in this field up until 1977 are reviewed by Hopfinger [155].

In discussing more recent results it must be mentioned, that there are three kinds of water molecules around the DNA:

The first one consists of 11–12 water molecules per nucleotide unit, which are coordinated directly to sites of the DNA double helix. Two of these water molecules are bound very tightly to the ionic phosphate residue and cannot be removed without completely destroying the structure of DNA. There are four other water molecules

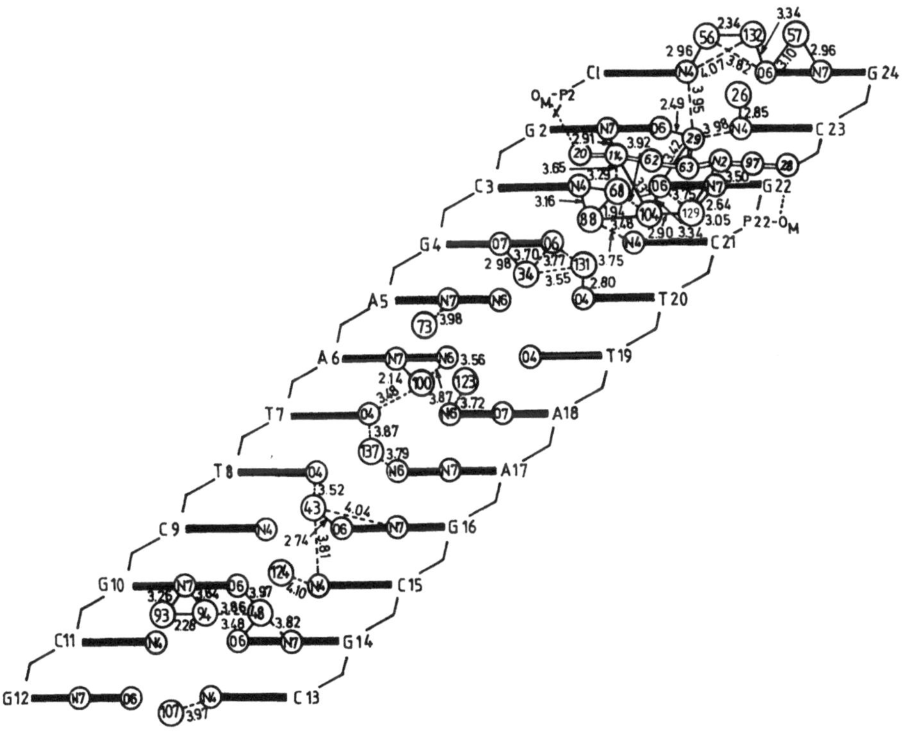

Fig. 20. Schematic representation of the unrolled major groove of the MPD 7 helix showing the first hydration shell, consisting of all solvent molecules that are directly associated with base edge N and O atoms. Base atoms are labeled N4, O4, N6, O6 and N7; solvent peaks are numbered. Interatomic distances are given in Å: up to 3,5 Å represented by unbroken lines, between 3,5–4,1 Å by dotted lines. The eight circles connected by double-lines represent the image of a spermine molecule bound to phosphate groups P2 and P22. There are 20 solvent molecules in a first hydration layer associated with N- and O-atoms [158]

bound at the pentose O-atoms and the phosphodiester linkages, whereas five H_2O are found in the grooves of the double helix near the base sites.

Above these water molecules are six to seven water molecules bound to what is usually known as the primary hydration water. In 1981 Drew and Dickerson investigated the crystal structure of a B-DNA dodecamer with the sequence C—G—C—G—A—A—T—T—C—G—C—G by X-ray measurements, however, the water molecules which should be located at the phosphate groups could not be detected [156]. It was assumed that this invisibility of the H_2O molecules was due to a blurring of images of these water peaks in the electron density map as an effect of the high crystallographic B values of atoms in the phosphate back bone. This indicates some overlapping of thermal vibration and statistic disorder from one molecule to the next in the crystal. These crystallographic B-values can be lowered by cooling the native DNA crystals to 16 °K. This was also found by investigating the 9-bromoderivate of the dodecamer mentioned above in 60% (v/v) 2-methyl-2,4-pentanediol at 7 °C (abbreviated MPD 7) according to Fratini et al. (1982) [157]. As was recently shown by Kopka et al. [158], the water molecules bound to the phosphate backbone are visible under these conditions. There are sixty-five solvent peaks associated with the phosphate groups. This corresponds to an average of three H_2O per phosphate residue.

Moreover there are nineteen water molecules forming the socalled primary hydration

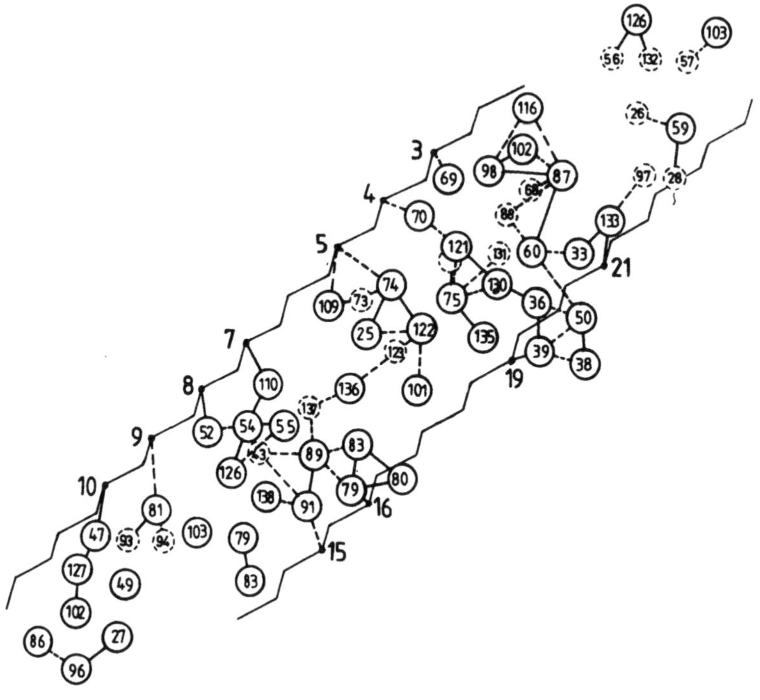

Fig. 21. Schematic representation of the second hydration shell in the major groove of MPD7 including all the water molecules interacting with those of the first shell (broken circles). Interacting phosphate groups are marked by numbered dots [158]

shell bound to base edge N and O-atoms within the major groove. Upper-hydration layers are formed by more than 36 H_2O molecules which are arranged in strings or clusters spanning the major groove from one phosphate to another. In addition a single spermine molecule spans the major groove.

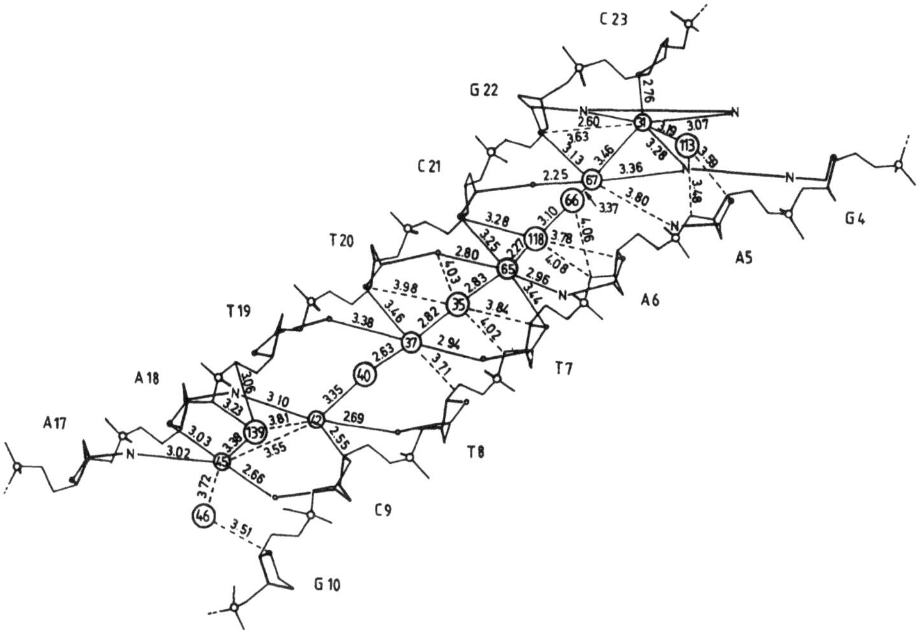

Fig. 22. Schematic representation of the unrolled minor groove of the MPD $^{-}$ helix showing the interaction with the spine of hydration. The sugar-phosphate back-bones of the two strands are depicted accurately in projection onto an 8 Å radius cylinder. Phosphorus atoms along the backbone are shown by small spheres, sugar O-1′ atoms by tiny letters O, and other atoms by intersections between bond lines. The C-3′—C-4′ bonds, closer to the viewer, are given heavy emphasis. Positions of N-3 atoms of adenine. O-2 of cytosine and thymine, and N-2 and N-3 of guanine residues are shown accurately in cylindrical projection, but base planes are represented only by thick lines connecting these atoms to the sugar C-1′. Solvent peaks are numbered circles whose diameters indicate relative distances from the viewer looking down into the groove from outside the helix. Thin unbroken lines mark hydrogen bonds of distances 3.5 Å or less, and broken lines indicate more tenuous interactions out to 4.1 Å. Note the frequent interaction between sugar O-1′ atoms and water molecules in the bottom of the groove (e.g. 45, 42, 37, 65 and 67), and the interactions between phosphate O-3′ or O-5′ atoms and water peaks in the second hydration shell (e.g. 139, 35, 118, 66 and 113) [158]

In Fig. 20 the hydration pattern of N and O-atoms is shown schematically [158]. Some of the solvent molecules underly interaction only with one base atom whereas others will bridge two atoms of the nucleotide (bidentate bridging).

Three types of bidentate bridging exist:

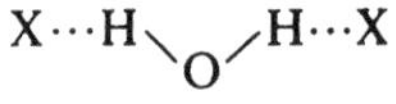
(1)

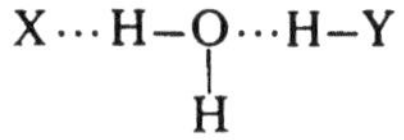
(2)

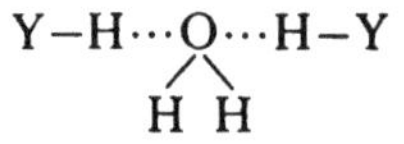
(3)

X is a proton acceptor (e.g. carbonyl-O, ring-N),
YH is a proton donor (e.g. —OH, $—NH_2$)

The three bidendate hydration mentioned above are all involved in hydration of the major groove, but only type 1 is encountered in the first shell hydration of the minor groove.

Figure 21 shows the arrangement of the water molecules in the second hydration shell in the major groove of MPD 7. It contains all the water molecules associated with those of the primary layer indicated by broken circles, and those interacting with one another in a network.

As one can see from the Figure, clusters of water molecules across the groove are formed.

By these studies half of the ten water molecules predicted by Wolf and Hanlon [159] for the primary shell could be found probably because these correspond to the number of water molecules at fixed positions under the experimental conditions used. In the second layer of water molecules only two water molecules on average were found whereas eight are predicted by Wolf and Hanlon [159]. Experimental reasons are probably responsible for these deviations.

Of course the interaction between water molecules and the hydration sites depends on their chemical nature. According to the results of IR-measurements the following order of decreasing affinity was given by Falk et al. [160] and was accepted as being quite reasonable [161]:

free phosphate O-atoms > esterified O-3′ and O-5′ phosphate O-atoms
> sugar O-1′ atoms > base edge N and C=O > base edge NH_2

On the other hand, according to Kopka et al. [158], a zig-zag spine of hydration in the minor groove of the B-DNA [156] is assumed to be mainly responsible for stabilizing its conformation and a 100% occupancy of the hydration sites in the A + T region has been found. Each first water molecule of this hydration-spine is close to two O-1′ atoms of desoxyribose-rings (Fig. 22).

From the results of their X-ray studies of the B-DNA dodecamer the following order of decreasing interaction has been obtained by the authors.

Minor groove spine (A + T rich region) > free phosphate O-atoms
> pentose O-1′ atoms > major groove base edge N and O-atoms
> esterified O-3′ and O-5′ phosphate O-atoms.

5 References

1. Luck, W. A. P.: Water in Biologic Systems, in: Topics in Current Chemistry (ed. Boschke, F. L.), p. 113, Berlin—Heidelberg—New York, Springer 1976
2. Luck, W. A. P.: Angew. Chem. *91*, 408 (1979), Int. ed. *18*, 350 (1979)
3. Stuart, H. A.: Molekülstruktur, p. 40, Berlin—Heidelberg—New York, Springer 1967
4. Rao, C. N. R.: Theory of Hydrogen Bonding in Water, in: Water — a Comprehensive Treatise, (ed. Franks, F.), Vol. 1, chapter 3, New York, Plenum Press 1972
5. Némethy, G.: The Structure of Water and the Thermodynamic Properties of Aqueous Solutions, in: Annali dell'Instituto Superiore die Sanita (ed. Marini-Bettôlo, Vd.), VI Roma, Instituto Superiore dell Sanita, 1970
6. Luck, W.: Über die Assoziation des flüssigen Wassers, in: Fortschr. chem. Forschg. *4*, 653 (1964)
7. Luck, W. A. P.: Angew. Chem. *92*, 29 (1980), Int. ed. *19*, 28 (1980)
8. Morgan, J., Warren, J.: J. Chem. Physics *6*, 666 (1938)
9. Danford, M. D., Levy, H. A.: J. Am. Chem. Soc. *84*, 3965 (1962)
10. Narten, A. H., Levy, H. A.: Liquid Water: Scattering of X-Rays, in: Water — a Comprehensive Treatise, (ed. Franks, F.), Vol. 1, chapter 8, New York, Plenum Press 1972
11. Luck, W. A. P.: Infrared Studies of Hydrogen Bonding in Pure Liquids and Solutions in: Water — a Comprehensive Treatise, (ed. Franks, F.), Vol. 2, chapter 4, New York, Plenum Press 1973
12. Walrafen, G. E.: Raman and Infrared Spectral Investigations of Water Structure, in: Water — a Comprehensive Treatise, (ed. Franks, F.), Vol. 1, chapter 5, New York, Plenum Press 1972
13. Glasel, J. A.: Nuclear Magnetic Resonance Studies on Water and Ice, in: Water — a Comprehensive Treatise, Vol. 1, chapter 6, New York, Plenum Press 1972
14. Hasted, J. B.: Liquid Water: Dielectric Properties, in: Water — a Comprehensive Treatise, (ed. Franks, F.), Vol. 1, chapter 7, New York, Plenum Press 1972
15. Fox, J., Martin, A. E.: Proc. Roy. Soc. *A 174*, 23 (1940)
16. Frank, H. S.: Structural Models, in: Water — a Comprehensive Treatise, (ed. Franks, F.), Vol. 1, chapter 14, New York, Plenum Press 1972
17. Némethy, G., Scherga, H.: J. Chem. Phys. *36*, 3382 (1962); Hagler, A. T., Scheraga, H., Némethy, G.: Ann. New York Acad. *204*, 51 (1973)
18. Angell, C. A.: Supercooled Water, in: Water — a Comprehensive Treatise (ed. Franks, F.), Vol. 7, chapter 1, New York, Plenum Press 1982
19. Franks, F.: The Properties of Aqueous Solutions at Subzero Temperatures, in: Water — a Comprehensive Treatise, (ed. Franks, F.), Vol. 7, chapter 3, New York, Plenum Press 1982
20. Speedy, R. J.: J. Phys. Chem. *88*, 3364 (1984)
21. Davidson, D. W.: Clathrate Hydrates, in: Water — a Comprehensive Treatise (ed. Franks, F.), Vol. 2, chapter 3, New York, Plenum Press 1973
22. a) Stackelberg, M. v., Müller, H. R.: J. chem. Physics *19*, 1319 (1951)
    b) Clausen, W. F.: J. Chem. Physics *19*, 259, 662, 1425 (1951)
    c) Pauling, L., Marsh, R. E.: Proc. Nat. Acad. Sci. *38*, 112 (1952)
23. Bernal, J. D., Fowler, R. H.: J. Chem. Physics *1*, 516 (1933)
24. Franks, F.: The Solvent Properties of Water, in: Water — a Comprehensive Treatise (ed. Franks, F.), Vol. 2, chapter 1, New York, Plenum Press 1973
25. Franks, F., Reid, D. S.: Thermodynamic properties, in: Water — a Comprehensive Treatise (ed. Franks, F.), Vol. 2, chapter 5, New York, Plenum Press 1973
26. Kaminsky, M.: Disçuss. Faraday Soc. *24*, 171 (1957)
27. Friedman, H. L., Krishnan, C. V.: Thermodynamics of Ion Hydration, in: Water — a Comprehensive Treatise (ed. Franks, F.), Vol. 3, chapter 1, New York, Plenum Press 1973
28. Verell, R. E.: Infrared Spectroscopy of Aqueous Electrolyte Solutions, in: Water — a Comprehensive Treatise (ed. Franks, F.), Vol. 3, chapter 5, New York, Plenum Press 1973
29. Lilley, T. H.: Raman Spectroscopy of Aqueous Electrolyte Solutions, in: Water — a Comprehensive Treatise (ed. Franks, F.), Vol. 3, chapter 6, New York, Plenum Press 1973
30. Hertz, H. G.: Nuclear Magnetic Relaxation, in: Water — a Comprehensive Treatise, (ed. Franks, F.), Vol. 3, chapter 7, New York, Plenum Press 1973

31. Zeidler, M. D.: NMR-Studies, in: Water — a Comprehensive Treatise, (ed. Franks, F.), Vol. 2, chapter 10, New York, Plenum Press 1973
32. Pottel, R.: Dielectric Properties, in: Water — a Comprehensive Treatise (ed. Franks, F.), Vol. 3, chapter 8, New York, Plenum Press 1973
33. Hilgenfeld, R., Saenger, W.: Structural Chemistry of Natural and Synthetic Ionophores and their Complexes with Cations, in: Topics in Current Chemistry (ed. Boschke, F. L.), p. 8, Berlin—Heidelberg—New York, Springer 1982
34. Luck, W. A. P.: Progr. Colloid & Polymer Sci. *65*, 6 (1978)
35. Lang, E. W.: Dissertat. Regensburg 1980
36. Lang, E. W., Lüdemann, H. D.: Angew. Chem. *94*, 351 (1982)
37. Überreiter, K.: Colloid & Polymer Sci. *260*, 37 (1982)
38. Miyajima, K., Sawada, M., Nakagaki, M.: Bull. Chem. Soc. Jpn. *56*, 1954 (1983)
39. Miyajima, K., Sawada, M., Nakagaki, M.: ibid. *56*, 1620 (1983)
40. Miyajima, K., Sawada, M., Nakagaki, M.: ibid. *56*, 2905 (1983)
41. Frank, H. S., Evans, M. W.: J. Chem. Phys. *13*, 507 (1945)
42. Chang, D. Y. C., Mitchell, D. J., Ninhann, B. W., Pailthorpe, B. A.: Solvent Structure and Hydrophobic Solutions, in: Water — a Comprehensive Treatise (ed. Franks, F.), Vol. 6, Chapter 5, New York, Plenum Press 1979
43. Tanford, Ch.: The Hydrophobic Effect: Formation of Micelles and Biological Membranes, New York, Wiley Interscienc 1980
44. Ben-Naim, A.: Water and Aqueous Solution, Introduction to a Molecular Theory, chapter 6, New York, Plenum Press 1974
45. Ben Naim, A.: J. Phys. Chem. *82*, 792 (1978)
46. Pratt, L. R., Chandler, D.: J. Chem. Phys. *67*, 3683 (1977)
47. Greco, F. A.: J. Phys. Chem. *88*, 3132 (1984)
48. Lindman, B., Wennerström, H.: Micelles. Amphiphile Aggregation in Aqueous Solution, in: Topics in Current Chemistry (ed. Boschke, F. L.), p. 3, Berlin—Heidelberg—New York, Springer 1980
49. Eicke, F. H.: Surfactants in Nonpolar Solvents. Aggregation and Micellization, in: Topics in Current Chemistry (ed. Boschke, F. L.), p. 85, Berlin—Heidelberg—New York, Springer 1980
50. Kitahara, A., Kon-No, K.: Micelle formation in non aqueous media, in: Colloidal dispersions and micellar behaviour. ACS-Symposia Series *9*, 225 (1975)
51. Eicke, H. F., Christen, H.: Helv. Chim. Acta *61*, 2258 (1978)
52. Zulauf, M., Eicke, H. F.: J. phys. Chem. *83*, 480 (1979)
53. Senō, M., Shiraishi, Sh., Araki, K., Kise, H.: Bull. Chem. Soc. Jpn. *48*, 3678 (1975)
54. Senō, M., Araki, K., Shiraishi, S.: ibid. *49*, 899 (1976)
55. Senō, M., Sawada, K., Araki, K., Iwamoto, K., Kise, H.: J. Colloid Interf. Sci. *78*, 57 (1980)
56. Kise, K., Iwamoto, K., Senō, M.: Bull. Chem. Soc. Jpn. *55*, 3856 (1982)
57. Tsujii, K., Sunamoto, J., Fendler, J. H.: ibid. *56*, 2889 (1983)
58. Boicelli, C. A., Giomini, M., Giuliani, A. M.: Spectrochimica Acta *37A*, 559 (1981)
59. Boicelli, C. A., Conti, F., Giomini, M., Giuliani, A. M.: Spectrochimica Acta, *38A*, 299 (1982)
60. Boicelli, C. A., Conti, F., Giomini, M., Giuliani, A. M.: Chem. Phys. Letters *89*, 490 (1982)
61. Boicelli, C. A., Giomini, M., Guiliani, A. M., Trotta, E.: Internat. Symposium on Colloid & Surface Science: Modern Trends of Colloid Science in Chemistry and Biology, Interlaken, 1984
62. Tamamushi, B., Watanabe, N.: Colloid & Polymer Sci. *258*, 174 (1980)
63. Kunieda, H., Shinoda, K.: J. Coll. Interf. Sci. *70*, 577 (1979)
64. Fujii, H., Kawai, T., Nishikawa, H., Ebert, G.: Colloid & Polymer Sci. *261*, 340 (1983)
65. Senō, M., Noritomi, H., Kuroyanagi, Y., Iwamoto, K., Ebert, G.: ibid. *262*, 727 (1984)
66. Ebert, G., Kuroyanagi, Y.: Polymer *23*, 1147 (1982)
67. Waks, M., Delahodde, A., Vacher, M., Nicot, C.: Internat. Symposium on Colloid & Surface Science: Modern Trends of Colloid Science in Chemistry and Biology, Interlaken 1984
68. Martenson, R. E.: J. biol. Chem. *253*, 8887 (1978)
69. Delahodde, A., Vacher, M., Nicot, C., Waks, M.: FEBS Lett. *172*, 343 (1984)
70. Luisi, P. L., Meier, P., Wolf, R.: European Sci. Foundation Workshop, Mainz 1981
(a) Luisi, P. L., Wolf, R.: Micellar Solubilization of Enzymes in Hydrocarbon Solvents, in:

Solution Behaviour of Surfactants (eds. Mittal, K. L., Fendler, E. J.), Vol. 2, New York, Plenum Publishing Corp. 1982
(b) Barbaric, S., Luisi, P. L.: J. Am. Chem. Soc. *103*, 4329 (1981)
(c) Grandi, C., Smith, R. E., Luisi, P. L.: J. Biol. Chem. *256*, 837 (1981)
71. Meier, P., Fleschar, M., Wolf, R., Smith, R., Luisi, P. L.: ibid. Mainz 1981
72. Steinmann, P., Zampieri, G., Jäckle, H., Luisi, P. L.: Internat. Symposium on Colloid & Surface Science: Modern Trends of Colloid Science in Chemistry and Biology, Interlaken 1984
73. Bangham, A. D., Standish, M. M., Watkins, J. C.: J. Mol. Biol. *13*, 238 (1965)
74. Huang, Ch.-H.: Biochemistry *8*, 344 (1969)
75. Tyrrell, D. A., Heath, T. D., Colley, C. M., Ryman, B. E.: Biochim. Biophys. Acta *457*, 259 (1976)
76. Nakagaki, M.: Maku Gaku Nyumon, Tokyo, Kitami Shobo 1978
77. Nakagaki, M., Handa, M., Shakutsui, Sh., Nakayama, M.: Yakugaku Zasshi *102*, 17 (1982)
78. Hauser, H., Philipps, M. C., Stubbs, M.: Nature *239*, 342 (1972)
79. Okumura, O., Ohbu, K., Yokoi, K., Yamada, K., Saika, D.: J. Am. Oil Chem. Soc. *60*, 1699 (1983)
80. Eckert, Th., Fischer, W.: Colloid & Polymer Sci. *259*, 533 (1981)
81. Fischer, W., Eckert, Th.: ibid. *260*, 880 (1982)
82. Ebert, G.: Biopolymere, UTB 673, Darmstadt, Steinkopff 1980
83. Tiffany, M. L., Krimm, S.: Biopolymers *8*, 34 (1969)
84. Tiffany, M. L., Krimm, S.: Biopolymers *11*, 2309 (1972)
85. Barteri, M., Pispisa, B.: Biopolymers *12*, 2309 (1973)
86. Makino, S., Wakabayashi, K., Sugai, Sh.: Biopolymers *6*, 551 (1968)
87. Rifkind, J. M.: Biopolymers *8*, 685 (1969)
88. Peggion, E., Cosani, A., Terbojevich, M., Borin, G.: Biopolymers *11*, 633 (1972)
89. Werner, W.: Diplomarbeit, Marburg 1972
90. Ebert, Ch., Ebert, G., Werner, W.: Kolloid Z. u. Z. Polymere *251*, 504 (1973)
91. Ebert, Ch., Ebert, G.: Progr. Colloid & Polymer Sci. *57*, 100 (1975)
92. Murai, N., Miyazaki, M., Sugai, Sh.: Nippon Kagaku Kaishi *1976*, 659
93. Ebert, Ch., Ebert, G.: Colloid & Polymer Sci. *255*, 1041 (1977)
94. Miyazaki, M., Yoneyama, M. Sugai, Sh.: Polymer *19*, 995 (1978)
95. Conio, G., Patrone, E., Rialdi, G., Ciferri, A.: Macromolecules *7*, 654 (1974)
96. Miyajima, K., Inari, K., Hamaguchi, N., Yoshida, H., Nakagaki, M.: Nippon Kagaku Kaishi *1975*, 1447
97. Miyajima, K., Yoshida, H., Nakagaki, M.: ibid. *1976*, 366
98. Ebert, G., Ebert, Ch.: Colloid & Polymer Sci. *254*, 25 (1976)
99. Ebert, G., Ebert, Ch., Paudjojo, L.: Progr. Colloid & Polymer Sci. *65*, 60 (1978)
100. Paudjojo, L.: Thesis, Marburg (L) 1979
101. Ebert, G., Paudjojo, J.: in: Dynamic Aspects of Biopolyelectrolytes and Biomembranes (ed. Oosawa, F.), p. 63, Tokyo, Kōdansha 1982
102. v. Hippel, P. H., Wong, K. Y.: J. biol. Chem. *240*, 3909 (1960)
(a) v. Hippel, P. H., Schleich, Th.: The Effect of Neutral salts on the Structure and Conformational Stability of Macromolecules in Solution, in: Structure and Stability of Biological Macromolecules (ed. Timasheff, S. N., Fasman, G. D.), p. 417, New York, Dekker 1969
103. Ebert, G., Ebert, Ch., Wendorff, J.: Kolloid Z. u. Polymere *237*, 229 (1970)
104. Friehmelt, V.: Thesis, Marburg (L) 1981
105. Ebert, G., Friehmelt, V.: Bunsen-Tagung, Kaiserslautern 1984
106. Ichimura, S., Mita, K., Zama, M.: Biopolymers *17*, 2769 (1978)
107. Kim, Y. H.: Thesis, Marburg (L) 1978
108. Ebert, G., Kim, Y. H.: Progr. Colloid & Polymer Sci. *68*, 113 (1983)
109. Ebert, G., Köhler, J.: Int. J. Biol. Macromol. *6*, 45 (1984)
110. Zembala, M., Czarnecki, J.: J. Coll. Interf. Sci. *89*, 1 (1982)
111. Nakagaki, M., Ebert, G.: Colloid & Polymer Sci. *260*, 781 (1982)
112. Kuroyanagi, Y.: Thesis, Marburg (L) 1981
113. Ebert, G., Kuroyanagi, Y.: Polymer *23*, 1147 (1982)
114. Ebert, G., Kuroyanagi, Y.: Colloid & Polymer Sci. *260*, 131 (1982)
115. Ebert, G., Kuroyanagi, Y.: Polymer *23*, 1154 (1982)

116. Brack, H., Orgel, L. E.: Nature *256*, 383 (1975)
117. Brack, H., Caille, A.: Int. J. Peptide Protein Res. *11*, 128 (1981)
118. Brack, H., Spach, G.: J. Am. Chem. Soc. *103*, 6319 (1981)
119. Ebert, G., Kuroyanagi, Y.: Int. J. Biol. Macromol. *5*, 109 (1983)
120. Epand, R. F., Scheraga, H. A.: Biopolymers *6*, 1383 (1968)
121. Arfmann, H. A., Labitzke, R., Wagner, K. G.: Biopolymers *14*, 1381 (1975)
122. Zander, R.: Thesis, Marburg (L) 1978
123. Ebert, G., Zander, R.: Schriftenreihe des Deutschen Wollforschungs-Instituts, Sonderband 1979, p. 103, Aachen 1979
124. Conio, G., Patrone, E., Brighetti, S.: J. biol. Chem. *245*, 3335 (1970)
125. Komiyama, J., Mori, T., Yamamoto, K., Iijima, T.: J. C. S. Faraday I *73*, 203 (1977)
126. Mori, T., Komiyama, M., Iijima, T.: J. C. S. Faraday I *74*, 2583 (1978)
127. Mori, T., Hagiwara, K., Komiyama, Y., Iijima, T.: J. Chem. Thermodynamics *12*, 41 (1980)
128. Timasheff, S. N.: Acc. Chem. Res. *3*, 62 (1970)
129. Inoue, H., Timasheff, S. N.: J. Am. Chem. Soc. *90*, 1890 (1968)
130. Timasheff, S. N., Inoue, H.: Biochemistry *7*, 2501 (1968)
131. Kuntz, I. D. Jr.: J. Am. Chem. Soc. *93*, 514 (1971)
132. Derbyshire, W.: The Dynamics of Water in Heterogeneous Systems with Emphasis to Subzero Temperatures, in: Water — a Comprehensive Treatise (ed. Franks, F.), Vol. 7, p. 368, New York, Plenum Press 1982
133. Kuntz, I. D., Kauzmann, W.: Adv. Protein Chem. *28*, 239 (1974)
134. Almassy, R., Zill, J. S. V., Lum, L. G., Ifft, J. B.: Biopolymers *12*, 2713 (1973)
135. Rochester, C. H., Westermann, A. V.: J. C. S. Faraday I *72*, 2753 (1976)
136. Zahn, H.: Kolloid Z. *197*, 14 (1964)
137. Schrier, E. E., Ingwall, R. T., Scheraga, H. A.: J. phys. Chem. *69*, 298 (1965)
138. Ebina, S., Nagai, Y., Uedaira, H.: Fukushima J. Med. Sci. *28*, 83 (1982)
139. Ebina, S., Suzuki, M., Naitoh, I., Yamauchi, K., Nagai, Y.: Int. J. Biol. Macromol. *4*, 406 (1982)
140. Cordone, L., Cupane, A., San Biagio, P. L., Vitrano, E.: Biopolymers *20*, 39 (1981)
141. Cordone, L., Cupane, A., San Biagio, P. L., Vitrano, E.: Biopolymers *20*, 53 (1981)
142. Cordone, L., Cupane, A.: Biopolymers *20*, 2137 (1981)
143. Cupane, A., Giacomazza, D., Cordone, L.: Biopolymers *21*, 1081 (1982)
144. Cordone, L., Cupane, A., D'Alia, F., De Stefano, M. G.: Faraday Symp. Chem. Soc. *17*, 205 (1982)
145. Monod, J., Wyman, J., Changeux, J. P.: J. Mol. Biol. *12*, 88 (1965)
146. Cordone, L., Cupane, A., Fornili, S. L.: Biopolymers *22*, 1677 (1983)
147. Cordone, L., Cupane, A., San Biagio, P. L., Vitrano, E.: Biopolymers *18*, 1975 (1979)
148. Roux, G., Roberts, D., Perron, G., Desnoyers, J. E.: J. Solution Chem. *9*, 629 (1980)
149. Matthews, B. W., Weaver, L. H.: Biochem. *13*, 1719 (1974)
150. Matthews, B. W.: Ann. Rev. Phys. Chem. *27*, 493 (1976)
151. Baker, E. N.: J. Mol. Biol. *141*, 441 (1980)
152. James, M. N. G., Sielecki, A. R.: J. Mol. Biol. *163*, 299 (1983)
153. Blake, C. C. F., Pulford, W. C. A., Artymiuk, P. J.: J. Mol. Biol. *167*, 693 (1983)
154. Finney, J. L.: The Organization and Function of Water in Protein Crystals, in: Water — a Comprehensive Treatise (ed. Franks, F.), Vol. 6, chapter 2, New York, Plenum Press 1979
155. Hopfinger, A. J.: Intermolecular Interactions and Biomolecular Organization, p. 159, New York, Wiley 1977
156. Drew, H. R., Dickerson, R. E.: J. Mol. Biol. *151*, 535 (1981)
157. Fratini, A. V., Kopka, M. L., Drew, H. R., Dickerson, E. R.: J. biol. Chem. *257*, 14686 (1982)
158. Kopka, M. L., Fratini, A. V., Drew, R. W., Dickerson, R. E.: J. Mol. Biol. *163*, 129 (1983)
159. Wolf, B., Hanlon, S.: Biochemistry *14*, 1661 (1975)
160. Falk, M., Hartman, K. A., Lord, R. C.: J. Am. Chem. Soc. *85*, 387 (1963)
161. Texter, J.: Progr. Biophys. Mol. Biol. *33*, 83 (1978)

Active Transport of Ions Using Synthetic Ionophores Derived from Macrocyclic Polyethers and the Related Compounds

Mitsuo Okahara and Yohji Nakatsuji

Department of Applied Chemistry, Faculty of Engineering, Osaka University, Yamadaoka 2-1, Suita, Osaka, Japan 565

Table of Contents

1 Introduction

Certain natural antibiotics selectively complex with specific cations, which are considered to participate in the maintenance of organisms in vivo, and transfer them across lipophilic membranes against the concentration gradient. This selective transport is an important biological phenomenon. The modification of structures is expected for the understanding of the interaction between ionophores and cations, but it is very difficult because the number of such natural ionophores so far known is limited and the structures are very complicated [1,2]. Thus, many efforts have been devoted to the design of synthetic host molecules which are able to complex with specific cations. Synthetic multidentate ligands such as crown ethers, cryptands, or the corresponding open-chain analogues, which have simple structures compared with antibiotics, can selectively complex with specific cations based on ion-dipole interaction between cations and their coordination sites and are considered to be promising models for such natural ionophores [3-7]. In the case of active transport systems, in addition to the selective complexation property, it is necessary to change the properties between two interfaces by adding a certain secondary function. This review surveys the characteristics of synthetic ionophores, which are potentially useful for active transport systems.

2 Proton-driven Transport Systems

2.1 Carboxylic Ionophores

Monensin, which is one of the natural antibiotics, selectively transports Na^+ across an artificial liquid membrane (organic solvent) from the basic aqueous phase (IN) to the acidic aqueous phase (OUT), driven by the proton gradient [8]. (Fig. 1, 2)

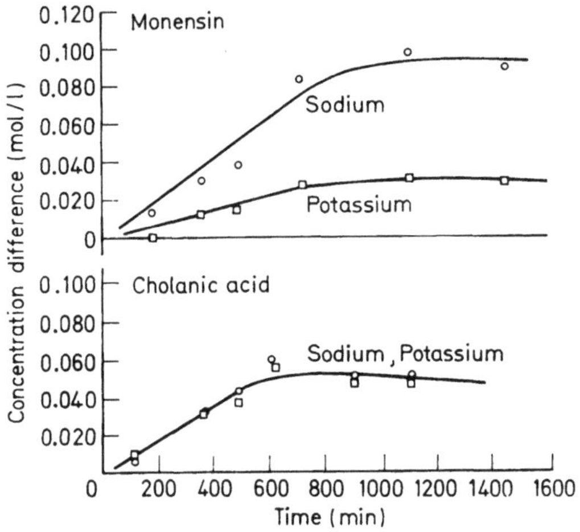

Fig. 1. The relative selectivity of two mobile carriers. In both parts of the figure, the circles and squares represent the sodium and potassium ions transported, respectively. (Cited from Ref. [8])

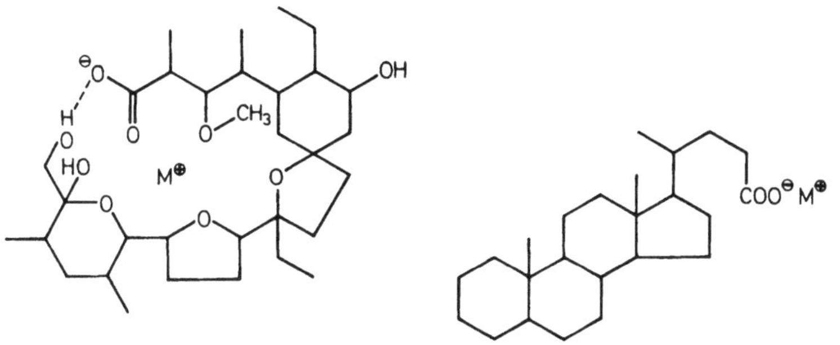

Monensin complex　　Cholanic acid complex

Fig. 2. Chemical structures of monensin (**a**) and cholanic acid (**b**). (Cited from Ref. [8])

Cholanic acid also possesses the ability of transporting cations across a lipophilic membrane but the selectivity is not observed because it contains no recognition sites for specific cations. In the basic region, monensin forms a lipophilic complex with Na^+, which is the counter ion of the carboxylate, by taking a pseudo-cyclic structure based on the effective coordination of the polyether moiety. The lipophilic complex taken up in the liquid membrane is transferred to the active region by diffusion. In the acidic region, the sodium cation is released by the neutralization reaction. The cycle is completed by the reverse transport of the free carboxylic ionophore.

In mimicking this type of function, noncyclic artificial carboxylic ionophores having two terminal groups of hydroxyl and carboxylic acid moieties were synthesized and the selective transport of alkali metal cations were examined by Yamazaki et al. [9,10]. Noncyclic polyethers take on a pseudo-cyclic structure when coordinating cations and so it is possible to achieve the desired selectivity for specific cations by adjusting the length of the polyether chain [2]. However, they were not able to observe any relationship between the selectivity and the structure of the host molecules in an active transport system using ionophores 1–3 [10]. (Table 1)

Table 1. Active and selective transport of sodium, potassium and cesium ions with synthetic ionophores[a]

Synthetic ionophore	Transported ion (%)			
	Na^+	K^+	Cs^+	Total
Control	12	11	9	32
1	26	42	21	89
2	29	42	17	88
3	25	42	22	89

[a] Temperature, 35 ± 2 °C; time, 5 days.

0.1 *N* NaCl	Synthetic ionophore	0.1 *N* NaCl
0.1 *N* KCl	10^{-4} mole/20 ml 1-hexanol	0.1 *N* KCl
0.1 *N* CsCl		0.1 *N* CsOH
0.1 *N* HCl		

(Cited from Ref. [10])

Table 2. Active transport of Na^+ and K^+ with ionophores 1, 4–6

Ionophore	Transported cation (%)		Total (%)
	Na^+	K^+	
1	13	60	73
4	0	0	0
5	0	0	0
6	28	41	69
control	0	0	0

Active transport was carried out at 35 ± 1 °C for 5 days. A 10 ml donor aqueous phase I (0.1 N NaOH, 0.1 N KCl) is separated from the 10 ml acceptor phase II (0.1 N HCl, 0.1 N NaCl, 0.1 N KCl) by dichloroethane liquid membrane (20 ml) containing the ionophore (10^{-4} mol). Each phase was stirred at 200 Hz. The amount of the transported cation was determined by flame analysis.
(Cited from Ref. [11])

Ionophore *1* selectively transports K^+ but ionophores *4* and *5* do not. This finding strongly suggests that the relative position of carboxylic acid group and other binding sites, or the ease of taking a pseudo-cyclic structure is important for attaining effective transport [11]. (Table 2) Ionophore *6*, the methyl ether of *1*, also shows active transport ability, though the selectivity is slightly decreased.

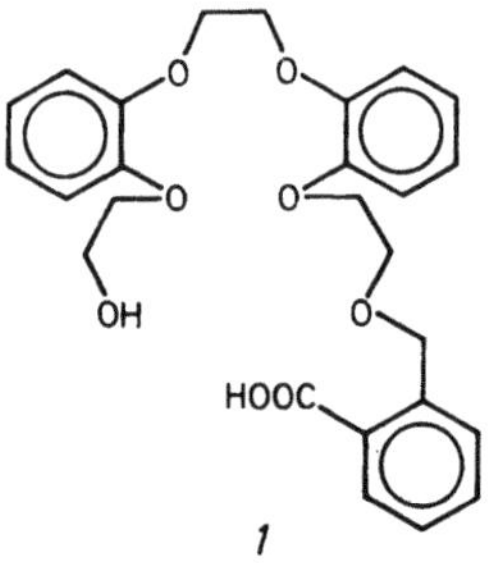

1

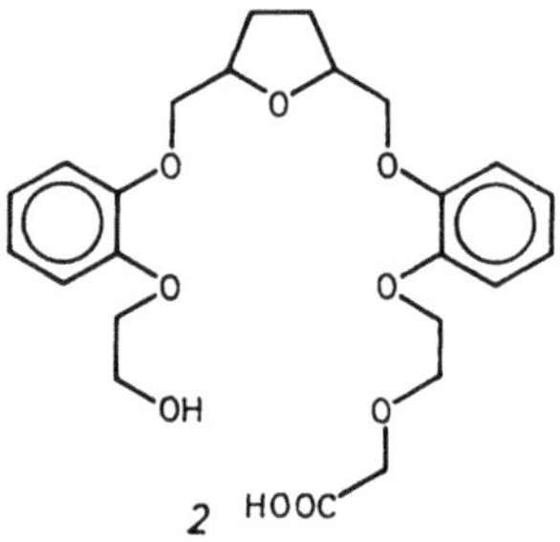

2

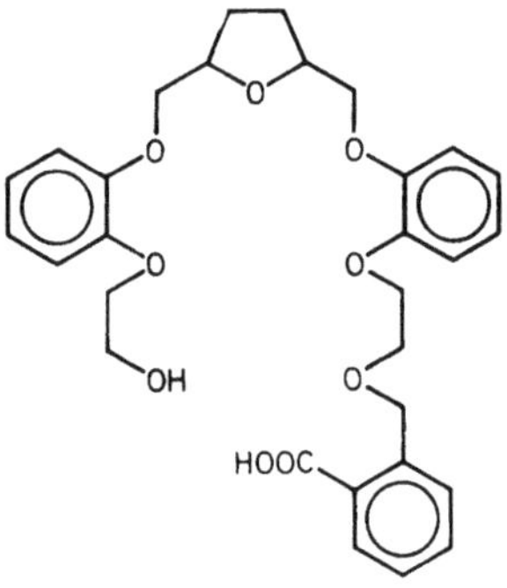

3

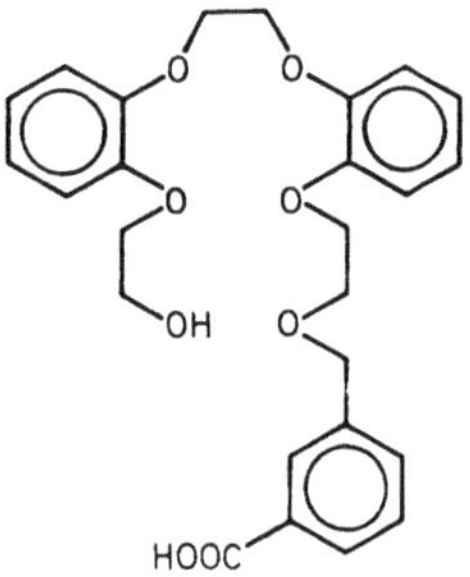

4

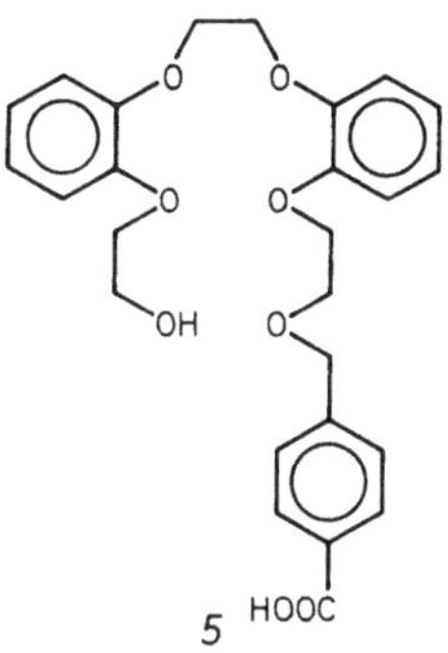

5

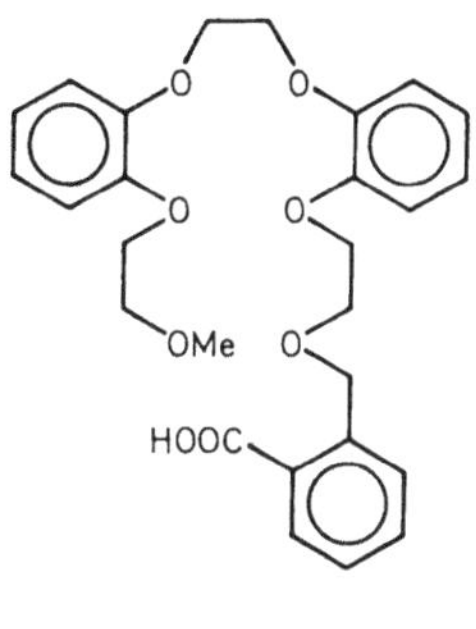

6

The modification of the terminal groups of the polyether chain enables a subtle adjustment of the complexation property toward cations. Above all, derivatives having a quinoline moiety as a terminal group show larger stability constants and higher selectivities in comparison with other analogues [12]. From this point of view, Hiratani et al. prepared new artificial ionophores having two terminal groups consisting of quinoline and carboxylic acid in mimicking the structure of *A 23187* ionophore [13] and systematically examined the selective and active transport ability toward alkali metal cations [14]. In accordance with the length of oxyethylene chain, *7*, *8* and *9* display interesting selectivities toward Li^+, Na^+, and K^+, respectively. The effect of the concentrations of the cations or anions on the transport properties is examined using ionophore *11*, which has an efficient selectivity toward K^+ [15,16].

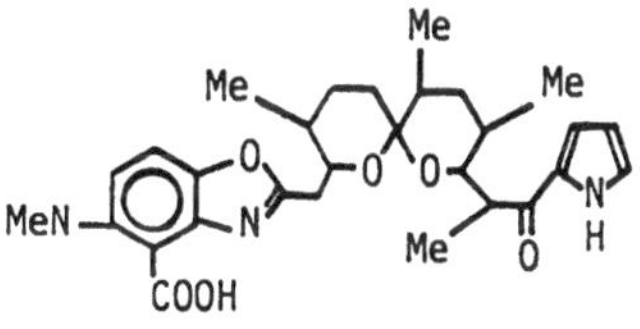

A 23187

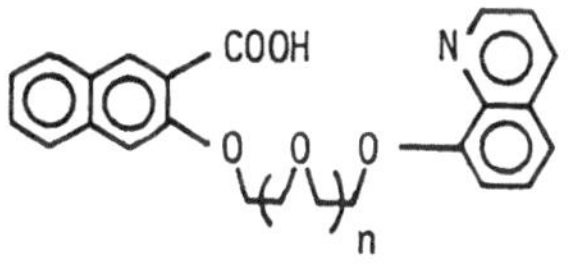

7; n=1
8; n=2
9; n=3

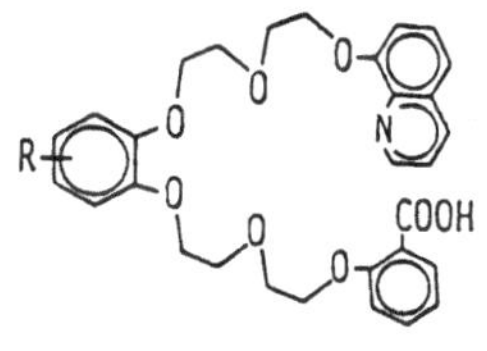

10; R = H
11; R = t-Bu

When changing the kind of acids with relatively higher lipophilicity, the efficiency of transport is lowered by the leak reaction based on their ease of transfer across the liquid membrane [16]. (Table 3) It is noteworthy that *12*, which has trimethylene chains, shows selectivity for Li^+ much better than *13*, which has ethylene chains [17]. (Table 4)

The result is explained by considering the stacking structure between the quinoline moiety and the benzene ring linked to the carboxylic acid, which gives the cavity size adequate for Li^+. (Fig. 3) Several selective host molecules for Li^+ such as [13]crown-4 [18], [14]crown-4 [19], [16]crown-4 [20], or noncyclic polyether amide derivatives [21] also possess trimethylene moiety, and this is an interesting finding from the point of view of molecular design of new host molecules for Li^+.

Table 3. Effect of acid in receiving phase on the K^+ transport

Initial conditions

Source phase	Chloroform membrane	Receiving phase
Run No. 1–9 15 ml of aq. solution containing 0.1 *M* KOH	1.5 × 10^{-4} mol of ionophore in 30 ml of chloroform	15 ml of aq. solution containing 0.1 *M* KOH + *n M* acid
Run No. 10 0.02 *M* KOH, 15 ml	0.3 × 10^{-4} mol of (II) in 30 ml of chloroform	0.02 *M* KOH + 0.028 *M* picric acid, 15 ml

Run No.	Ionophore	Acid in receiving phase	Initial rate[a] of transport (μmol/hr)	K^+ transported[a] after 2 days	
				%	mmol
1	*11*	0.2 *M* HCl	20	50	0.75
2	*11*	0.2 *M* HNO_3	19	50	0.75
3	*11*	0.1 *M* H_2SO_4	24	61	0.92
4	*11*	0.15 *M* H_2SO_4	25	62	0.93
5	*10*	0.1 *M* H_2SO_4	26	59	0.89
6	*10*	0.1 *M* (COOH)	26	60	0.90
7	*11*	0.2 *M* $(1/x)\,(HPO_3)_x$	24	61	0.92
8	*11*	1/15 *M* H_3PO_4	20	35	0.53
9	*11*	2/15 *M* H_3PO_4	26	56	0.84
10	*11*	0.028 *M* picric acid	0.75	9	0.027
				(55	0.21)[b]

[a] Reproducibility: ±10% or better.
[b] The amount of picric acid countertransported after 2 days.
(Cited from Ref. [16)])

Table 4. Amounts of cation transported by the synthetic ionophores through chloroform liquid membrane after 2 days

Ionophore	Amounts of ion transported, %[a]				Selectivity	
	Li	Na	K	Total	Li/Na	Li/K
12	41	12	8	61	3.4	5.1
13	20	16	9	45	1.3	2.2
14	10	7	6	23	1.4	1.7

[a] Reproducibility, ±10%.
(Cited from Ref. [17)])

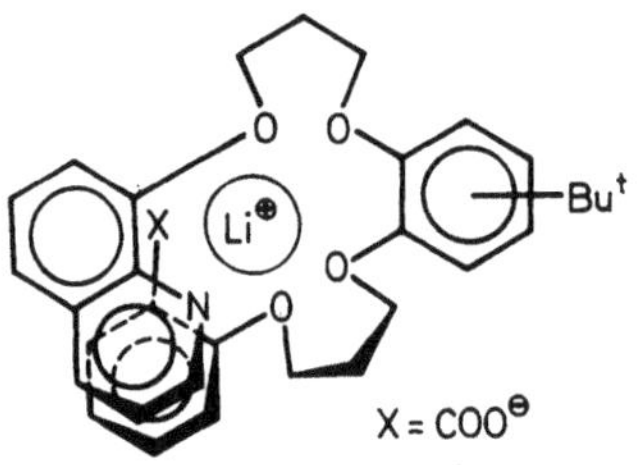

Fig. 3. Pseudocyclic structure of 12-Li^+ salt on the basis of the CPK model. (Cited from Ref. [17)])

Although the transport system is not active, rather complicated synthetic monocarboxylic ionophores such as *15* successfully transport Ca^{2+}, facilitated by protons [22]. Selective transport of Ca^{2+} is achieved using noncyclic dicarboxylic ionophore 16; Ba^{2+} is selectively transported using ionophores *17* and *18* [23]. Interestingly, ionophore *16* shows an Mg^{2+}-selectivity in the range of higher OH^- concentrations of IN aqueous phase (Fig. 4). A drastic increase in transport velocity of Mg^{2+} is observed when $Mg(OH)_2$ begins to precipitate. In addition, this peculiar phenomenon is specific for *16* and not observed for *X 537A*, *17*, or *18*.

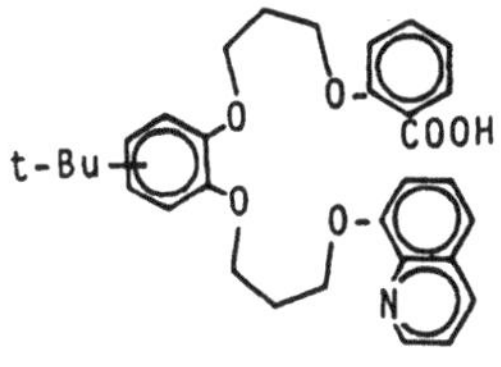

12

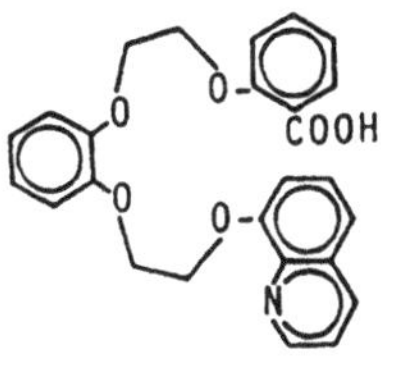

13

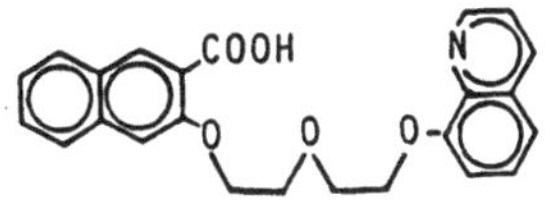

14

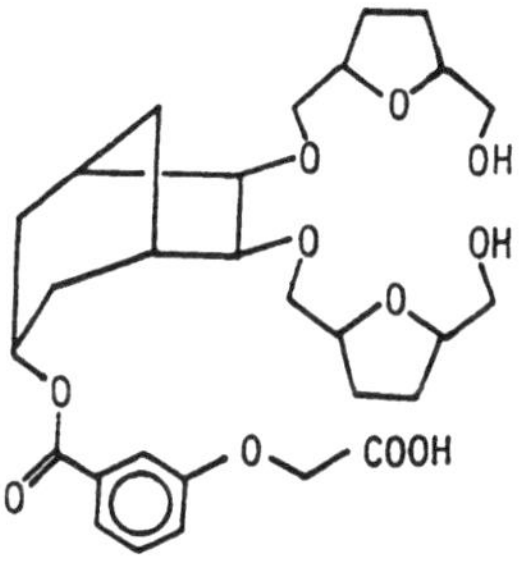

15

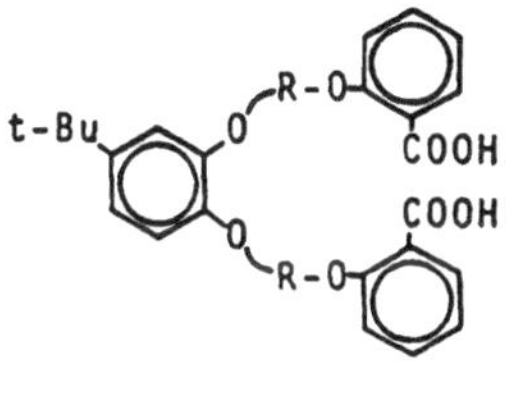

R

16; $(CH_2)_3$

17; $(CH_2)_2$

18; $(CH_2)_2O-(CH_2)_2$

Other types of carboxylic ionophores which have a crown ether moiety as a recognition site for specific cations are known, though they also mimick the function of monensin or nigericin. Generally speaking, crown ether shows a stronger complexing ability for a variety of cations and a higher selectivity compared with the corresponding open-chain analogs. As for transport systems, strong complexation between host and guest molecules is not necessarily desired for obtaining an optimum result, different from extraction experiments [20,24,25]. In passive transport systems, Izatt et al. observe a good correlation between the stability constant in methanol (log K_{MeOH}) and the transport velocity for alkali metal and alkaline earth metal cations by using a veriety of carriers. For example, the ranges of stability constants (log K_{MeOH}) for obtaining optimum transport velocity are 5.5–6.0 for K^+ or Rb^+ and 6.5–7.0 for Ba^{2+} or Sr^{2+} [26].

By considering the stability constant and the lipophilicity of host molecules, Fyles et al. synthesized a series of carboxylic ionophores having a crown ether moiety and energetically developed the active transport of alkali metal cations [27–32]. Ionophores *19–21* possess appropriate stability constants for K^+ and show effective K^+-selective transports (Fig. 5). Although all of the corresponding [15]crown-5 derivatives (*22–24*) selectively transport Na^+, their transport rates are rather slow compared with

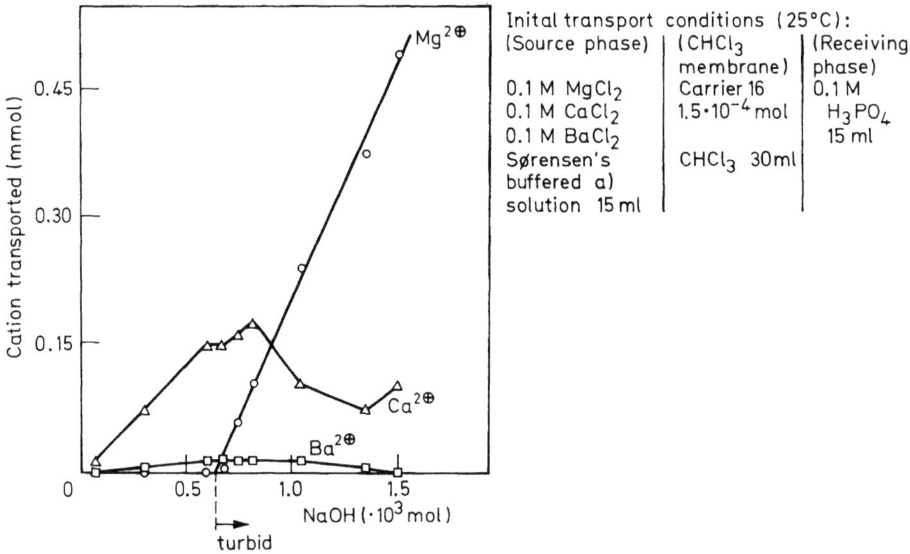

Fig. 4. Relationship between the amounts of cation transported after 1 d and the concentration of NaOH added into the source phase.[a] (Cited from Ref. [23])

[a] Sørensen's buffered solutions were adjusted by the mixture of 0.1 M NaCl, 0.1 M glycine, and 0.1 M NaOH solutions. Amount of Na^+ transported was about zero in the all range

monensin. In principle, the carrier of an ion and the carrier of a proton need not be in the same molecule. However, the result that the combination of stearic acid and dicyclohexano[18]crown-6 is less successful in this active transport system shows the disadvantage of the separated carriers. When there are lipophilic anions such as PF_6^- or ClO_4^- in the system, the transport efficiency is lowered because the leak reaction becomes larger. This is why ionophores *19–21* maintain higher stability

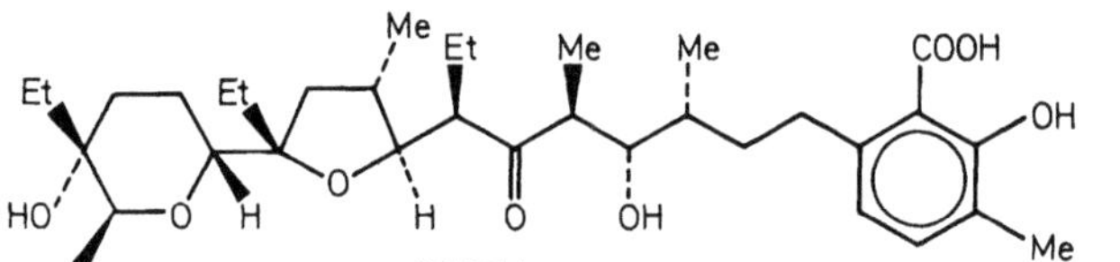

X 537 A

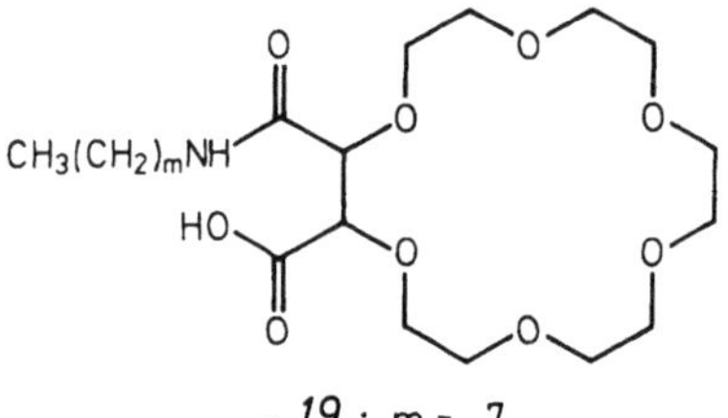

19 ; m = 7
20 ; m = 13
21 ; m = 17

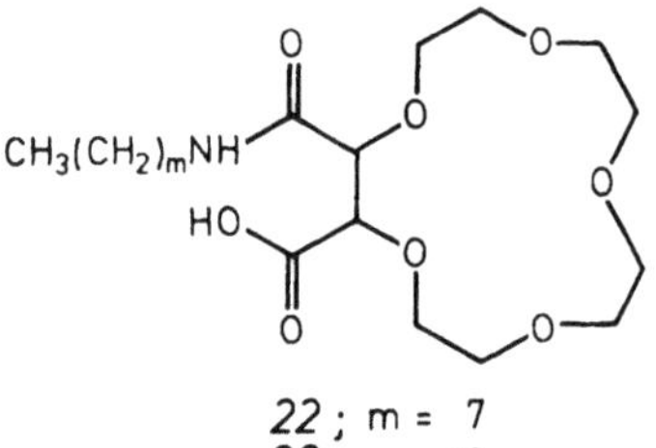

22 ; m = 7
23 ; m = 13
24 ; m = 17

constants for K^+ based on the [18]crown-6 ring even in the acidic phase (OUT phase), and thus can passively transport K^+ from the OUT phase to the IN phase as lipophilic complexes.

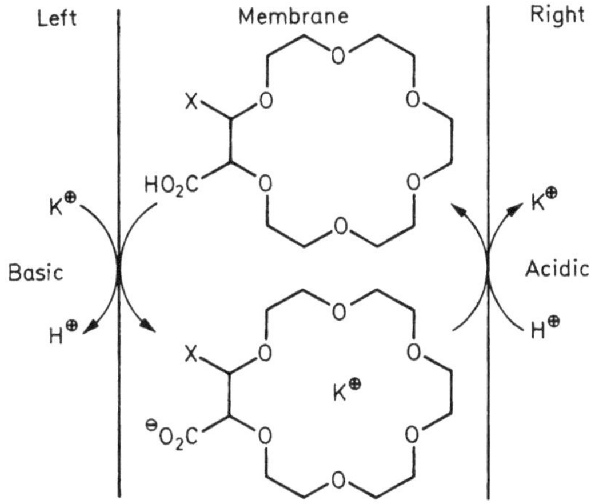

Fig. 5. A schematic mechanism for the carrier mediated countertransport of a metal ion and a proton. (Cited from Ref. [29])

The same investigators recently synthesized [18]crown-6 derivatives having two carboxyl groups (*25–30*) and investigated their transport abilities for alkaline earth cations [31,32]. Despite the complexing ability of the complexes formed, the rate of transport attained by these cations is comparable to that of cations with single charge. Some carriers show selectivity to Sr^{2+} and Ba^{2+} (Table 5).

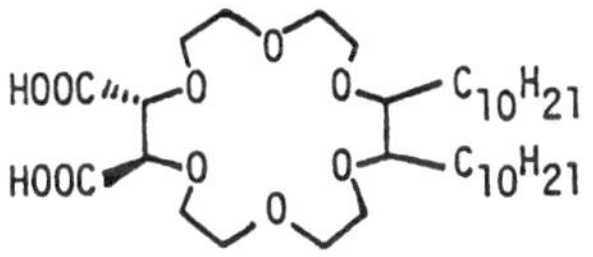

25

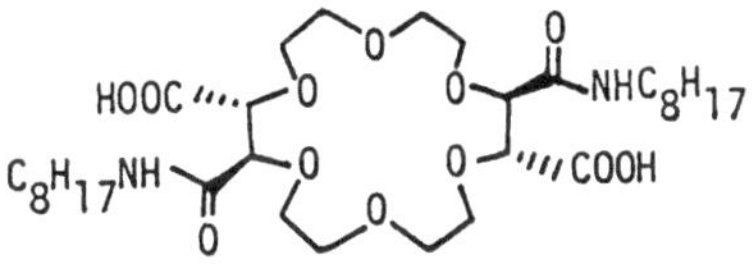

26

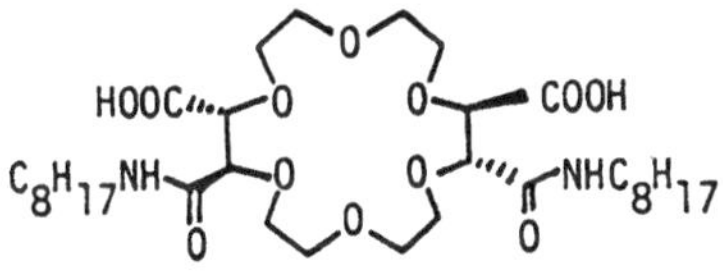

27

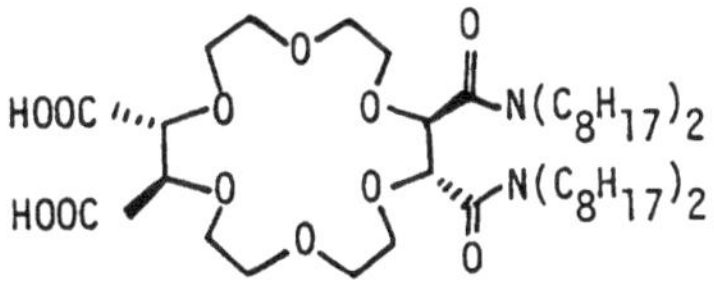

28

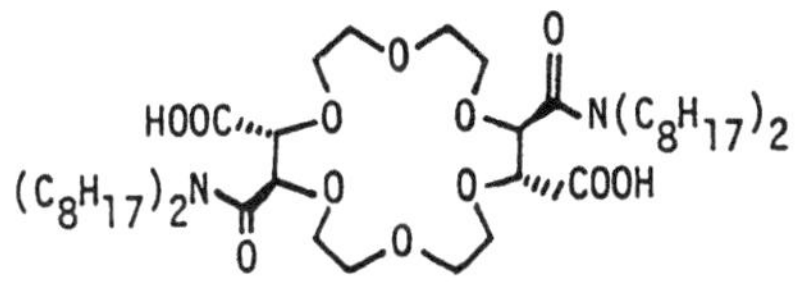

29

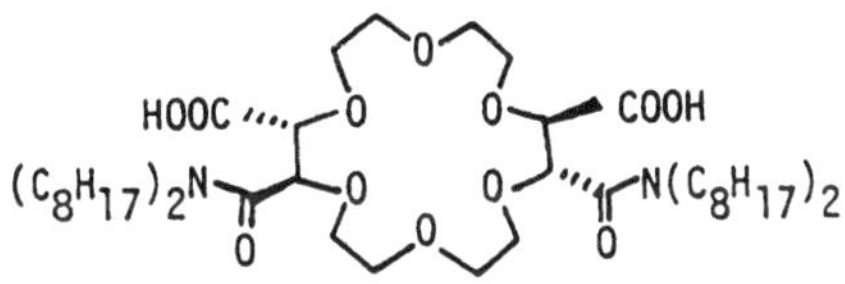

30

Table 5. Transport rates for various carriers

Carrier	Metal ion [Rate, μmol/h (selectivity relative to Ca^{2+})]			
	Ca^{2+}	Sr^{2+}	Ba^{2+}	K^{+}
25[a]	3.02 (1.0)	2.57 (0.85)	2.22 (0.74)	0.31 (0.10)
26[a]	2.68 (1.0)	3.82 (1.42)	3.70 (1.38)	<0.1 (<0.03)
27[a]	4.29 (1.0)	5.15 (1.20)	4.55 (1.06)	<0.1 (<0.02)
28[b]	0.61 (1.0)	1.07 (1.75)	0.27 (0.44)	0.21 (0.35)
29[b]	0.30 (1.0)	0.27 (0.90)	0.57 (1.90)	0.22 (0.73)
30[b]	0.70 (1.0)	0.75 (1.07)	1.04 (1.48)	0.77 (1.10)

[a] carrier 10^{-3} M, [b] carrier 5×10^{-4} M
(Cited from Ref. [32])

On the other hand, Bartsch et al. have studied cation transports using crown ether carboxylic acids, which are ascertained to be effective and selective extractants for alkali metal and alkaline earth metal cations [33–42]. In a proton-driven passive transport system (HCl) using a chloroform liquid membrane, ionophore *31* selectively transports Li^+, whereas *32–36* and *37* are effective for selective transport of Na^+ and K^+, respectively, corresponding to the compatible sizes of the ring cavity and the cation. By increasing the lipophilicity from *33* to *36*, the transport rate is gradually

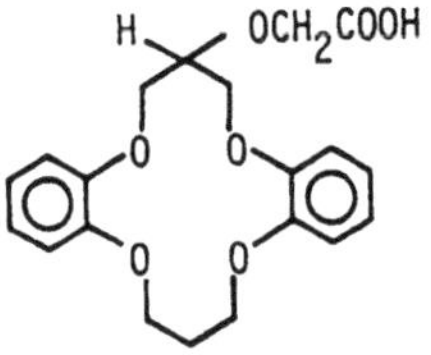

31

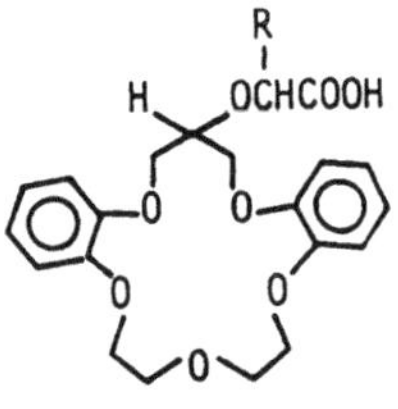

32; R= H
33; R= CH_2CH_3
34; R= $(CH_2)_3CH_3$
35; R= $(CH_2)_5CH_3$
36; R= $(CH_2)_7CH_3$

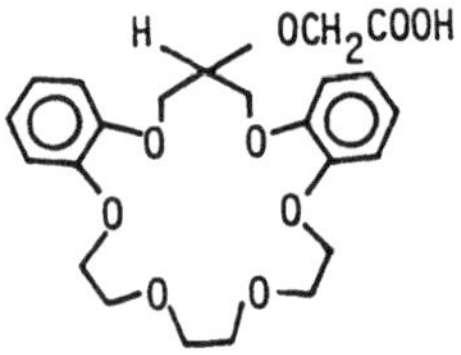

37

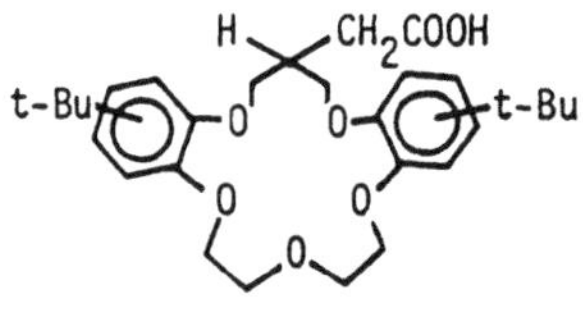

38

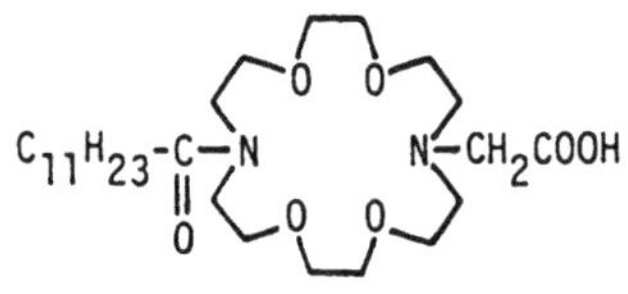

39

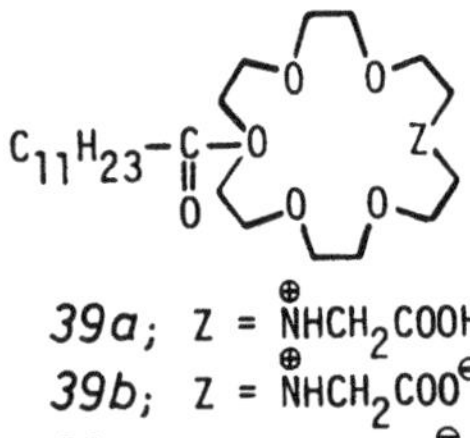

39a; Z = $\overset{\oplus}{N}HCH_2COOH$
39b; Z = $\overset{\oplus}{N}HCH_2COO^{\ominus}$
39c; Z = $NCH_2COO^{\ominus}$

increased without changing the cation selectivity. This shows that the uptake process is rate-determining in this range of lipophilicity [38]. However, a quantitative estimate of the effect of the lipophilicity on the transport property seems to be rather difficult [29].

When the chloroform membrane is replaced by toluene, ionophore *38* loses the Na^+/K^+ selectivity observed in the $CHCl_3$ system [41]. This strongly demonstrates the importance of the selection of the liquid membrane [10]. In addition, Bartsch et al. compare the results obtained in extraction experiments, liquid membrane transports, and liquid surfactant membrane transport (w/o/w emulsion) using ionophores *36* and *38* [37,42] (Table 6, 7). The liquid surfactant membrane is very thin and has by far a larger contact area with aqueous phases in comparison with the liquid membrane. It is interesting that the cation selectivity of both systems is rather similar.

Recently, Shinkai and Manabe achieved the active transport of K^+ using a new type of carrier *39* derived from diaza crown ether [43,44]. The ionophore forms the zwitter-ionic species *39b*, which is most lipophilic among other species (*39a*, *39c*), at about neutral pH region, and it acts as effective ion carrier in the active transport

Table 6. Maximum molar ratios of transported alkali metal ions to crown ether carrier for several separation techniques

Separation technique	Time	Organic phase vol: source aqueous phase vol	pH of source phase	Transported metal ions: crown ether carrier 36	38	Figure no.
Liquid-liquid extraction	few min	1.0	8.9	0.76	0.56	2a, c
			10.5	0.12[a]	0.56	2a, c
Bulk liquid membrane transport	60–80 h	0.5	10.0	2.4	9.0	3a, b
Liquid surfactant membrane transport	25 min	0.06	8.5	9.3	8.9	4a, b

[a] Third phase has separated.
(Cited from Ref. [37])

Table 7. Selectivity orders for transport of alkali metal ions into toluene by crown ether carboxylic acids for several separation techniques

Technique	Selectivity order for 36	38
Liquid-liquid extraction	$Na^+ > K^+ \approx Li^+ > Rb^+ > Cs^+$	$Na^+ > K^+ \approx Li^+ > Rb^+ > Cs^+$
Bulk liquid membrane transport	$Na^+ > K^+ > Rb^+ > Cs^+ > Li^+$	$Na^+ > K^+ > Li^+ > Rb^+ > Cs^+$
Liquid surfactant membrane transport	$Na^+ > K^+ > Rb^+ > Li^+$ [a]	$Na^+ \approx K^+ > Rb^+ > Li^+$ [a]

[a] Transport of Cs^+ was not studied.
(Cited from Ref. [37])

system in a slightly acidic solution, since the release process is accelerated by the protonation to the ring nitrogen. However, the introduction of an amide group for the purpose of increasing the lipophilicity lowers the stability constant for alkali metal cations[45] and may decrease the recognition ability for such guest ions compared with normal crown ether derivatives.

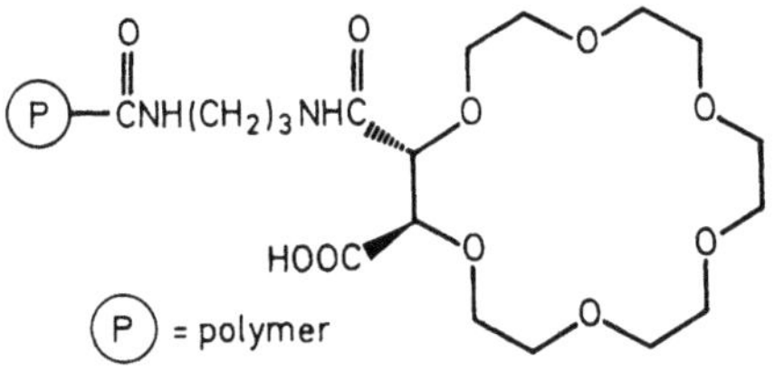

40; (P) = from acryloyl chloride

41; (P) = from styrene–acryloyl chloride (1:1)

42; (P) = from styrene–acryloyl chloride (5:1)

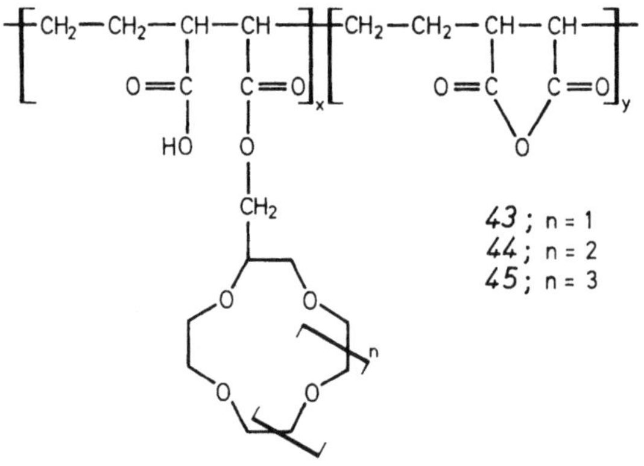

43; n = 1
44; n = 2
45; n = 3

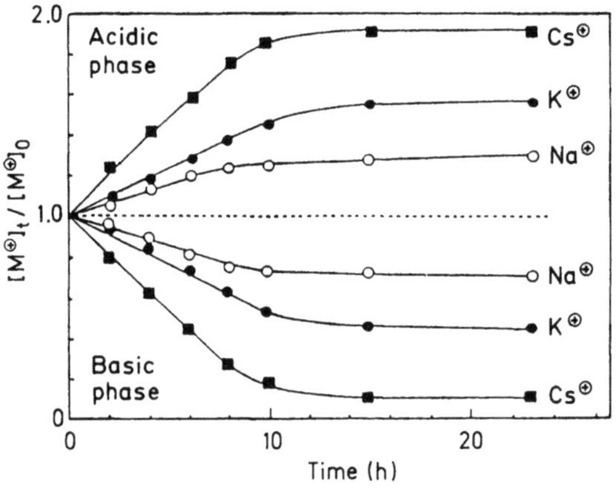

Fig. 6. Proton-driven transport of alkali metal ions through a membrane formed from 12-crown-4 polymer (43; n = 1) (crown ether content of about 30%). $[M^+]_t$ and $[M^+]_0$ refer to metal ion concentrations at time $= t$ and 0, respectively. (Cited from Ref.[47])

In contrast to the cation transport system using the mobile carriers mentioned above, a new transport system using crown polymers with carboxyl group has been developed [46]. In this system, cations are transported by hopping among binding sites immobilized on the polymer chain. The membrane, consisting of polyamide derivatives containing pendent [18]crown-6 carboxylic acid (*40–42*), works as an effective proton-driven pump for alkali metal cations and selectively transports K^+. Shono et al. also achieved the active transport of alkali metal cations using the membrane containing *43–45* and in producing the selectivity by retarding the transport velocity of the specified cation by means of complexation between the cation and the crown ring [47] (Fig. 6)

2.2 Crown Ether Derivatives Having Amino Functions

Carboxylic ionophores selectively transport cations by using intramolecular complexation in the uptake process of cations (basic region). A new ion transport system has been developed which incorporates a structural device which assists in the release process by using intramolecular complexation of an [18]crown-6 ring and a primary ammonium ion [48]. The experimental conditions are shown in Fig. 7. All these com-

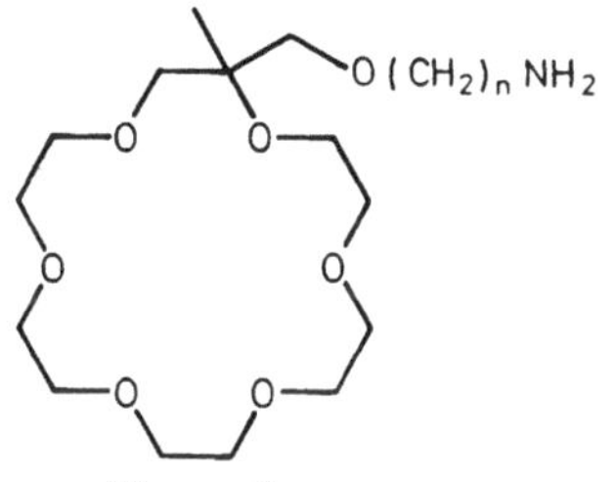

46; n = 5
47; n = 2

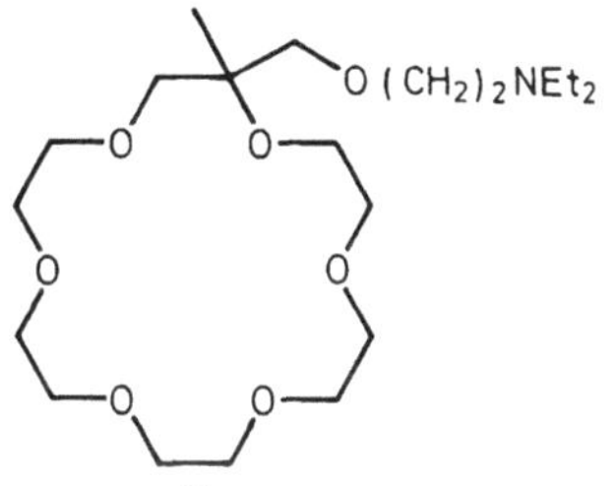

48

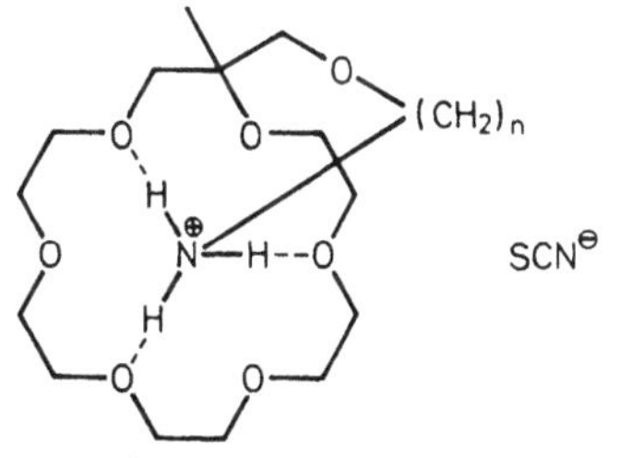

46a; n = 5
47a; n = 2

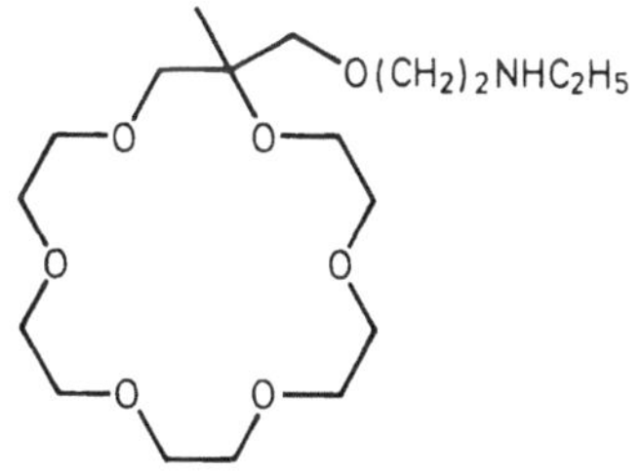

49

pounds *46–48* shows similar stability constants (in methanol, 25 °C) for K^+ in the range of about 5.4 to 5.8 and for Na^+ in the range of about 3.6 to 4.1. Ionophores *46* and *47*, which both have a primary amino group, display an effective transport of K^+ different from ionophore *48* (Fig. 8). It is well known that [18]crown-6 derivatives can effectively complex with primary ammonium ions, but hardly complex at all with tertiary ones [49–51]. Judging from the above facts, ionophores *46* and *47* may form intramolecular complexes such as *46a* and *47a* in the acidic region, and these are soluble in the liquid membrane. Consequently, the marked difference in the transport ability among these ionophores seems to be reasonably explained by considering intramolecular complexation between the [18]crown-6 ring and the primary ammonium ion. The transport of Na^+ using synthetic ionophores which possess an [18]crown-6 ring is rather diffcult, as expected, since the characteristic complexation property is based on the relative sizes of the ring and the cation. As a result, ionophores *46* and *47* display an efficient K^+/Na^+ selectivity. Ionophore *49* having a secondary amino group is found to possess transport ability similar to ionophore *48*.

Phase 1 (H_2O, 10 ml)	Liquid membrane (CH_2Cl_2, 20 ml)	Phase 2 (H_2O, 10 ml)
		Transport system I
Transport system I and II		Me_4NOH, 0.1 M
KSCN 0.1 M	Ionophore: 5×10^{-5} mol	Transport system II
NaSCN 0.1 M		KSCN 0.1 M
Me_4NOH 0.1 M		NaSCN 0.1 M
		HCl 0.1 M

Fig. 7. Conditions for competitive transport of K^+ and Na^+. (Cited from Ref. [48])

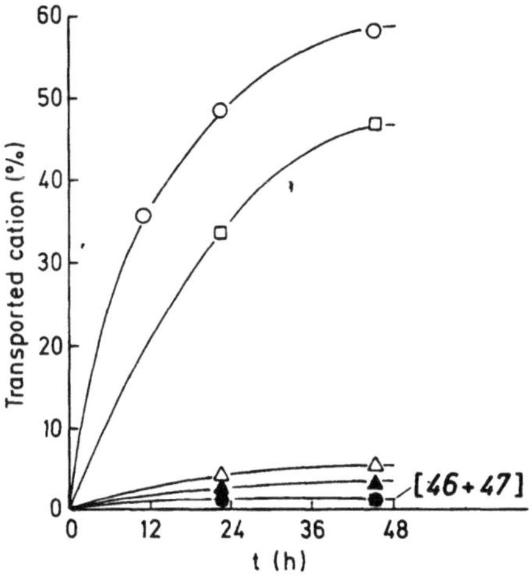

Figure 8. Competitive transport of K^+ and Na^+ in transport system II. Open symbols, K^+; filled symbols, N^+; ○, ●, 46; □, ■, 47; △, ▲, 48, (Cited from Ref. [48])

A simple active transport system using long-chain alkylsubstituted monoaza coronand derivatives has been reported [52]. Those ligands show a relatively high complexing ability under basic conditions, but under acidic conditions, in which the nitrogen atom is protonated to form the ammonium ion, its complexing ability is remarkably reduced. This active transport system should be complete if the tertiary ammonium ion accompanying the thiocyanate anion formed in the acidic phase is transferred through the membrane. The large change in the complexing ability between two interfaces and the sufficient lipophilicity of the carrier are pointed out to be important factors dominating this system. Monoaza coronands having an electron-donat-

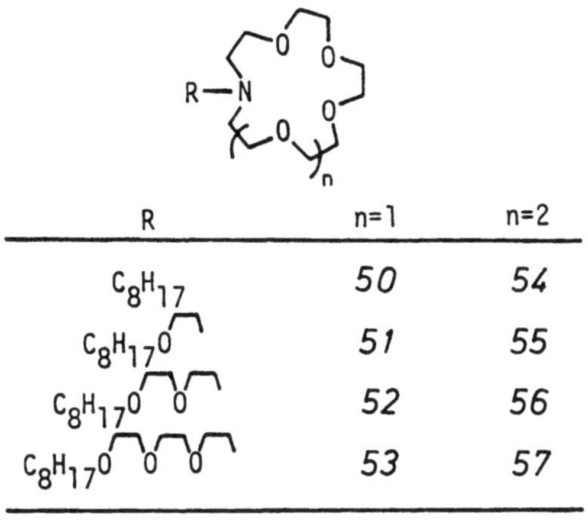

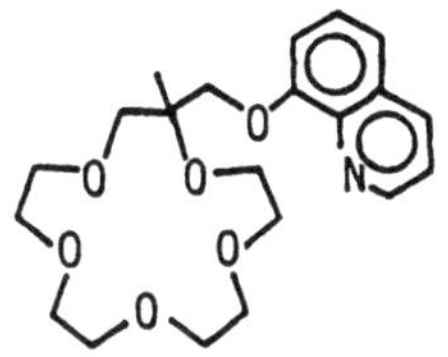

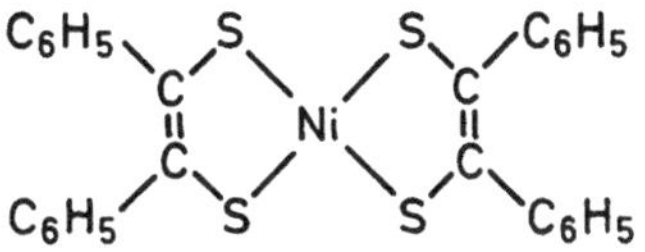

58 59

ing sidearm (nitrogen lariat ethers) (*51–53* and *55–57*) are known to possess a higher complexing ability than the corresponding alkyl monoaza coronand having no electron-donating sidearm (*50*, *54*).[53–56] It is notable that the log K_s values (in methanol, 25 °C) of these compounds for K^+ are just within the optimum range from 5.5 to 6.0[26], giving the maximum transport velocity observed in the passive

Table 8. Stability constants[a] and competitive transport ability of ionophores (50–57) for potassium and sodium cations

Ionophore	log K_s		Transported cation (%)	
	Na^+	K^+	Na^+	K^+
50	3.08	2.82	4	6
51	3.83	3.58	12	8
52	4.26	4.57	13	18
53	4.36	4.74	15	22
54	3.59	4.87	5	26
55	4.21	5.73	11	56
56	4.27	5.74	12	57
57	4.33	5.75	14	61

[a] in MeOH at 25 °C.
(Cited from Ref. [58])

transport system (Table 8). In a similar transport system, Na^+ selective transport is achieved by using methyl lariat ethers [57] having a quinoline moiety (*58*), which show excellent Na^+/K^+ selectivity (about 20 times) based on potentiometric titration [58] (Fig. 9). This transport system needs the presence of lipophilic anions in contrast to the system using carboxylic ionophores [16,29]. For instance, in the presence of an equimolar amount of picric acid, ionophore *58* is found to actively transport about 80% of K^+ from the basic region (initial conditions: 0.1 N KOH) to the acidic region (initial conditions: 0.1 N HCl, 0.1 N KCl) [58]. From this finding, we can regard the salt of the monoaza coronand and the picric acid as the carrier corresponding to carboxylic ionophores.

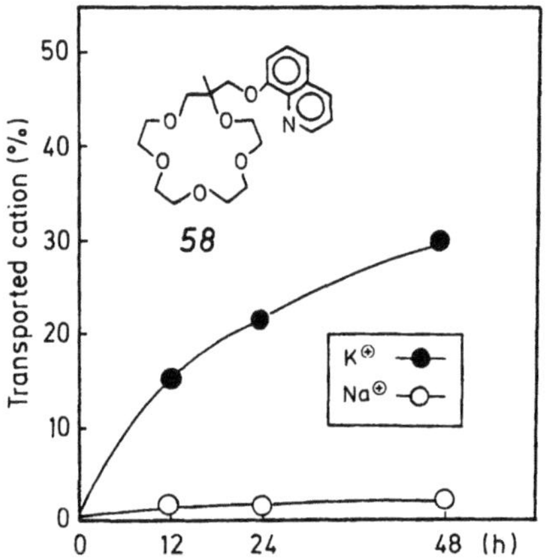

Fig. 9. Competitive transport of K^+ and Na^+ using ionophore 58 (Active transport system). (Cited from Ref. [58])

The carrier used in this transport system is only required to possess the nitrogen atom as a structural unit, which participates in the complexation. From the viewpoint of the simplicity in the structure, the development of new ionophores are expected.

3 Electron-driven Transport Systems

Grimaldi and Lehn succeeded in developing a new active transport system which performs selective transport of potassium cations coupled with the simultaneous flow of electrons in the same direction (symport) across a membrane containing an electron carrier and a cation carrier [59]. As shown in Fig. 10, one of the aqueous phase, RED, contains a reducing agent (sodium dithionite) and the other, OX phase, an oxidizing agent (potassium ferricyanide). Dicyclohexyl-18-crown-6 and the nickel bisdithiolene complex *59*, which is negative in the RED phase and neutral in the OX phase, are used as the cation and the electron carriers. The macrocyclic ligand complexes with K^+ in the RED phase and transfer it with the counter anion (reduced *59*) to the OX phase across the membrane. In the OX phase, *59* is oxidized to neutral molecule and then the potassium cation is released, and thus active transport of potassium cation occurs driven by the redox gradient and the electron flow. It goes without saying that the presence of other lipophilic anions are not desirable in the OX phase for this active transport system.

This system has characteristic in ingeneous combination of a redox pump and the selective complexation of cation by the macrocyclic ligand. It must be noted that this new system is very promising from the point of view of extending the scope of the selection of the cation carriers, since any carrier can be employed so long as it has selectivity for a special cation and has enough stability toward redox system.

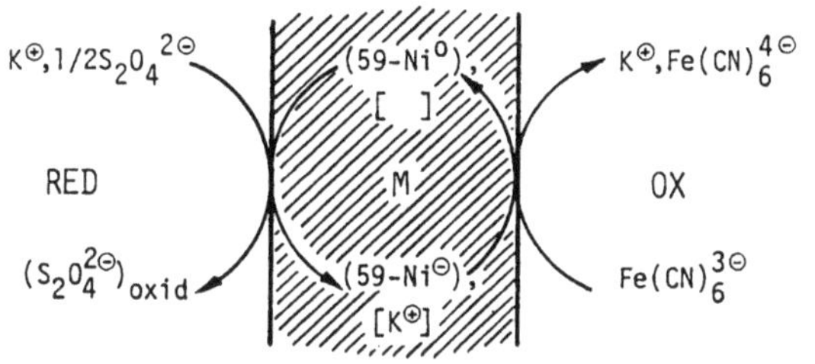

Fig. 10. Representation of the mechanism of redox driven K^+ transport using an electron and a cation carrier. (59-Ni^0) and (59-Ni^-) are the oxidized and reduced form of the electron carrier, the nickel bis-dithiolene complex 59; [] and [K^+] are dicyclohexyl-18-crown-6 and its K^+ complex. (Cited from Ref. [59])

4 Light-driven Transport Systems

Azobenzene derivatives are well known for their reversible interconversion of *E* and *Z* isomers by light irradiation. Shinkai and Manabe et al. synthesized a series of photoresponsive bis(crown ether)s [60] or cryptands [61], which have an azobenzene moiety. Detailed results on the transport system using them were reviewed recently [62]. Since these compounds can change their complexation properties on the bases of reversible isomerization, it is possible to control the uptake or the release velocity when applied to the transport system [63–65]. An acceleration of the transport rate of alkali metal [63,64] and alkaline earth metal cations [65] is observed in systems using ionophores *60*, *61* and the anion-capped ionophores *62–64*, respectively. Although the findings obtained are mainly restricted to passive transport systems, it is noteworthy that this system should be applied to the active transport system (Fig. *11*).

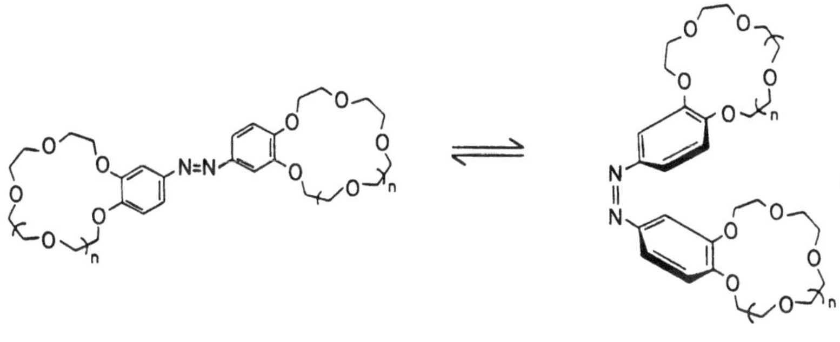

60; n = 1
61; n = 2

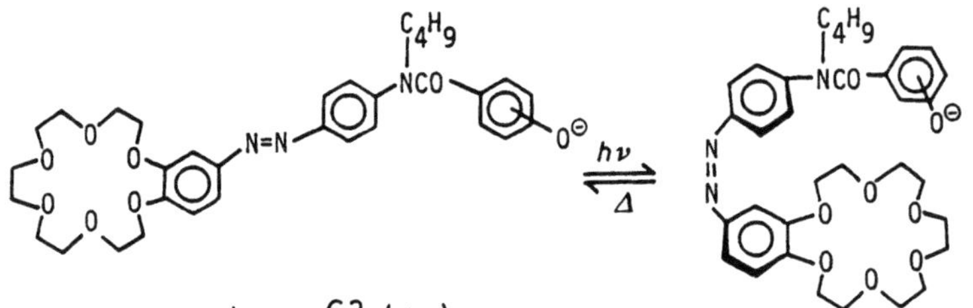

trans-62 (*p*-)
trans-63 (*m*-)

cis-62 (*p*-)
cis-63 (*m*-)

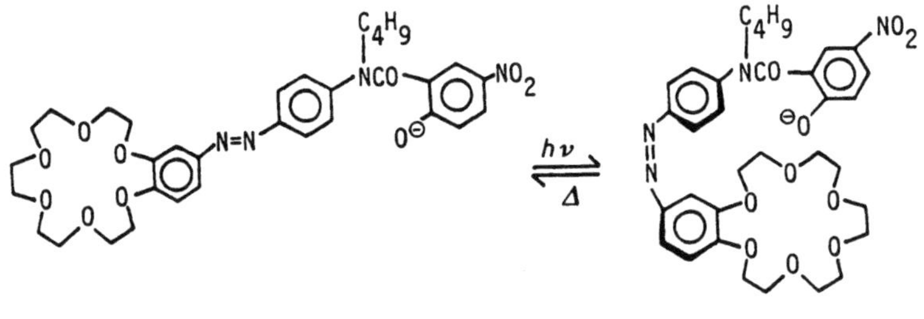

trans-64 (*o*-)

cis-64 (*o*-)

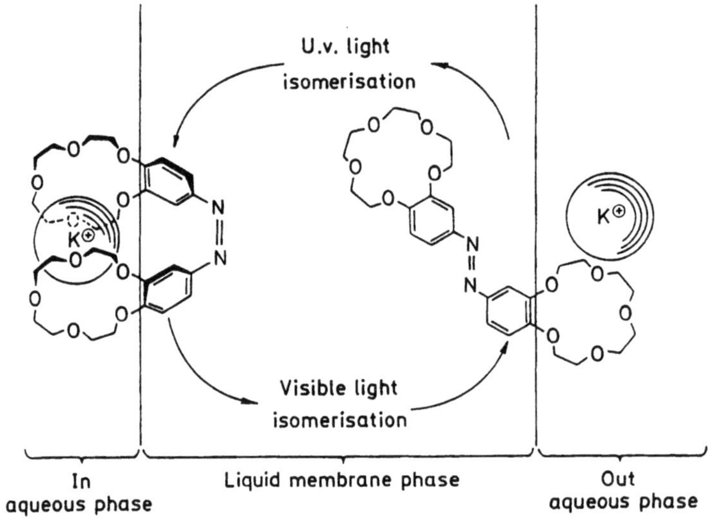

Fig. 11. Schematic representation of ion-transport mediated by alternative irradiation of u.v. and visible light. (Cited from Ref. [64])

5 Other Transport Systems

Although much attention has been devoted to the transport of inorganic cations in the previous chapters, the membrane transport of a variety of organic ions, which play an important role in biological systems, is also another interesting problem. A lot of important findings have been accumulated concerning this problem, though they have not necessarily been obtained in active transport systems. For example, in the case of amine salts, primary ammonium ions are much more effectively transported than secondary ones by using dicyclohexyl-18-crown-6 as the cation carriers[49]. Azacrown ethers are better suited for the transport of ammonium ions, on account of the N–H hydrogen bonding, than normal crown ethers[66]. Enantiomers of amine salts[67] or the anion of mandelic acid[25] can be successfully separated by using chiral crown ether derivatives. In addition, Tabushi et al. have achieved the selective transport of nucleotides by using a lipophilic diammonium cation containing a rigid structure (*65*)[68,69].

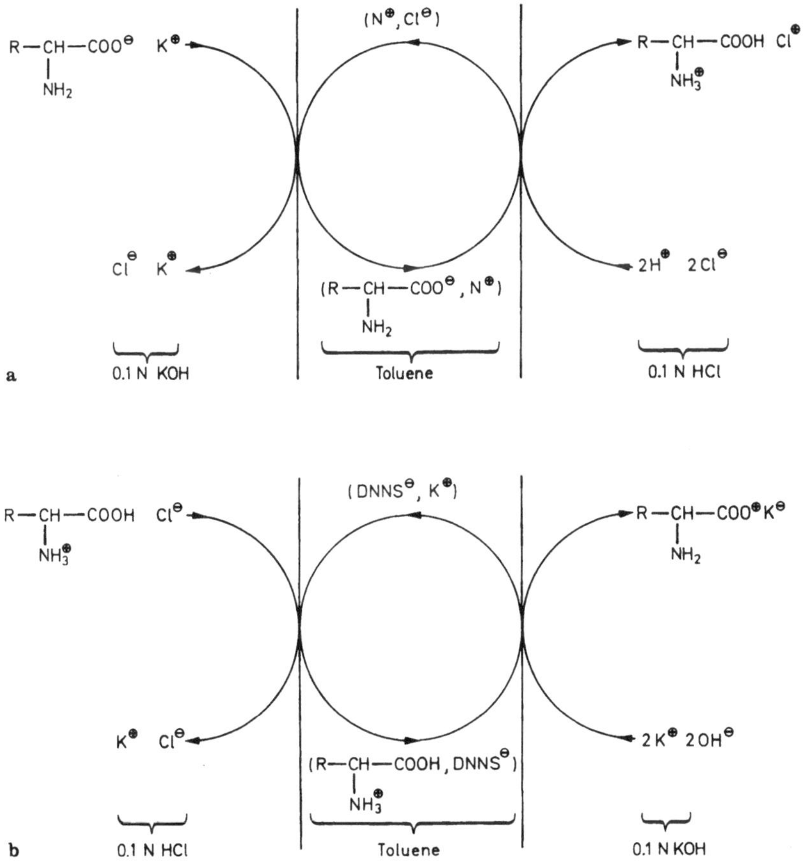

Fig. 12a and b. Transport of amino acids through a toluene barrier: **a** from basic to acid aqueous phases using a positively charged carrier (N^+, tricaprylylmethylammonium chloride, Aliquat 336); **b** from acid to basic aqueous phases using a negatively charged carrier ($DNNS^-$, dinonylnaphthalenesulfonate). (Cited from Ref. [70])

Within active transport systems, Behr and Lehn report the active transport of amino acids by using surfactants as carriers. Quaternary ammonium salt, *Aliquat 336*, is useful for the transport of amino acids from the basic phase to the acidic phase across the toluene membrane, whereas anionic surfactant, DNNS (dinonyl naphthalene sufonate) transfer it in the reverse direction [70] (Fig. 12). The amino acid behaves as a cation in the acidic region and as an anion in the basic region and this suggests that there are two ways of active transport of such compounds. Sugihara and Sinbo have achieved the active transport of picrate anion through the dichloroethane membrane against its concentration gradient [71]. This method attains the active transport of anions using the concentration gradient of potassium cation, and dicyclohexyl-18-crown-6 (*66*) has been shown to be most effective among the carriers examined. Recently, Tsukube applied this method to the transport of N-benzyloxyamino acid and oligopeptides [72,73]. Ionophore *66–69* are used as the K^+ carrier and the selectivity toward K^+ is known to be controlled by the lipophilicity of the anion and the complexation between K^+ and the selective carrier [73]. In addition,

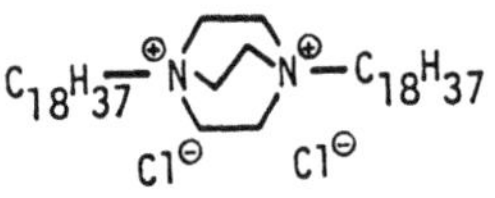

65

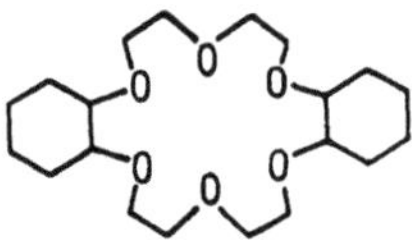

66

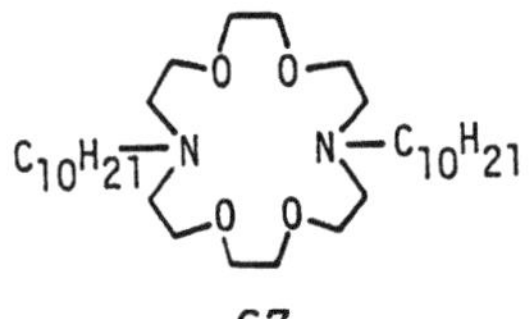

67

68

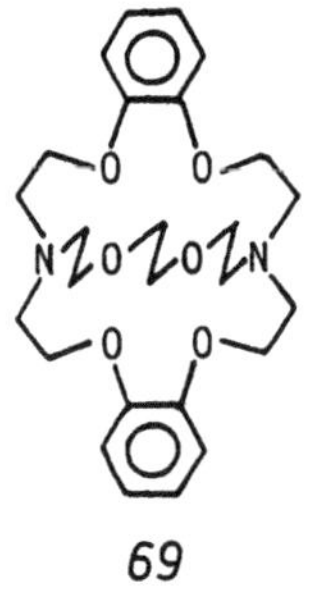

69

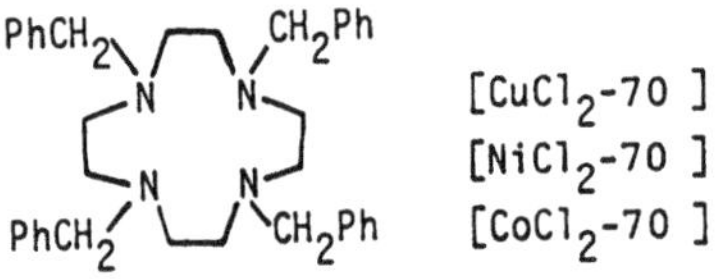

70

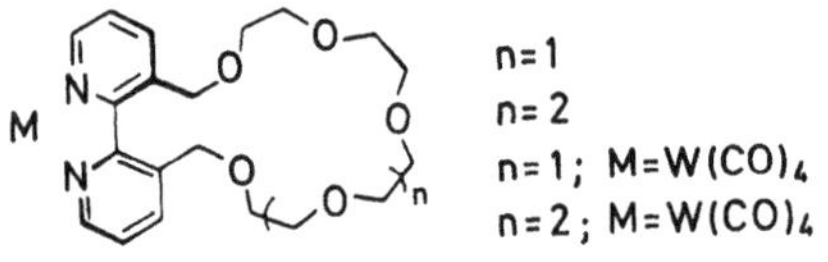

71

the anion transport system using a variety of transition metal complexes as the carrier has been developed [74,75]. The complex formed between tetracyclene and a transition metal salt *70* is especially noted because this type of complex can control the transport ability and the selectivity only by changing the central metal [75].

A certain crown ether having additional coordination sites for a trasition metal cation (*71*) changes the transport property for alkali metal cations when it complexes with the transition metal cation [76] (Fig. 13). The fact that a carrier can be developed which has a reversible complexation property for a transition metal cation strongly suggests that this type of ionophore can be applied to the active transport system.

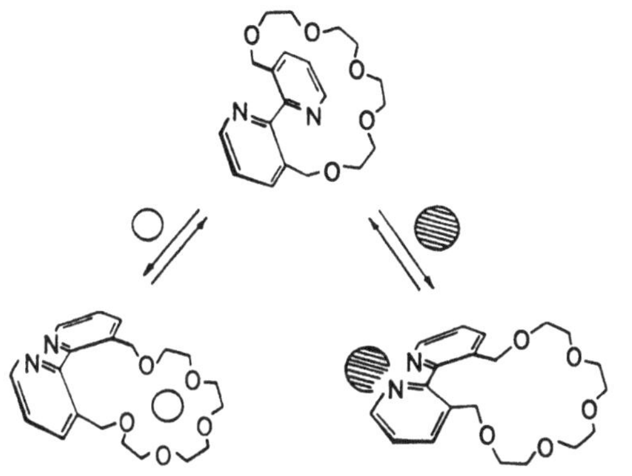

Fig. 13. Binding-induced conformational charges. (Cited from Ref. [76])

6 Conclusion

This review surveys the types of host molecules that are applicable to the active transport system. It need scarcely be said that these results, which are based on selective transport in passive transport systems (see the Chaps. 3 and 5), strongly supports this consideration. From this point of view, a systematic investigation into the passive transport system as that by Izatt et al. is noted as one of the best approaches for clarifying the question of membrane transport [77].

Anyway, it is clear that the findings obtained in these artificial transport systems do contribute to the understanding of biological phenomena and point the way to possible practical applications, such as the separation of ions. Accordingly, the development of synthetic ionophores which possess high selectivity for specific cations is expected to gain importance in the future.

7 Summary

In the biological field, much attention has been directed toward the transport phenomena through membrane. Although the function of some natural ionophores has been known, the investigation of active and selective transport of ions using the artificial ionophores in the simple model systems may be important to simulate the biological systems and clarify the transport behaviour of natural membranes.

To achieve the transport of ions against their concentration gradients, the reversible change in the nature of ionophores at the both interfaces of a membrane is necessary, and for this object, many ingenious devices in the structure of ionophores and the transport systems have recently been developed.

In this review, recent development of active transport of ions accross the liquid membranes using the synthetic ionophores such as crown ethers and other acyclic ligands, which selectively complex with cations based on the ion-dipole interaction, was surveyed.

8 References

1. Hilgenfeld, R., Saenger, W.: Structural Chemistry of Natural and Synthetic Ionophores and their Complexes with Cations, in: Host Guest Complex Chemistry II (ed. Vögtle, F.), Top. Curr. Chem. *101*, p. 1–82 (1982), Berlin Heidelberg New York, Springer-Verlag 1982
2. Painter, G. R., Pressman, B. C.: Dynamic Aspects of Ionophore Mediated Membrane Transport, ibid., p. 83–110
3. Izatt, R. M., Christensen, J. J. (eds.): Synthetic Multidentate Macrocyclic Compounds, New York, Academic Press 1978
4. Lamb, J. D., Izatt, R. M., Christensen, J. J., Eatough, D. J.: Thermodynamics and Kinetics of Cation-Macrocycle Interaction, in: Coordination Chemistry of Macrocyclic Compounds (ed. Melson, G. A.), p. 145–217, New York, Plenum Press 1979
5. Vögtle, F. (ed.): Host Guest Complex Chemistry I, Top. Curr. Chem. *98* (1981), Berlin Heidelberg New York, Springer-Verlag 1982
6. Gokel, G. W., Korzeniowski, S. H.: Macrocyclic Polyether Syntheses, Berlin Heidelberg New York, Springer-Verlag 1982
7. Reviews: Christensen, J. J., Eatough, D. J., Izatt, R. M.: Chem. Rev. *74*, 351 (1974);
   Gokel, G. W., Durst, H. D.: Synthesis *1976*, 168;
   Bradshaw, J. S., Stott, P. E.: Tetrahedron *36*, 461 (1980);
   Jolly, S. T., Bradshaw, J. S., Izatt, R. M.: J. Heterocyclic Chem. *19*, 3 (1982);
   Gokel, G. W., Dishong, D. M., Shultz, R. A., Gatto, V. I.: Synthesis *1982*, 997;
   Weber, E.: Kontakte (Merck) 1984 (1), 26, and preceding contributions in this Journal
8. Choy, E. M., Evans, D. F., Cussler, E. L.: J. Am. Chem. Soc. *96*, 7085 (1974)
9. Yamazaki, N., Nakahama, S., Hirao, A., Negi, S.: Tetrahedron Lett. *1978*, 2429
10. Yamazaki, N., Hirao, A., Nakahama, S.: J. Macromol. Sci.-Chem. *A 13*, 321 (1979)
11. Kuboniwa, H., Yamaguchi, K., Hirao, A., Nakahama, S., Yamazaki, N.: Chem. Lett. *1982*, 1937
12. Vögtle, F., Weber, E.: Angew. Chem. Int. Ed. Engl. *18*, 753 (1979)
13. Hiratani, K., Aiba, S., Nakagawa, T.: Chem. Lett. *1980*, 477
14. Hiratani, K.: ibid. *1981*, 21
15. Hiratani, K.: Bull. Chem. Soc. Jpn. *55*, 1963 (1982)
16. Hiratani, K., Nozawa, I., Nakagawa, T., Yamada, S.: J. Membr. Sci. *12*, 207 (1982)
17. Hiratani, K.: Chem. Lett. *1982*, 1021
18. Olsher, U., Grodzinski, J. J.: J. Chem. Soc. Dalton *1981*, 501
19. Olsher, U.: J. Am. Chem. Soc. *104*, 4006 (1982)
20. Kobuke, Y., Hanji, K., Horiguchi, K., Asada, M., Nakayama, Y., Furukawa, J.: ibid. *98*, 7414 (1976)
21. Shanzer, A., Samuel, D., Korenstein, R.: ibid. *105*, 3815 (1983)
22. Wierenga, W. A., Evans, B. R., Woltersom, J. A.: ibid. *101*, 1334 (1979)
23. Hiratani, K., Sugihara, H., Taguchi, K., Iio, K.: Chem. Lett. *1983*, 1657
24. Kirch, M., Lehn, J.-M.: Angew. Chem. Int. Ed. Engl. *14*, 555 (1975)
25. Lehn, J.-M.: Pure & Appl. Chem. *51*, 979 (1979)
26. Lamb, J. D., Christensen, J. J., Oscarson, J. L., Nielsen, B. L., Asay, B. W., Izatt, R. M.: J. Am. Chem. Soc. *102*, 6820 (1980)
27. Frederick, L. A., Fyles, T. M., M.-Diemer, V. A., Whitfield, D. M.: J. Chem. Soc., Chem. Commun. *1980*, 1211
28. Frederick, L. A., Fyles, T. M., Gruprasad, N. P., Whitfield, D. M.: Can. J. Chem. *59*, 1724 (1981)
29. Fyles, T. M., M.-Diemer, V. A., Whitfield, D. M.: ibid. *59*, 1734 (1981)

30. Fyles, T. M., M.-Diemer, V. A., McGavin, C. A., Whitfield, D. M.: ibid. *60*, 2259 (1982)
31. Fyles, T. M., McGavin, C. A., Whitfield, D. M.: J. Org. Chem. *49*, 753 (1984)
32. Dulyea, L. M., Fyles, T. M., Whitfield, D. M.: Can. J. Chem. *62*, 498 (1984)
33. Strzelbicki, J., Bartsch, R. A.: Anal. Chem. *53*, 1894 (1981)
34. Strzelbicki, J., Bartsch, R. A.: ibid. *53*, 2247 (1981)
35. Strzelbicki, J., Bartsch, R. A.: ibid. *53*, 2251 (1981)
36. Charewicz, W. A., Heo, G. S., Bartsch, R. A.: ibid. *54*, 2094 (1982)
37. Charewicz, W. A., Bartsch, R. A.: ibid. *54*, 2300 (1982)
38. Strzelbicki, J., Bartsch, R. A.: J. Membr. Sci. *10*, 35 (1982)
39. Bartsch, R. A., Heo, G. S., Kang, S. I., Liu, Y., Strzelbicki, J.: J. Org. Chem. *47*, 457 (1982)
40. Bartsch, R. A., Liu, Y., Kang, S. I., Son, B., Heo, G. S., Hipes, P. G., Bills, L. J.: ibid. *48*, 4864 (1983)
41. Charewicz, W. A., Bartsch, R. A.: J. Membr. Sci. *12*, 323 (1983)
42. Bartsch, R. A., Charewicz, W. A., Kang, S. I.: ibid. *17*, 97 (1984)
43. Shinkai, S., Kinda, S., Araragi, Y., Kang, S. I.: J. Chem. Soc., Chem. Commun. *1982*, 125
44. Shinkai, S., Kinda, H., Araragi, Y., Manabe, O.: Bull. Chem. Soc. Jpn. *56*, 559 (1983)
45. Nakatsuji, Y., Kobayashi, H., Okahara, M., Matsushima, K.: Chem. Lett. *1982*, 1571
46. Fyles, T. M., McGavin, C. A., Thompson, D. E.: J. Chem. Soc., Chem. Commun. *1982*, 924
47. Kimura, K., Sakamoto, H., Yoshinaga, M., Shono, T.: ibid. *1983*, 978
48. Nakatsuji, Y., Kobayashi, H., Okahara, M.: ibid. *1983*, 800
49. Bacon, E., Jung, L., Lehn, J.-M.: J. Chem. Res. (S) *1980*, 136
50. Izatt, R. M., Izatt, N. E., Rossiter, B. E., Christensen, J. J., Haymore, B. L.: Science *199*, 994 (1978)
51. Izatt, R. M., Lamb, J. D., Izatt, N. E., Rossiter, Jr., B. E., Christensen, J. J., Haymore, B. L.: J. Am. Chem. Soc. *101*, 6273 (1979)
52. Matsushima, K., Kobayashi, H., Nakatsuji, Y., Okahara, M.: Chem. Lett. *1983*, 701
53. Schultz, R. A., Dishong, D. M., Gokel, G. W.: Tetrahedron Lett. *1981*, 2623
54. Schultz, R. A., Dishong, D. M., Gokel, G. W.: J. Am. Chem. Soc. *104*, 625 (1982)
55. Masuyama, A., Nakatsuji, Y., Ikeda, I., Okahara, M.: Tetrahedron Lett. *1981*, 4665
56. Masuyama, A., Kuo, P.-L., Ikeda, I., Okahara, M.: Nippon Kagaku Kaishi *1983*, 249
57. Nakatsuji, Y., Nakamura, T., Okahara, M.: Chem. Lett. *1982*, 1207
58. Okahara, M.: unpublished results
59. Grimaldi, J. J., Lehn, J.-M.: J. Am. Chem. Soc. *101*, 1333 (1979)
60. Shinkai, S., Ogawa, T., Kusano, Y., Manabe, O.: Chem. Lett. *1980*, 283
61. Shinkai, S., Ogawa, T., Nakaji, T., Kusano, Y., Manabe, O.: Tetrahedron Lett. *1979*, 4569
62. Shinkai, S., Manabe, O.: Topics in Current Chem. *121*, 67–104 (1984)
63. Shinkai, S., Nakaji, T., Ogawa, T., Shigematsu, K., Manabe, O.: J. Am. Chem. Soc. *103*, 111 (1981)
64. Shinkai, S., Shigematsu, K., Sato, M., Manabe, O.: J. Chem. Soc., Perkin Trans. 1, *1982*, 2735
65. Shinkai, S., Minami, T., Kusano, Y., Manabe, O.: J. Am. Chem. Soc. *104*, 1967 (1982)
66. Tsukube, H.: Bull. Chem. Soc. Jpn. *55*, 3882 (1982)
67. Newkomb, M., Toner, J. L., Helgeson, R. C., Cram, D. J.: J. Am. Chem. Soc. *101*, 4941 (1979)
68. Tabushi, I., Kobuke, Y., Imuta, J.: ibid. *102*, 1744 (1980)
69. Tabushi, I., Kobuke, Y., Imuta, J.: ibid. *103*, 6152 (1981)
70. Behr, J.-P., Lehn, J.-M.: ibid. *95*, 6108 (1973)
71. Sugiura, M., Shinbo, T.: Bull. Chem. Soc. Jpn. *52*, 684 (1979)
72. Tsukube, H.: Tetrahedron Lett. *1981*, 3981
73. Tsukube, H.: J. Chem. Soc., Perkin Trans. 1, *1982*, 2359
74. Maruyama, K., Tsukube, H., Araki, T.: J. Am. Chem. Soc. *104*, 5197
75. Tsukube, H.: J. Chem. Soc., Perkin Trans. 1, *1983*, 29
76. Rebek, Jr., R., Wattley, R. V.: J. Am. Chem. Soc. *102*, 4853 (1980)
77. Izatt, R. M., Dearden, D. V., Brown, P. R., Bradshaw, J. S., Lamb, J. D., Christensen, J. J.: J. Am. Chem. Soc. *105*, 1785 (1983), and references cited therein

Quantitative Structure-Reactivity Analysis of the Inclusion Mechanism by Cyclodextrins

Yoshihisa Matsui[1], Takaaki Nishioka[2], and Toshio Fujita[2]

1 Department of Agricultural Chemistry, Shimane University, Matsue 690, Japan
2 Department of Agricultural Chemistry, Kyoto University, Kyoto 606, Japan

Table of Contents

The applications of quantitative structure-reactivity analysis to cyclodextrin complexation and cyclodextrin catalysis, mostly from our laboratories, as well as the experimental and theoretical backgrounds of these approaches, are reviewed. These approaches enable us to separate several intermolecular interactions, acting simultaneously, from one another in terms of physicochemical parameters, to evaluate the extent to which each interaction contributes, and to predict thermodynamic stabilities and/or kinetic rate constants experimentally undetermined. Conclusions obtained are mostly consistent with those deduced from experimental measurements.

1 Introduction

Quantitative structure-reactivity analysis involves the formulation of a multiple linear free-energy relationship existing between the structural parameters and the biological activities of organic compounds by means of the multivariate technique of statistics [1-5]. This method has been widely and successfully utilized for elucidating the mechanisms of interactions between biologically active substances and biopolymers *in vivo* or *in vitro*. The correlations obtained have also served well as guides for drug design. This approach, however, mostly does not give information on the structures of action sites. Both the living organism and the biopolymer remain "black boxes": Only inputs to and outputs from the "black box" are dealt with in the correlation analysis. No one can observe what is really happening *in vivo* or at the action sites. Thus, discussions relating to actual interactions *in vivo* are admittedly quite speculative.

Cyclodextrin is a cyclic oligomer composed of six (α-cyclodextrin), seven (β-cyclodextrin), eight (γ-cyclodextrin), or more α-D-glucopyranose units linked 1 → 4 as in amylose (Fig. 1). The interior cavity of the doughnut-shaped molecule provides a space within which a variety of molecules can be included [6-12]. The rate and stereochemistry of an organic reaction are significantly changed by the inclusion complexation of cyclodextrin with a substrate. These modes of action are very similar to those in enzymes or biological receptors, so that cyclodextrins have been regarded as a good model compound for such biopolymers. Since the structure of cyclodextrin is well defined, one can observe what is really happening at the action site more unequivocally than in the cases of biopolymers.

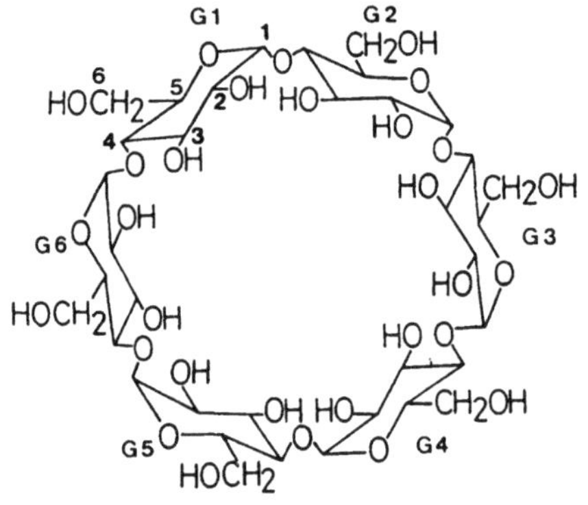

Fig. 1. Chemical structure of α-cyclodextrin. Six glucopyranose units are numbered G1 to G6. The numbers on the G1 glucopyranose refer to those of the carbon atoms

It is known that several intermolecular interactions are responsible for cyclodextrin complexation, acting simultaneously. These interactions are separable from one another by quantitative structure-reactivity analysis. Furthermore, correlations obtained by the analysis can be discussed in direct connection with actual interactions already elucidated experimentally for the action site of cyclodextrin. Thus, the results must serve to make the background of the correlation analysis more concrete.

The present review is concerned with the applications of the quantitative structure-reactivity analysis to cyclodextrin complexation and cyclodextrin catalysis, mostly from our laboratories, as well as the experimental and theoretical backgrounds of these approaches.

2 Binding Forces Contributing to the Formation of Cyclodextrin Inclusion Complexes

Several intermolecular interactions have been proposed and discussed as being responsible for the formation of cyclodextrin inclusion complexes in an aqueous solution [6-10]. They are

1) hydrophobic interaction,
2) van der Waals interaction,
3) hydrogen bonding,
4) the relief of high energy water from the cyclodextrin cavity upon substrate inclusion, and
5) the relief of conformational strain in a cyclodextrin-water adduct, together with the formation of a hydrogen bonding network around the O(2), O(3) side of the cyclodextrin macrocycle, upon substrate inclusion.

Water plays a crucial role in the inclusion process. Although cyclodextrin does form inclusion complexes in such nonaqueous solvents as dimethyl sulfoxide, the binding is very weak compared with that in water [13]. Recently, it has been shown that the thermodynamic stabilities of some inclusion complexes in aqueous solutions decrease markedly with the addition of dimethyl sulfoxide to the solutions [14,15]. Kinetic parameters determined for inclusion reactions also revealed that the rate-determining step of the reactions is the breakdown of the water structure around a substrate molecule and/or within the cyclodextrin cavity [16,17].

Hydrophobic interaction involves a process in which the structure of water around nonpolar molecules is broken upon association and the liberated water molecules are taken up by bulk water [18]. This process is accompanied by an increase in the degree of freedom of water molecules, so that hydrophobic interaction essentially involves a favorable positive entropy change ($\Delta S^\circ > 0$), together with a slightly positive enthalpy change ($\Delta H^\circ \geqq 0$). However, thermodynamic parameters determined for the formation of cyclodextrin complexes showed that the inclusion process is mostly governed by a negative enthalpy change ($\Delta H^\circ < 0$) rather than by a positive entropy change. Bender et al. [6,7] explained the thermodynamic data in terms of the relief of high energy water from the cyclodextrin cavity. Water molecules enclosed within the uncomplexed cyclodextrin cavity cannot have a full complement of hydrogen bonds owing to interference from the glucopyranose ring of cyclodextrin, so that they are enthalpy-rich. The expulsion of these enthalpy-rich molecules into bulk water upon substrate inclusion results in a negative enthalpy change, together with a negative entropy change. This explanation was supported by an X-ray crystallographic study of β-cyclodextrin dodecahydrate [19], which showed that 6.5 water molecules within the cavity are disordered over 8 sites and display extensive thermal motion.

However, the importance of this binding force is still controversial [14,20-22]. The uncomplexed β-cyclodextrin includes a larger number of high energy water molecules than the uncomplexed α-cyclodextrin [19,20]. If the major part of the binding energy is derived from the relief of high energy water, β-cyclodextrin should give more negative enthalpy of complexation than α-cyclodextrin. In contradiction to this presumption, the values of ΔH° and ΔS° for the formation of β-cyclodextrin complexes with p-nitrophenol [22], p-nitrophenyl glycosides [22], and m- and p-disubstituted benzenes [23] are considerably less negative than those of the corresponding α-cyclodextrin complexes.

Similar results have been presented for the association of cyclodextrins with alcohols (Table 1) [24]. Interestingly, both ΔH° and ΔS° for β-cyclodextrin-1-alkanol systems are positive. This fact indicates that the complexation is governed by entropy rather than by enthalpy. It has rarely been reported that the inclusion process of cyclodextrin is accompanied by a positive entropy change. It had been reported that an α-cyclodextrin-1-adamantanecarboxylate system falls under this category [25]. However, the reinvestigation of this system using a different methodology [26] has shown that the value of ΔS° is slightly negative [—5.4 J/(K · mol)]. Uekama et al. [27] have shown that the complexation of β-cyclodextrin with sulfamonomethoxine is apparently entropy-driven. Among the binding forces responsible for cyclodextrin complexation, only hydrophobic interaction is attended by a positive entropy change. Thus, it is obvious that hydrophobic interaction plays a primary role in the association of β-cyclodextrin with 1-alkanols. In the cases of β-cyclodextrin complexes with such bulky alcohols as 2,2-dimethyl-1-propanol and cyclohexanol, the ΔH° term is negative and the contribution of ΔS° to the binding is less significant than in the complexes with 1-alkanols. Binding force(s) other than hydrophobic interaction must be responsible for the complex formation. It is reasonable to consider that the bulky alcohols are in close 'van der Waals' contact with the β-cyclodextrin cavity. The 'van der Waals' interaction is accompanied by a negative enthalpy change as well as a negative entropy change.

Table 1. Thermodynamic parameters for the association of cyclodextrin with alcohol in an aqueous solution at 25 °C

Alcohol	α-Cyclodextrin			β-Cyclodextrin		
	ΔG°	ΔH°	ΔS°	ΔG°	ΔH°	ΔS°
	kJ/mol	kJ/mol	kJ/mol · K	kJ/mol	kJ/mol	kJ/mol · K
1-Butanol	−11.1	−12	− 2	− 7.0	2.9	33
1-Pentanol	−14.3	−16	− 5	−10.3	4.6	50
1-Hexanol	−16.8	−19	− 8	−13.3	0.4	46
2,2-Dimethyl-1-propanol	− 8.4	−12	−12	−15.5	− 8.8	21
Cyclohexanol	−10.3	−14	−14	−15.3	−10.0	17

On the other hand, the values of ΔH° and ΔS° for α-cyclodextrin-1-alkanol systems are significantly more negative than those for the corresponding β-cyclodextrin systems. 1-Alkanols must fit closely into the cavity of α-cyclodextrin, so that the complexation is governed by 'van der Waals' interaction rather than by hydrophobic interaction.

The van der Waals interaction generally consists of dipole-dipole and dipole-induced dipole interactions and London dispersion force. The importance of these forces in the cyclodextrin complexation has been emphasized through thermodynamic [14, 21–23, 28, 29] and theoretical [30–32] investigations. A remarkable feature of cyclodextrin complexation is that the van der Waals interaction plays a decisive role in regulating the geometry of cyclodextrin-guest inclusion complexes, especially when the guest molecules are large in size. The fact that bulky guest molecules are in close van der Waals contact with the cavity of cyclodextrin has been shown by means of X-ray crystallography on α-cyclodextrin complexes with p-iodoaniline [33, 34], sodium benzenesulfonate [30], p-nitrophenol [35], p-hydroxybenzoic acid [35], m-nitrophenol [31], 2-pyrrolidone [36], N,N-dimethylformamide [36], m-nitroaniline [37], benzaldehyde [38], γ-aminobutyric acid [39], and 1-phenylethanol [40], and on β-cyclodextrin complexes with p-nitroacetanilide [41], p-ethylaniline [42], and 1-adamantanecarboxylic acid [43].

Matsui [44] has computed energies (E_{vdw}) due to the van der Waals interaction between α-cyclodextrin and some guest molecules by the use of Hill's potential equation [45]:

$$E_{vdw} = a \cdot \sum_i \sum_j \varepsilon_{ij}(r_{ij}/d_{ij})^6 + b \cdot \sum_i \sum_j \varepsilon_{ij} \cdot \exp(d_{ij}/c \cdot r_{ij}) \qquad (1)$$

In this equation, the subscripts i and j denote interacting atoms belonging to α-cyclodextrin and a guest molecule respectively, d_{ij}, the interatomic distance, and r_{ij} and ε_{ij}, the sum of van der Waals radii (r) and a geometrical mean of energetic parameters (ε) respectively. The first and second terms of the right-hand side of this equation correspond to attractive and repulsive van der Waals interactions, respectively. The relative position of a guest molecule at which the E_{vdw} value becomes minimum was searched for by the steepest descent method for nonlinear optimization programming. The results of calculation showed that a geometric relationship at which the E_{vdw} value becomes minimal is in good agreement with that observed by X-ray crystallography for such guest molecules as 1-propanol [46] and p-nitrophenol [35] (Fig. 2 and 3). However, a significant large deviation was found between the observed and the calculated geometries of the inclusion complex with such polar and/or small guest molecules as methanol [47] and krypton [48].

The relief of conformational strain energy in the α-cyclodextrin-water adduct upon substrate inclusion is regarded as the main binding force of complexation by Saenger et al. [8, 9, 47, 49–53] They demonstrated by X-ray crystallography that the macrocycle of α-cyclodextrin bears an unstrained hexagonal conformation in most of α-cyclodextrin adducts except for the α-cyclodextrin-water adduct, in which the macrocycle of α-cyclodextrin is unsymmetrically distorted and energetically unstable. It is thought that this binding force can occur only with α-cyclodextrin, since uncomplexed β- and γ-cyclodextrins are not strained [19, 54–58]. However, the contribution

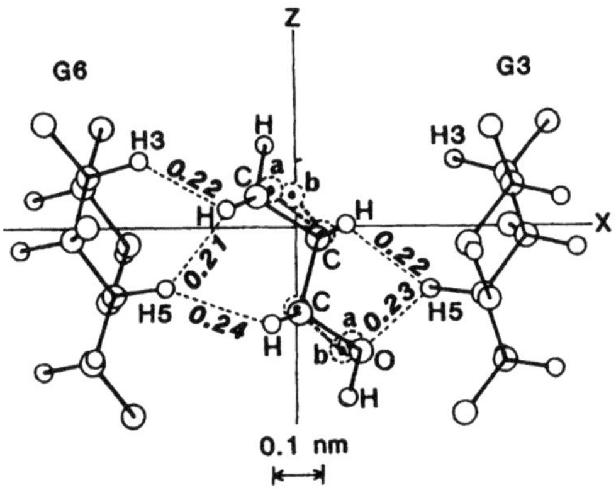

Fig. 2. Geometries calculated (solid lines) and observed (bold dashed lines) for 1-propanol in its α-cyclodextrin adduct. G3 and G6 denote the numbers of glucopyranose units of α-cyclodextrin. H3 and H5 refer to the hydrogen atoms located inside of the cyclodextrin cavity. The hydrogen atoms for the observed geometry of 1-propanol are not shown, since their atomic coordinates have not been determined. The observed 1-propanol is twofold disordered, with site a occupied 80%, site b 20%. Interatomic distances are shown in bold italics on fine dashed lines (nm). Reproduced with permission from the Chemical Society of Japan

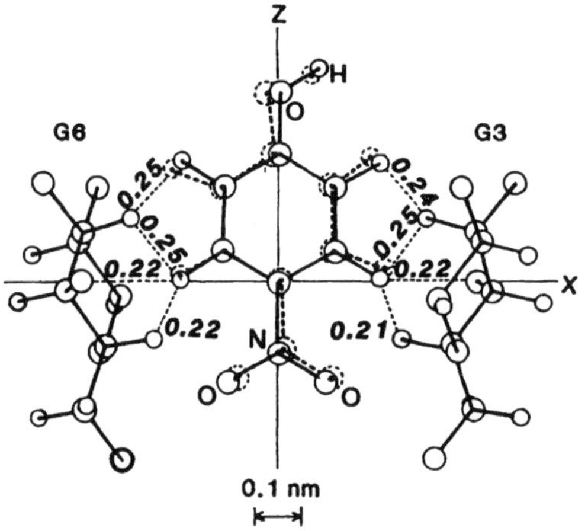

Fig. 3. Geometries calculated (solid lines) and observed (bold dashed lines) for p-nitrophenol in its α-cyclodextrin adduct. G3 and G6 denote the numbers of glucopyranose units of α-cyclodextrin. Interatomic distances are shown in bold italics on fine dashed lines (nm). Reproduced with permission from the Chemical Society of Japan

of this binding force to the formation of α-cyclodextrin inclusion complexes is still controversial. Structural [59)], thermodynamic [14, 24)], and theoretical [31, 32)] considerations have suggested that this binding force is of minor importance.

It has been thought that hydrogen bonding is also one of the binding forces in cyclodextrin complexation [60, 61)]. Several X-ray crystallographic investigations have shown that the C-6 primary hydroxyl groups on cyclodextrin are hydrogen bonded to polar guest molecules, such as potassium acetate [62)], methanol [47, 58)], ethanol [63)], 1-propanol [46)], Methyl Orange [64)], sodium 1-propanesulfonate [65)], p-nitrophenol [35)], and p-hydroxybenzoic acid [35)]. Connors et al. [66)] have suggested that the p-substituted groups capable of functioning as hydrogen-bond donors increase the stability of α-cyclodextrin complexes with the benzoate anions. However, no decisive evidence which substantiates that the hydrogen bonding also acts as a major driving force in an aqueous solution, has been presented. Griffith and Bender [6] regarded the hydrogen bonding as a minor binding force in water.

Finally, we can conclude that the various intermolecular interactions described above act simultaneously. The extent to which these interactions contribute may depend on the nature of host and guest molecules. Therefore it is reasonable to postulate that the total free energy change (ΔG_t°) upon complexation is the sum of free energy changes due to hydrophobic interaction (ΔG_{hp}°), van der Waals interaction (ΔG_{vdw}°), hydrogen bonding (ΔG_{hb}), the relief of high energy water (ΔG_{rhw}°), and the relief of conformational strain (ΔG_{rcs}°):

$$\Delta G_t^\circ = \Delta G_{hp}^\circ + \Delta G_{vdw}^\circ + \Delta G_{hb}^\circ + \Delta G_{rhw}^\circ + \Delta G_{rcs}^\circ \qquad (2)$$

In this equation, ΔG_{rcs}° is taken to be negligible for β- and γ-cyclodextrin systems and to be constant, if there is any, for the α-cyclodextrin system. The ΔG_{rhw}° term is virtually independent of the kind of guest molecules, though it is dependent on the size of the cyclodextrin cavity. The ΔG_{vdw}° term is divided into two terms, ΔG_{elec}° and ΔG_{ster}°, which correspond to polar (dipole-dipole or dipole-induced dipole) interactions and London dispersion forces, respectively. The former is mainly governed by the electronic factor, the latter by the steric factor, of a guest molecule. Thus, Eq. 2 is converted to Eq. 3 for the complexation of a particular cyclodextrin with a homogeneous series of guest molecules:

$$\Delta G_t^\circ = \Delta G_{hp}^\circ + \Delta G_{elec}^\circ + \Delta G_{ster}^\circ + \Delta G_{hb}^\circ + \text{Const.} \qquad (3)$$

The ΔG° terms on the right-hand side of Eq. 3 can be reasonably correlated to the physicochemical properties of guest molecules. Thus, the ΔG_{hp}°, ΔG_{elec}°, and ΔG_{ster}° terms are closely associated with the hydrophobicity, polarity, and bulkiness of guest molecules, respectively. The hydrophobicity of a molecule is generally represented by an indication, log P [2,67], where P is a partition coefficient between an organic solvent and water. A parameter, π [2,67], is also widely used as a measure of the hydrophobicity of a substituent group. The dipole moment and the Hammett σ constant are good parameters for representing polarities of a molecule and a substituent, respectively. A number of parameters, such as the Taft E_s constant, the Hancock's steric E_s^c constant, STERIMOL parameters, van der Waals molecular volume, molecular refraction (MR), and parachor, have been thought to be effective as indications of the steric factor of a molecule or a substituent [68]. The ΔG_{hb}° term can be rationalized with the polarity of a molecule, so that the contribution of hydrogen bonding to total binding energy is represented by such a parameter as σ [69], when the effect of variable substituents on a specified hydrogen bonding group is considered. In such cases, no independent variable is required for hydrogen bonding in regression analysis. When the site of hydrogen bonding is located at the variable substituent position, an indicator variable, I_{hb}, is available for representing hydrogen bonding ability [69]. When the substituent can act as a hydrogen bonding donor or acceptor,

$I_{hb} = 1$, whereas $I_{hb} = 0$ when it is inert to hydrogen bonding. Since $-\Delta G_t^\circ$ is proportional to $\log 1/K_d$, where K_d is the dissociation constant of a cyclodextrin complex with a guest molecule, we can derive a quantitative structure-reactivity relationship as shown, for example, in Eq. 4:

$$\log 1/K_d = a\pi + b\sigma + cE_s + dI_{hb} + e \tag{4}$$

In this equation, a, b, c, d, and e are regression coefficients. The quantitative structure-reactivity analyses of cyclodextrin inclusion processes are essentially based on this or a similar equation.

3 Quantitative Structure-Reactivity Analyses of the Inclusion Processes of Cyclodextrin Complexes

The thermodynamic stability of a cyclodextrin inclusion complex is generally expressed in terms of $\log 1/K_d$. The K_d is the dissociation constant of the complex. The $\log 1/K_d$ values determined for a variety of cyclodextrin-guest systems have frequently been correlated to the physicochemical properties of guest molecules. Thus, the $\log 1/K_d$ values were correlated to Hammett σ constants [70] for α-cyclodextrin-substituted benzoic acid complexes, to parachors [71] of guest molecules for α-cyclodextrin-substituted phenyl acetate complexes, and to $\log P_c$ [27,72] of various drug molecules (P_c: the partition coefficient of a molecule between an organic solvent and water) for their α- and β-cyclodextrin complexes. Rosanske and Connors [73] showed that the $\log 1/K_d$ values for α-cyclodextrin complexes with cinnamic acid and its related compounds decrease linearly with an increase in the dipole moment of substrates. This finding was ascribed to the hydrophobic environment of the cyclodextrin cavity. Connors et al. [66] showed that virtually linear correlations exist between $\log 1/K_d$ and the Hammett σ value for α-cyclodextrin complexes with para-substituted benzoic acids and their conjugate anions. The slopes (ϱ) of the relations were negative (−0.31) for the acids and positive (0.77) for the conjugate bases. This observation was explained in terms of the difference in geometry of inclusion complexes between the undissociated and dissociated forms of the substrates: The carboxyl group enters the cyclodextrin cavity in the acid form, whereas the substituent group enters the cavity in the base form. A linear correlation was found to exist between the thermodynamic parameters ΔH° and ΔS° and the volumes of the substitutents of guest molecules for the complexation of β-cyclodextrin with 2-substituted naphthalenes [74]. Furthermore, the ΔH° values for α-cyclodextrin complexes with organic acids and phenols were linearly correlated to the displacement upon substrate inclusion of ^{13}C chemical shift for the anomeric C-1 atom of the cyclodextrin [29]. All the above-mentioned correlations served to elucidate the main driving forces contributing to the formation of cyclodextrin complexes. However, as described in the previous section, actual inclusion processes are governed by several simultaneous intermolecular interactions. Therefore, it is a primary problem to evaluate quantitatively the extent to which each binding force contributes. The quantitative structure-reactivity analysis developed by Hansch and Fujita [1-5] is a promising method for this purpose.

3.1 Inclusion of Alcohols and Bicyclic Phosphates

Matsui and Mochida [24] have determined the thermodynamic stabilities ($\log 1/K_d$) for α- and β-cyclodextrin complexes with a variety of alcohols (Table 2) and analyzed the results in connection with the physicochemical properties of the guest molecules by the multivariate technique. The $\log 1/K_d$ values were plotted against $\log P_e$, where P_e is the partition coefficient of alcohol in a diethyl ether-water system. The plots for the α- and β-cyclodextrin complexes with eight 1-alkanols gave approximately straight lines with slopes of around one.

For the α-cyclodextrin system,

$$\log 1/K_d = 0.91 \log P_e + 1.25 \qquad (5)$$

$$n = 8\,, \quad r = 0.994\,, \quad s = 0.153$$

and for the β-cyclodextrin system,

$$\log 1/K_d = 0.94 \log P_e + 0.58 \qquad (6)$$

$$n = 8\,, \quad r = 0.994\,, \quad s = 0.099$$

Figures 4 and 5 show the plots of $\log 1/K_d$ vs. $\log P_e$ for branched or cyclic alcohol-cyclodextrin systems. Both of the plots showed considerable scatter in contrast to the plots for 1-alkanol systems (solid lines). However, a remarkable trend was found by comparing both plots. Most of the plots for an α-cyclodextrin system (Fig. 4) are located below the straight line due to Eq. 5, whereas those for a β-cyclodextrin system (Fig. 5) are located above the straight line given by Eq. 6. This shows that it is general for a bulky alcohol to associate with α-cyclodextrin less strongly and with β-cyclodextrin more strongly than a rod-like 1-alkanol if the $\log P_e$ values are the

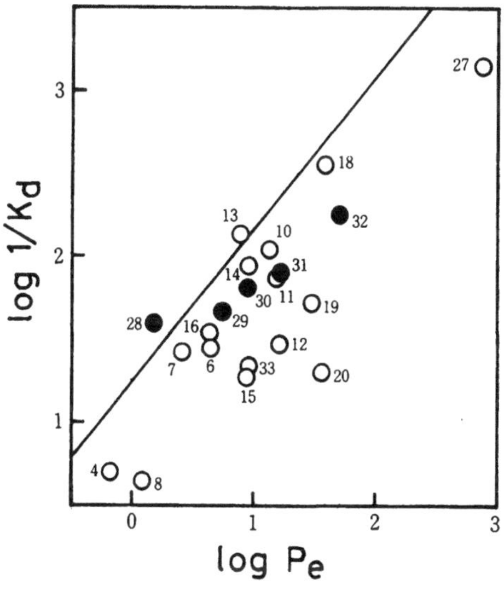

Fig. 4. Plots of $\log 1/K_d$ vs. $\log P_e$ for complexes of α-cyclodextrin with branched alkanols (○) and cycloalkanols (●). The solid line was given by the plots for an α-cyclodextrin-1-alkanol system. Numbers shown refer to the numbers in the first column of Table 2. Reproduced with permission from the Chemical Society of Japan

Table 2. The values of log $1/K_d$ and log P_e at 25 °C, together with E_s and E_{min}, for cyclodextrin-alcohol systems

No.	Alcohol	log $1/K_d$		log P_e	E_s	$-E_{min}$ (kJ/mol)	
		α-CD	β-CD			α-CD	β-CD
1	Methanol	−0.03	−0.49	−1.15	0.00	29.2	23.1
2	Ethanol	0.75	−0.03	−0.50	−0.07	41.7	31.4
3	1-Propanol	1.37	0.57	−0.02	−0.36	57.4	38.7
4	2-Propanol	0.69	0.58	−0.18	−0.47	39.4	40.7
5	1-Butanol	1.95	1.22	0.61	−0.39	65.5	47.7
6	2-Methyl-1-propanol	1.44	1.62	0.65	−0.93	42.7	49.4
7	2-Butanol	1.42	1.19	0.41	−1.13	49.0	46.1
8	2-Methyl-2-propanol	0.64	1.68	0.09	−1.54	32.8	51.3
9	1-Pentanol	2.51	1.80	1.20	−0.40	72.8	53.8
10	2-Methyl-1-butanol	2.04	2.08	1.13	—	45.0	53.4
11	3-Methyl-1-butanol	1.87	2.25	1.19	−0.35	61.2	60.9
12	2,2-Dimethyl-1-propanol	1.47	2.76	1.21	−1.74	48.9	58.9
13	2-Pentanol	2.13	1.49	0.89	—	47.9	53.7
14	3-Pentanol	1.94	1.35	0.96	—	—	—
15	3-Methyl-2-butanol	1.27	1.92	0.94	—	—	59.9
16	2-Methyl-2-butanol	1.53	1.91	0.64	—	34.9	62.2
17	1-Hexanol	2.95	2.34	1.80	—	79.5	58.8
18	2-Hexanol	2.55	1.98	1.58	—	—	—
19	4-Methyl-2-pentanol	1.72	2.04	1.47	—	—	—
20	3,3-Dimethyl-2-butanol	1.30	2.75	1.56	—	—	—
21	2-Methyl-2-pentanol	—	1.99	1.16	—	—	—
22	3-Methyl-3-pentanol	—	2.15	1.10	—	—	—
23	1-Heptanol	3.36	2.85	2.40	—	85.4	61.5
24	2-Methyl-2-hexanol	—	2.33	1.79	—	—	—
25	3-Ethyl-3-pentanol	—	2.28	1.63	—	—	
26	1-Octanol	3.80	3.17	2.91	−0.33	87.8	63.9
27	2-Octanol	3.15	3.13	2.87	—	—	—
28	Cyclobutanol	1.59	1.18	0.18	−0.06	—	—
29	Cyclopentanol	1.66	2.08	0.74	−0.51	—	—
30	Cyclohexanol	1.81	2.70	0.95	−0.79	—	59.5
31	Cycloheptanol	1.90	3.23	1.22	−1.10	—	—
32	Cyclooctanol	2.25	3.30	1.70	—	—	—
33	Benzyl alcohol	1.33	1.70	0.96	−0.38	—	45.4
34	2-Phenylethanol	—	2.15	1.18	−0.38	—	—
35	1,2-Ethanediol	−0.10	−0.19	−2.27	—	46.4	33.2
36	1,3-Propanediol	0.53	0.67	−2.00	—	55.4	42.3
37	1,4-Butanediol	0.91	0.64	−1.72	—	69.3	49.7
38	1,5-Pentanediol	1.50	1.22	−1.26	—	75.9	55.6
39	1,6-Hexanediol	1.97	1.69	−0.92	—	82.6	59.9
40	2-Methoxyethanol	0.33	0.22	−1.28	—	57.7	41.5
41	2-Ethoxyethanol	0.89	0.49	−0.86	—	67.1	50.4
42	2-Propoxyethanol	1.57	—	−0.11	—	74.7	—
43	2-Chloroethanol	1.18	—	0.00	—	60.6	—
44	2-Bromoethanol	1.73	—	0.29	—	66.4	—

same. Taft's steric substituent constant (E_s) was used for a comparative index of molecular bulkiness in the quantitative study of structure-activity relationship. The correlation analysis with the known values of E_s gave Eqs. 7 and 8.

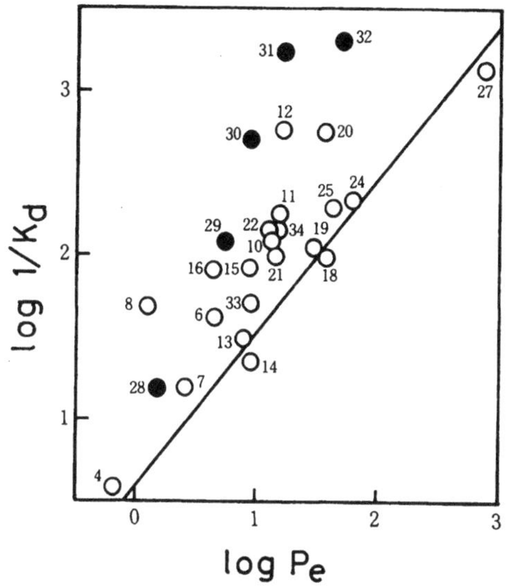

Fig. 5. Plots of log $1/K_d$ vs. log P_e for complexes of β-cyclodextrin with branched alkanols (○) and cycloalkanols (●). The solid line was given by the plots for a β-cyclodextrin-1-alkanol system. Numbers shown refer to the numbers in the first column of Table 2. Reproduced with permission from the Chemical Society of Japan

For the α-cyclodextrin system,

$$\log 1/K_d = 0.92 \log P_e + 0.42E_s + 1.24 \quad (7)$$

$$n = 17, \quad r = 0.953, \quad s = 0.271$$

and for the β-cyclodextrin system,

$$\log 1/K_d = 0.97 \log P_e - 0.65E_s + 0.60 \quad (8)$$

$$n = 18, \quad r = 0.938, \quad s = 0.383$$

These equations show that hydrophobic and steric (van der Waals) interactions are of prime importance in the inclusion processes of cyclodextrin-alcohol systems. The coefficient of E_s was positive in sign for an α-cyclodextrin system and negative for a β-cyclodextrin system. These clear-cut differences in sign reflect the fact that a bulky alcohol is subject to van der Waals repulsion by the α-cyclodextrin cavity and to van der Waals attraction by the β-cyclodextrin cavity.

Matsui [75] has computed energies (E_{min}) which correspond to the minimal values of E_{vdw} in Eq. 1 for cyclodextrin-alcohol systems (Table 2). Besides normal and branched alkanols, some diols, cellosolves, and haloalkanols were involved in the calculations. The E_{min} values obtained were adopted as a parameter representing the London dispersion force in place of E_s. Regression analysis gave Eqs. 9 and 10 for α- and β-cyclodextrin systems respectively.

$$\log 1/K_d = 0.53 \log P_e - 0.026E_{min} - 0.10 \quad (9)$$

$$n = 27, \quad r = 0.967, \quad s = 0.25$$

$$\log 1/K_d = 0.34 \log P_e - 0.056E_{min} - 1.47 \quad (10)$$

$$n = 27, \quad r = 0.963, \quad s = 0.27$$

Again, these equations show that both hydrophobic and steric interactions are of primary importance in the inclusion processes.

Ozoe et al.[76] have determined the K_d values for complexes of α- and β-cyclodextrin with a variety of 4-substituted bicyclic phosphates (4-substituted 2,6,7-trioxa-1-phosphabicyclo[2.2.2]octane 1-oxides, *1*), which are highly toxic convulsants.

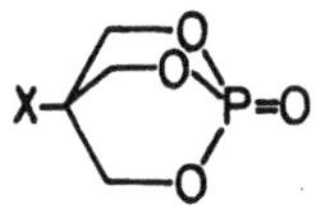

1

Regression analysis of the results gave Eqs. 11 and 12 for α- and β-cyclodextrin systems respectively.

$$\log 1/K_d = \underset{(0.090)}{0.167\pi} + \underset{(0.207)}{0.319E_s^c} + 2.036 \tag{11}$$

$$n = 17, \quad r = 0.762, \quad s = 0.224, \quad F = 9.68$$

$$\log 1/K_d = \underset{(0.068)}{0.068\pi} - \underset{(0.135)}{0.189E_s^c} + 2.415 \tag{12}$$

$$n = 17, \quad r = 0.826, \quad s = 0.135, \quad F = 15.04$$

In these equations, π is the hydrophobic parameter of a bridgehead substituent and E_s^c, Hancock's steric parameter. Although the correlations were not good in terms of r, both equations were statistically significant at the 95% level of confidence. Interestingly, the regression coefficient of the E_s^c term was positive for the α-cyclodextrin system and negative for the β-cyclodextrin system. This is quite similar to the finding for the cyclodextrin-alcohol systems. Bulky substituents of *1* are favorable for β-cyclodextrin complexation, but unfavorable for α-cyclodextrin complexation. These steric effects were attributed to the attractive and repulsive van der Waals interactions respectively.

3.2 Inclusion of Phenols, Phenyl Acetates, and Related Compounds

Silipo and Hansch[77] have developed correlation equations for the formation of α-cyclodextrin-substituted phenyl acetate complexes (Eq. 13), α-cyclodextrin-$RCOO^-$ complexes (Eq. 14), and β-cyclodextrin-substituted phenylcyanoacetic acid anion complexes (Eq. 15).

$$\log 1/K_d = \underset{(0.17)}{0.41MR_{3,4}} + \underset{(0.19)}{1.53} \tag{13}$$

$$n = 12, \quad r = 0.858, \quad s = 0.154$$

$$\log 1/K_d = \underset{(0.22)}{0.92MR} - \underset{(0.60)}{1.79} \log (\beta \cdot 10^{MR} + 1) - \underset{(0.57)}{0.82} \quad (14)$$

$n = 12$, $r = 0.962$, $s = 0.225$
optimum $MR = 3.96$, $\log \beta = -3.90$

$$\log 1/K_d = \underset{(0.36)}{0.76MR_4} + \underset{(0.18)}{1.30} \quad (15)$$

$n = 8$, $r = 0.905$, $s = 0.131$

In these equations, $MR_{3,4}$, MR, and MR_4 are the molar refractivities of 3- and 4-substituents, of R-, and of 4-substituents, respectively. All the equations exhibited positive coefficients of the MR terms. This suggests that the dispersion forces of substituents are actually responsible for the binding of ligands to cyclodextrin. Eq. 14 shows that the stability of α-cyclodextrin-$RCOO^-$ complexes increases linearly up to $MR = 4.0$ and then falls off linearly.

Nishioka and Fujita [78] have determined the K_d values for α- and β-cyclodextrin complexes with m- and p-substituted phenols at pH 7.0. Taking into account the directionality in inclusion of a guest molecule, they assumed three and two probable orientational isomers for the cyclodextrin complexes with m- and p-substituted phenols respectively (Fig. 6). Then the observed K_d values were divided into two or three terms corresponding to the dissociation of the orientational isomers involved (Eqs. 16, 17):

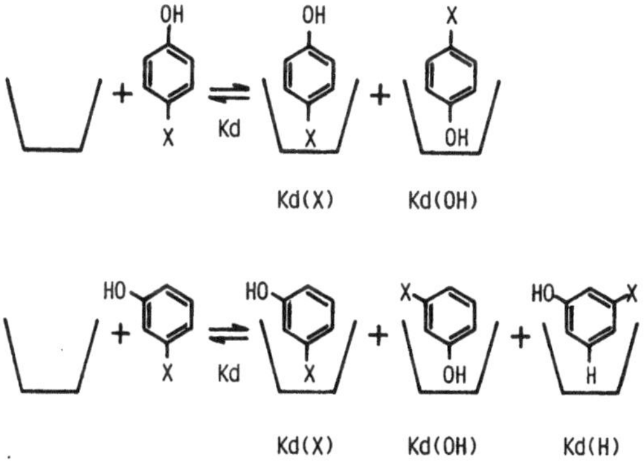

Fig. 6. Probable orientational isomers for cyclodextrin complexes with m- and p-substituted phenols

For m-substituted phenols,

$$1/K_d = 1/K_d(X) + 1/K^d(OH) + 1/K_d(H) \quad (16)$$

and for p-substituted phenols,

$$1/K_d = 1/K_d(X) + 1/K_d(OH) \quad (17)$$

The values of $K_d(H)$ were determined by the extrapolation to $\pi = 0$ of the virtually linear plots of $\log 1/K_d$ vs. π obtained for cyclodextrin complexes with methyl-, ethyl-, n-propyl-, and n-butylphenols (Table 3). The values of $K_d(OH)$ in Eqs. 16 and 17 were calculated from Eq. 18:

$$\log 1/K_d(OH/X) = 0.959\sigma^\circ + \log 1/K_d(OH/H) \qquad (18)$$

where $K_d(OH/X)$ and $K_d(OH/H)$ are the values of $K_d(OH)$ for X-substituted and unsubstituted phenols respectively. This equation was introduced on the assumption that the inclusion of the OH moiety of phenols within the cyclodextrin cavity is influenced by the σ° values of their substituents to the same extent as the transfer of the phenols from water to 1-octanol is influenced. For the latter process, Eq. 19 had been formulated on the basis of the regression analysis of partition coefficients for substituted phenols [79].

$$\pi(X/PhOH) = 1.028\pi + 0.959\sigma^\circ + 0.00$$
$$n = 13, \quad r = 0.993, \quad s = 0.06$$

Table 3. The log $1/K_d(X)$ values for cyclodextrin-substituted phenol systems at pH 7.0 and 25 °C and the parameters of substituents

Substituent	log $1/K_d(X)$						
	β-Cyclodextrin		α-Cyclodextrin		π	σ°	
	p-X-PhOH	m-X-PhOH	p-X-PhOH	m-X-PhOH		para	meta
H	1.89	1.66	1.08	1.16	0.00	0.00	0.00
Me	2.36	1.98	1.51	1.17	0.56	−0.12	−0.07
Et	2.86	2.60	2.13	1.90	1.02	−0.13	−0.07
n-Pr	3.55	3.28	2.75	2.51	1.55	−0.13	−0.07
iso-Pr	3.58	3.44	—	—	1.53	−0.16	−0.07
n-Bu	3.97	3.76	3.26	2.90	2.13	−0.16	−0.07
iso-Bu	—	4.21	—	—	2.06	—	−0.07
sec-Bu	4.18	4.06	—	—	2.04	−0.16	-0.07
tert-Bu	4.56	4.41	—	—	1.98	−0.17	−0.07
n-Amyl	4.19	—	3.36	—	2.65	−0.17	−0.07
neo-Pentyl	—	4.79	—	—	2.51	−0.18	−0.07
tert-Amyl	4.70	—	—	—	2.51	−0.17	—
OH	—	1.40	0.86	1.12	−0.67	−0.13	0.04
OMe	2.21	2.11	1.59	1.03	−0.02	−0.16	0.06
OEt	2.33	2.35	1.87	1.69	0.38	−0.14	0.04
O-iso-Pr	—	2.86	—	—	0.85	—	0.04
F	1.73	1.70	0.91	—	0.14	0.17	0.35
Cl	2.45	2.28	2.45	2.17	0.71	0.27	0.37
Br	2.65	2.51	2.88	2.46	0.86	0.26	0.38
I	3.05	2.93	3.33	3.12	1.12	0.27	0.35
CHO	1.75	1.34	1.75	—	−0.65	0.49	0.30
COMe	2.16	2.06	—	—	−0.55	0.46	0.34
COEt	2.63	2.61	—	—	0.06	0.46	0.34
CN	—	—	1.86	1.75	−0.57	0.69	0.62
Nitro	—	1.61	2.27	1.88	−0.28	0.82	0.70
3,5-Di-Cl	—	2.07	—	2.53	0.71	—	0.74
3,5-Di-Br	—	2.56	—	3.12	0.86	—	0.76
3,5-Di-Me	—	—	—	1.20	0.56	—	−0.14
3,5-Di-OMe	—	2.34	—	1.19	−0.02	—	0.12

The values of $K_d(OH/H)$ in Eq. 18 were estimated from the K_d and $K_d(H)$ values for unsubstituted phenol and/or the $K_d(OH/OH)$ values for hydroxyphenols by use of Eq. 16 or 17.

The $K_d(X)$ values thus obtained (Table 3) were analyzed by the multivariant technique using such parameters as π, $\sigma°$, B_1, I_{brnch}, and I_{hb}, where B_1 is a STERIMOL parameter showing the minimum width of substituents from an axis connecting the α-atom of the substituents and the rest of molecule, and I_{brnch}, an indicator variable representing the number of branches in a substituent.

For β-cyclodextrin-p-X-phenols,

$$\log 1/K_d(X) = \underset{(0.116)}{0.922\pi} + \underset{(0.243)}{0.343 I_{hb}} + \underset{(0.126)}{0.355 I_{brnch}} + \underset{(0.165)}{1.88} \quad (20)$$

$$n = 19, \quad r = 0.989, \quad s = 0.153$$

for β-cyclodextrin-m-X-phenols,

$$\log 1/K_d(X) = \underset{(0.108)}{0.839\pi} + \underset{(0.154)}{0.454 I_{brnch}} + \underset{(0.125)}{1.85} \quad (21)$$

$$n = 25, \quad r = 0.976, \quad s = 0.219$$

for α-cyclodextrin-p-X-phenols,

$$\log 1/K_d(X) = \underset{(0.213)}{0.762\pi} + \underset{(0.559)}{1.64 B_1} + \underset{(0.411)}{0.534 I_{hb}} - \underset{(0.90)}{1.05} \quad (22)$$

$$n = 16, \quad r = 0.961, \quad s = 0.259$$

and for α-cyclodextrin-m-X-phenols,

$$\log 1/K_d(X) = \underset{(0.194)}{0.849\pi} + \underset{(0.45)}{1.52\sigma°} + \underset{(0.21)}{1.17} \quad (23)$$

$$n = 17, \quad r = 0.945, \quad s = 0.260$$

Upon formulating these relationships, phenols with branched alkyl substituents were not included in the data of α-cyclodextrin systems, though they were included in β-cyclodextrin systems. In all the above equations, the π term was statistically significant at the 99.5% level of confidence, indicating that the hydrophobic interaction plays a decisive role in the complexation of cyclodextrin with phenols. The I_{brnch} term was statistically significant at the 99.5% level of confidence for β-cyclodextrin complexes with m- and p-substituted phenols. The stability of the complexes increases with an increasing number of branches in substituents. This was ascribed to the attractive van der Waals interaction due to the close fitness of the branched substituents to the β-cyclodextrin cavity. The steric effect of substituents was also observed for α-cyclodextrin complexes with p-substituted phenols (Eq. 22). In this case, the B_1 parameter was used in place of I_{brnch}, since no phenol with a branched alkyl substituent was involved in data for this regression analysis. On the other hand, the B_1 term was negligible in an α-cyclodextrin-m-substituted phenol system, where the σ° term, as well as the π term, was statistically significant at the 99.5% level of confidence. This electronic effect of substituents was explained in terms of probable hydrogen bonding between the C-2 and C-3 hydroxyl groups of α-cyclodextrin and the phenol OH group of m-substituted phenols. An investigation with molecular models showed that these OH groups are located close to each other upon complexation. The contribution of hydrogen bonding to the stability of inclusion complexes was also recognized for α- and β-cyclodextrin complexes with p-substituted phenols (Eqs. 20 and 22) by the fact that the I_{hb} term is statistically significant. The I_{hb} term corresponds to the hydrogen bonding between the C-6 hydroxyl groups of cyclodextrin and the substituents of phenols. It seems that the p-substituents are able to penetrate into the cyclodextrin cavity deeply enough to form hydrogen bonding with the C-6 hydroxyl groups.

Nishioka and Fujita [78] concluded that the stabilities of cyclodextrin complexes with phenols are mainly governed by hydrophobic interaction, London dispersion forces, and hydrogen bonding, which act simultaneously. Furthermore, they suggested that the dominant orientation of a guest molecule within the cyclodextrin cavity is predictable by comparing the log $1/K_d(X)$ values, calculated by Eqs. 20 to 24, for the possible orientational isomers. For example, the log $1/K_d(X)$ value was always larger than the corresponding log $1/K_d(OH)$ value for each cyclodextrin-alkylphenol system, suggesting that the alkyl moiety of the phenol is included within the cyclodextrin cavity in preference to the OH moiety. On the other hand, the log $1/K_d(X)$ values for the complexes with such guests as nitro- and cyanophenols were either similar to or even smaller than the corresponding log $1/K_d(OH)$ values, suggesting that the inclusion of the OH moiety in place of the nitro or cyano moiety is possible. These predictions were substantiated by ^{13}C-NMR studies on the structures of cyclodextrin inclusion complexes in solution [80,81].

Matsui et al. [82] have analyzed the same data of log $1/K_d(X)$ for cyclodextrin-phenol systems from a somewhat different standpoint. They computed the minimal van der Waals interaction energies (E_{min}) for the systems by using the same method as described in a previous section (Table 4). The calculated E_{min} values were applied, in place of such steric parameters as I_{brnch} and B_1, to the correlation analysis. The correlations obtained are given in Eqs. 24 to 27.

For β-cyclodextrin-p-X-phenols,

$$\log 1/K_d(X) = 0.515\pi - 0.0381E_{min} + 0.573 \qquad (24)$$

$$n = 19, \quad r = 0.982, \quad s = 0.196$$

for β-cyclodextrin-m-X-phenols,

$$\log 1/K_d(X) = 0.686\pi - 0.0312E_{min} + 0.528 \qquad (25)$$

$$n = 25, \quad r = 0.951, \quad s = 0.311$$

for α-cyclodextrin-p-X-phenols,

$$\log 1/K_d(X) = 0.841\sigma^\circ - 0.0596E_{min} - 1.26 \qquad (26)$$

$$n = 16, \quad r = 0.975, \quad s = 0.200$$

Table 4. Van der Waals energy for cyclodextrin-phenol systems ($-E_{min}$: kJ/mol)

Substituent	β-Cyclodextrin		α-Cyclodextrin	
	p-X-PhOH	m-X-PhOH	p-X-PhOH	m-X-PhOH
H	34.77	34.77	35.35	35.35
Me	38.66	44.64	48.24	51.25
Et	43.68	50.88	55.94	55.69
n-Pr	52.67	53.35	69.33	69.12
iso-Pr	55.30	58.20	—	—
n-Bu	61.59	59.25	80.42	80.29
iso-Bu	—	59.66	—	—
sec-Bu	61.34	70.48	—	—
tert-Bu	70.75	73.26	—	—
n-Amyl	62.43	—	78.37	—
neo-Pentyl	—	72.38	—	—
tert-Amyl	79.58	—	—	—
OH	—	41.97	40.21	43.05
OMe	45.31	44.14	53.18	53.01
OEt	52.05	50.00	57.49	66.73
O-iso-Pr	—	54.45	—	—
F	35.69	38.66	35.65	—
Cl	40.92	46.19	52.30	55.44
Br	43.26	52.01	61.30	64.27
I	53.85	58.62	72.63	75.60
CHO	36.57	43.25	45.73	—
COMe	41.63	48.17	—	—
COEt	52.43	43.93	—	—
CN	—	—	40.88	43.97
Nitro	—	47.95	52.26	55.15
3,5-Di-Cl	—	53.69	—	64.48
3,5-Di-Br	—	62.72	—	75.52
3,5-Di-Me	—	—	—	58.16
3,5-Di-OMe	—	46.07	—	55.90

and for α-cyclodextrin-m-X-phenols,

$$\log 1/K_d(X) = 0.844\sigma^\circ - 0.0462E_{min} - 0.968 \quad (27)$$

$$n = 17\,, \quad r = 0.889\,, \quad s = 0.364$$

The results were simple and clear-cut: Only the two terms of σ° and E_{min} were involved for the α-cyclodextrin systems, and the two terms of π and E_{min}, for β-cyclodextrin systems. This means that the stabilities of the inclusion complexes are mainly governed by the electronic and steric interactions in α-cyclodextrin systems and by the hydrophobic and steric interactions in β-cyclodextrin systems, regardless of the position of the substituents in the phenols. These observations agree well with those by Harata [23], who showed that there is no appreciable difference in thermodynamic parameters between cyclodextrin complexes of m- and p-disubstituted benzenes and that the contribution of the enthalpy term to the complexation is more significant in α-cyclodextrin systems than in β-cyclodextrin systems, where the inhibitory effect of the entropy term is less significant. The electronic term observed for the α-cyclodextrin systems could be attributed to the hydrogen bonding between the C-2 and C-3 hydroxyl groups of the cyclodextrin and the OH groups of phenols. However, another explanation is possible and may be plausible: The electronic term corresponds to the dipole-dipole interactions between the host and guest molecules. The tight fitting of a guest molecule into the relatively small α-cyclodextrin cavity makes the role of the dipole-dipole interactions more important than in the case of β-cyclodextrin, with a larger cavity. The relatively poor correlation ($r = 0.889$) in the α-cyclodextrin-m-X-phenol system was explained in connection with the disagreement in direction of molecular dipole between the host and guest molecules. It might be thought that the absence of the π term in Eqs. 26 and 27 for α-cyclodextrin complexation is somewhat strange, since the π term is apparently taking part in the foregoing equations (Eqs. 22 and 23). However, statistically significant correlations were found to exist between E_{min} and π:

For an α-cyclodextrin-p-X-phenol system.

$$-E_{min} = 12.8\pi + 47.8 \tag{28}$$

$$n = 16, \quad r = 0.868, \quad s = 7.4$$

and for an α-cyclodextrin-m-X-phenol system,

$$-E_{min} = 13.2\pi + 52.1 \tag{29}$$

$$n = 17, \quad r = 0.795, \quad s = 7.7$$

Thus, it is reasonable to regard the π term as being partly involved in the E_{min} term of Eqs. 23 and 24.

Nishioka and Fujita [78] have also determined the K_d values for α- and β-cyclodextrin complexes with p- and/or m-substituted phenyl acetates through kinetic investigations on the alkaline hydrolysis of the complexes. The K_d values obtained were analyzed in the same manner as those for cyclodextrin-phenol complexes to give the $K_d(X)$ values (Table 5). The quantitative structure-activity relationships were formulated as Eqs. 30 to 32:

For a β-cyclodextrin-p-X-phenyl acetate system at pH 10.6,

$$\log 1/K_d(X) = \underset{(0.08)}{0.625\pi} + \underset{(0.12)}{0.192 I_{brnch}} + \underset{(0.10)}{2.20} \tag{30}$$

$$n = 16, \quad r = 0.981, \quad s = 0.123$$

for a β-cyclodextrin-m-X-phenyl acetate system at pH 9.7,

$$\log 1/K_d(X) = \underset{(0.10)}{0.740\pi} + \underset{(0.17)}{0.237 I_{brnch}} + \underset{(0.11)}{2.03} \tag{31}$$

$$n = 19, \quad r = 0.973, \quad s = 0.163$$

Table 5. The log $1/K_d(X)$ values for cyclodextrin-phenyl acetate systems

Substituent	β-Cyclodextrin		α-Cyclodextrin
	p-X-PhOAc	m-X-PhOAc	m-X-PhOAc
H	2.10	1.94	1.23
Me	2.49	2.21	1.47
Et	2.83	2.68	1.91
n-Pr	3.15	3.28	2.42
iso-Pr	2.88	3.36	—
n-Bu	3.62	3.66	2.85
iso-Bu	—	3.83	—
tert-Bu	3.85	—	—
n-Amyl	3.80	—	—
OMe	2.45	2.03	1.38
OEt	2.54	2.49	1.73
O-iso-Pr	—	2.78	—
F	2.11	1.91	1.15
Cl	2.50	2.44	2.26
Br	2.68	2.67	2.80
I	3.00	3.07	3.32
CHO	—	1.84	1.11
COMe	2.17	2.08	—
COEt	2.32	2.38	—
CN	—	1.49	1.74
Nitro	2.13	1.87	1.72
3,5-Di-Cl	—	—	2.48
3,5-Di-Me	—	—	1.54
3,5-Di-OMe	—	—	1.28

for an α-cyclodextrin-m-X-phenyl acetate system at pH 9.7,

$$\log 1/K_d(X) = \underset{(0.19)}{0.533\pi} + \underset{(0.53)}{1.461B_1} - \underset{(0.82)}{0.643} \quad (32)$$

$$n = 17, \quad r = 0.936, \quad s = 0.253$$

Only the hydrophobic and steric terms were involved in these equations. There are a few differences between these equations and the corresponding equations for cyclodextrin-substituted phenol systems. However, it is not necessarily required that the mechanism for complexation between cyclodextrin and phenyl acetates be the same as that for cyclodextrin-phenol systems. The kinetically determined K_d values are concerned only with productive forms of inclusion complexes. The productive forms may be similar in structure to the tetrahedral intermediates of the reactions. To attain such geometry, the penetration of substituents of phenyl acetates into the cyclodextrin cavity must be shallow, compared with the cases of the corresponding phenol systems, so that the hydrogen bonding between the substituents of phenyl acetates and the C-6 hydroxyl groups of cyclodextrin may be impossible.

4 Proximity Effect in Cyclodextrin Catalysis

Cyclodextrin catalyzes a number of organic reactions which occur in aqueous solutions [6,7,10–12]. The reaction which is most extensively investigated in connection with cyclodextrin catalysis is the hydrolytic cleavage of organic esters [6,7]. The catalytic process generally involves the prior formation of a cyclodextrin-substrate inclusion complex, followed by the nucleophilic attack of the cyclodextrin alkoxide ion on the reaction site of the included substrate, although Komiyama et al. [83,84] have recently presented a few exceptional cases in which the ester cleavage is subject to general base catalysis, rather than nucleophilic catalysis, by cyclodextrin.

The catalytic site of cyclodextrin is the secondary hydroxyl group aligned around one edge of the toroidal molecule. The hydroxyl group acts as a weak acid and dissociates into the corresponding alkoxide ion in a strongly alkaline solution. The pK_a value was first estimated to be 12.1 at 25 °C by analyzing the pH dependence of the cleavage of m-tolyl acetate in the presence of α-cyclodextrin [85]. Recently, Gelb et al. [86,87] have precisely determined the pK_a values for α-, β-, and γ-cyclodextrins to be 12.332, 12.202, and 12.081, respectively, at 25 °C by means of pH potentiometry. They also found that the ^{13}C-NMR peaks of both C-2 and C-3 carbons are substantially displaced upon acid dissociation, whereas the C-6 peak shows lesser displacement. This fact clearly indicates that the C-2 and C-3 secondary hydroxyl groups rather than the C-6 primary hydroxyl groups are involved in the ionization process. However, it is not yet known whether the C-2 or C-3 alkoxide ion is the one which attacks a substrate first in cyclodextrin-catalyzed cleavage of phenyl esters. Breslow et al. [88,89] have shown that the reactions of β-cyclodextrin with a few p-nitrophenyl esters give an acylated β-cyclodextrin which is, by NMR, a mixture of the C-2 acylated and C-3 acylated compounds. Since acyl migration in glycol esters is fast, it is not clear whether the acylation occurs randomly or it goes specifically to one of the carbons, followed by equilibration. Interestingly, the cleavage of 3-nitrophenyl tosylate in the presence of β-cyclodextrin yields exclusively the C-2 tosylated cyclodextrin [90,91]. The tosyl group does not migrate once attached to the C-2 hydroxyl. This suggests that the acylation of cyclodextrin also occurs initially on the C-2 hydroxyl. A recent ^{1}H-NMR study [92] has also suggested that the C-3 hydroxy hydrogens are hydrogen bonded to the C-2 hydroxyl oxygen and, therefore, the C-2 hydroxyl is more liable to dissociate into the conjugate base than the C-3 hydroxyl. However, Bergeron and Burton [93] have presented ^{1}H- and ^{13}C-NMR data which cast some doubt upon the premise that the C-3 hydroxyls are inherently unreactive and therefore uninvolved in cyclodextrin-catalyzed ester cleavage.

The catalytic effect of cyclodextrin on ester hydrolysis is dependent on the structures of substrate and cyclodextrin. The rate accelerations imposed by the cyclodextrin are always larger for the m-substituted phenyl acetates than for the corresponding p-substituted phenyl acetates [6,7,71]. This was explained by invoking a specific, rate-accelerating interaction between the secondary hydroxyl groups of cyclodextrin and the carbonyl group of the included ester: The carbonyl group is fixed in close proximity to the secondary hydroxyl groups of cyclodextrin in an m-substituted phenyl acetate complex, whereas it extends into the bulk solution, away from the secondary hydroxyl groups, in a p-substituted phenyl acetate complex. In the latter case, a considerable change in conformation of the inclusion complex is required for the access of the carbonyl group to the secondary hydroxyl groups to react with each other. The conformational change should accompany an increase in enthalpy. When the carbonyl group is located in close proximity to the secondary hydroxyl groups in an inclusion complex, the activation energy required is so small that the rate acceleration is large.

The concept that a geometric relationship between the catalytic site of cyclodextrin and the reaction site of a substrate is of primary importance has been substantiated by kinetic investigations for a variety of cyclodextrin-organic ester systems. Among them, a series of investigations by Breslow et al. [88,89,94,95] are most noticeable. At first, they modified β-cyclodextrin by building in an intrusive floor, so that the cavity was well defined and shallower [88]. The modified cyclodextrin improved the rate of the m-tert-butylphenyl acetate cleavage by an order of magnitude, compared with the unmodified cyclodextrin. Even more striking effects were obtained by modifying the substrate geometry itself. In the optimum case examined in their earliest study [88], the p-nitrophenyl ester of ferroceneacrylic acid showed a rate acceleration (k_2/k_{un}) of more than 10^5 in comparison with acyl transfer (k_2) to β-cyclodextrin with spontaneous hydrolysis (k_{un}) under the same conditions. Further improvement in the rate acceleration was realized by freezing out residual rotational degrees of freedom in the acylation transition state. Thus, (E)-3-(carboxymethylene)-1,2-ferrocenocyclohexene p-nitrophenyl ester (*2*) [89] and its cyclopentenoferrocene congener [94] showed rate accelerations of 3.2×10^6 and 5.9×10^6 respectively for one of their enantiomers. These large rate accelerations are comparable to those by such pro-

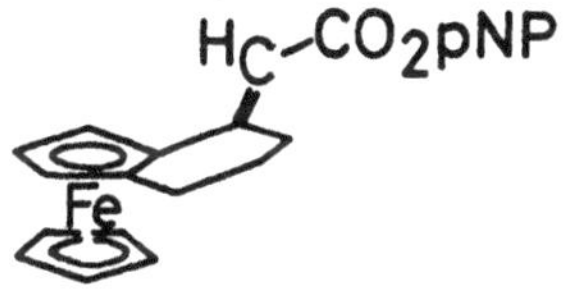

2

teolytic enzymes as chymotrypsin. Besides these investigations, the effect of substrate structure on the rate acceleration has recently been studied on phenyl salicylates [96], thiocarboxylic S-esters [97], nitrophenyl 1-adamantanecarboxylates [98], and p-nitrophenyl 1-adamantaneacetate [99]. Fujita et al. [100,101] have shown that the modification of the β-cyclodextrin C-6 hydroxyl group(s) results in a change in meta/para selectivity for the cyclodextrin catalyzed cleavage of substituted phenyl acetates. Similar effects of cyclodextrin modification have recently been reported for such modified cyclodextrins as α-cyclodextrin-N-(N,N′-dimethylaminoethyl)acetohydroxamic acid [102], α-cyclodextrin-N-(4-imidazolemethyl)acetohydroxamic acid [102], O-[2-(hydroxyimino)ethyl]-α-cyclodextrin [103], mono-(6-trimethylammonio-6-deoxy)-β-cyclodextrin [104], and β-cyclodextrin having a hydroxamate group at the C-6 position [105].

The importance of the proximity effect in cyclodextrin catalysis has been discussed on the basis of the structural data. Harata et al. [31,35] have determined the crystal structures of α-cyclodextrin complexes with m- and p-nitrophenols by the X-ray method. Upon the assumption that m- and p-nitrophenyl acetates form inclusion complexes in the same manner as the corresponding nitrophenols, they estimated the distances between the carbonyl carbon atoms of the acetates and the adjacent secondary hydroxyl group of α-cyclodextrin to be 3.78 Å in the m-nitrophenyl acetate complex and more than 5 Å in the p-nitrophenyl acetate complex. Evidently, the carbonyl carbon atom of m-nitrophenyl acetate can approach the secondary hydroxyl group more easily than that of the p-isomer. Komiyama and Hirai [80,106] have evaluated time-averaged distances (D_{C-O}) between the carbonyl carbon atoms of phenyl acetates and the C-2 hydroxyl oxygen atoms of cyclodextrin in the inclusion complexes by ^{1}H-NMR spectroscopy. They found that the magnitude (k_2/k_{un}) of the rate acceleration for the cleavage of the substrates by cyclodextrin increases with a decrease in the D_{C-O} value. A similar relationship has been demonstrated by the theoretical analysis of the structures of the cyclodextrin complexes with substituted phenyl acetates [107]. The geometries of the inclusion complexes were determined by the calculation of van der Waals energies, as have been described in previous sections. Consequently, the D_{C-O} values were computed. The D_{C-O} values thus obtained agreed approximately with those determined experimentally. The magnitudes ($\log k_2/k_{un}$) of the rate accelerations were plotted against the D_{C-O} values (Fig. 7). The plot gave a virtually hyperbolic curve, regardless of the size of the cyclodextrin cavity. These results clearly suggest that the rate acceleration is primarily governed by the proximity between the cyclodextrin catalytic site and the substrate reaction site in the inclusion complex.

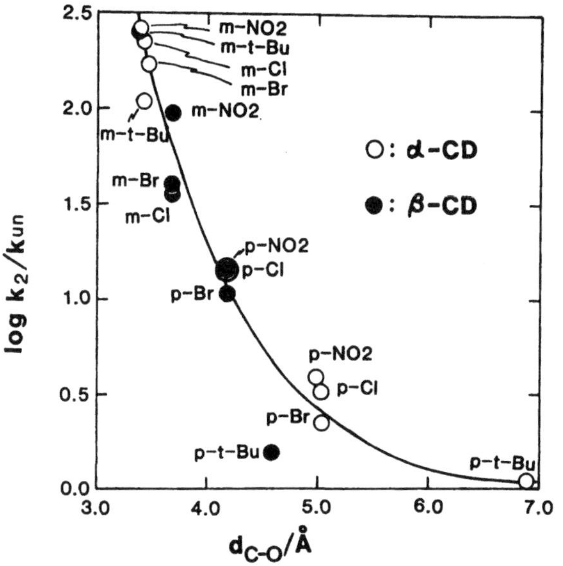

Fig. 7. Plots of $\log k_2/k_{un}$ vs. D_{C-O} determined by the calculation of van der Waals energies for cyclodextrin complexes with m- and p-substituted phenyl acetates

5 Quantitative Structure-Reactivity Analysis of Cyclodextrin Catalysis

Quantitative structure-reactivity analysis is one of the most powerful tools for elucidating the mechanisms of organic reactions. In the earliest study, Van Etten et al. [71] analyzed the pseudo-first-order rate constants for the alkaline hydrolysis of a variety of substituted phenyl acetates in the absence and in the presence of cyclodextrin. The hydrolysis in the absence of cyclodextrin followed a normal Hammett relationship between the logarithm of the rate constant (log k_{un}) and the appropriate Hammett substituent constant (σ or σ^*). However, when the hydrolysis solution contained cyclodextrin, very large and variable accelerations were found, and the logarithm of the rate constant (log k_2) did not follow a Hammett relationship. Later, these data were analyzed by Silipo and Hansch [77] to give such correlation equations as Eqs. 33 and 34:

For the α-cyclodextrin-catalyzed cleavage of phenyl acetates,

$$\log k_2 = \underset{(0.41)}{0.64}MR_3 + \underset{(0.61)}{0.96}\sigma + \underset{(0.27)}{0.51}E_{s4} - \underset{(0.39)}{1.75} \qquad (33)$$

$$n = 11, \quad r = 0.960, \quad s = 0.283$$

and for the uncatalyzed hydrolysis of phenyl acetates,

$$\log k_{un} = \underset{(0.11)}{0.79}\sigma^- - \underset{(0.06)}{3.11} \qquad (34)$$

$$n = 17, \quad r = 0.968, \quad s = 0.111$$

In these equations, MR_3 is the molar refractivity of the 3-substituent, σ and σ^-, Hammett substituent constants, and E_{s4}, a steric substituent constant of the 4-substituent. Equation 34 is a normal Hammett-type relationship. The parameter σ^- gave a better correlation than σ. On the other hand, Eq. 33 involves two steric parameters, MR_3 and E_{s4}, in addition to the Hammett substituent constant, σ. The positive coefficient of MR_3 indicates that a bulky substituent in the meta position facilitates the access of the acetyl moiety of the substrate to the secondary hydroxyl groups of cyclodextrin. The positive coefficient of E_{s4}, however, indicates a deleterious effect of bulky groups in the para position. The larger the group, the more negative its E_s constant. Bulky groups in the para position may displace the substrate so that it is not in a good position for nucleophilic attack by one of the α-cyclodextrin OH groups. The effect of the cyclodextrin cavity size on the catalytic activity was also examined by the analysis of a larger data set of the overall pseudo-first-order rate constants (k_{obsd}) for substituted phenyl acetate hydrolyses. Equations 35 and 36 were formulated for α- and β-cyclodextrin catalysts respectively.

$$\log k_{obsd} = \underset{(0.40)}{0.95}MR_3 + \underset{(0.46)}{0.99}\sigma + \underset{(0.27)}{0.31}E_{s4} - \underset{(0.37)}{2.26} \qquad (35)$$

$$n = 16, \quad r = 0.925, \quad s = 0.335$$

$$\log k_{obsd} = \underset{(0.20)}{0.89}MR_3 + \underset{(0.29)}{1.29}\sigma - \underset{(0.14)}{2.34} \qquad (36)$$

$$n = 15, \quad r = 0.963, \quad s = 0.183$$

Equation 35 is very similar to Eq. 33, although the correlation is not as sharp. The most striking difference between Eqs. 35 and 36 is that the latter does not contain a term in E_{s4}. This means that para substituents have no effect on hydrolysis through steric interactions. The obvious conclusion is that these substituents are not in contact with the cavity of β-cyclodextrin. Their only effect on hydrolysis is electron withdrawal.

Nishioka and Fujita [68, 78] have, independently of Van Etten et al. [71], determined and analyzed the rate constants, k_2, for the cyclodextrin-catalyzed cleavage of m- and p-substituted phenyl acetates (Table 6). They have derived Eqs. 37 to 40:

Table 6. The log k_2 values for cyclodextrin-phenyl acetate systems ($k_2 : s^{-1}$)

Substituent	β-Cyclodextrin		α-Cyclodextrin
	p-X-PhOAc (pH 10.6)	m-X-PhOAc (pH 9.7)	m-X-PhOAc (pH 9.7)
H	−1.987	−2.854	−2.374
Me	−2.188	−2.618	−1.848
Et	−2.495	−2.159	−1.710
n-Pr	−2.564	−2.067	−1.996
iso-Pr	−2.790	−1.914	—
n-Bu	−2.514	−2.177	−2.036
iso-Bu	—	−1.932	—
tert-Bu	−3.052	—	—
n-Amyl	−2.257	—	—
OMe	−2.386	−2.466	−1.631
OEt	−2.652	−2.324	−1.790
O-iso-Pr	—	−2.020	—
F	−1.680	−2.370	−1.644
Cl	−1.656	−2.156	−1.349
Br	−1.775	−2.024	−1.380
I	−2.062	−1.790	−1.434
CHO	—	−1.943	−1.064
COMe	−1.896	−1.629	—
COEt	−2.046	−1.410	—
CN	—	−1.622	−0.845
Nitro	−1.047	−1.181	−0.721
3,5-Di-Cl	—	—	−0.879
3,5-Di-Me	—	—	−1.745
3,5-Di-OMe	—	—	−1.199

For a β-cyclodextrin-p-X-phenyl acetate system at pH 10.6,

$$\log k_2 = \underset{(0.26)}{1.43\sigma^\circ} - \underset{(0.07)}{0.200D_{max}} - \underset{(0.32)}{1.41} \tag{37}$$

$$n = 16, \quad r = 0.965, \quad s = 0.142$$

for a β-cyclodextrin-m-X-phenyl acetate system at pH 9.7,

$$\log k_2 = \underset{(0.25)}{1.28\sigma^\circ} + \underset{(0.06)}{0.275D_{max}} - \underset{(0.28)}{3.47} \tag{38}$$

$$n = 19, \quad r = 0.959, \quad s = 0.123$$

for an α-cyclodextrin-m-X-phenyl acetate system at pH 9.7,

$$\log k_2 = \underset{(0.32)}{1.45\sigma^\circ} + \underset{(0.44)}{0.878D_{max}} - \underset{(0.05)}{0.103D_{max}^2} - \underset{(0.92)}{3.67} \tag{39}$$

$$n = 21, \quad r = 0.951, \quad s = 0.158$$

and for an uncatalyzed alkaline hydrolysis (k_{OH}, M^{-1} min^{-1}) of phenyl acetates,

$$\log k_{OH} = \underset{(0.06)}{1.17\sigma^\circ} + \underset{(0.02)}{2.26} \tag{40}$$

$$n = 18, \quad r = 0.996, \quad s = 0.037$$

In these equations, D_{max} is the larger of the summed values of STERIMOL parameters, B_i, for the opposite pair [68]. It expresses the "maximum" total width of substituents. The coefficients of the σ° terms in Eqs. 37 to 39 were virtually equal to that in Eq. 40. This means that the σ° terms essentially represent the hydrolytic reactivity of an ester itself and are virtually independent of cyclodextrin catalysis. The catalytic effect of cyclodextrin is only involved in the D_{max} term. Interestingly, the coefficient of D_{max} was negative in Eq. 37 and positive in Eq. 38. This fact indicates that bulky substituents at the meta position are favorable, while those at the para position unfavorable, for the rate acceleration in the β-cyclodextrin catalysis. Similar results have been obtained for α-cyclodextrin catalysis, but not for β-cyclodextrin catalysis, by Silipo and Hansch described above. Equation 39 suggests the existence of an optimum "diameter" for the proper fit of m-substituents in the cavity of α-cyclodextrin. The optimum D_{max} value was estimated from Eq. 39 as 4.4 Å, which is approximately equivalent to the diameter of the α-cyclodextrin cavity. The situation is shown in Fig. 8. A similar parabolic relationship would be obtained for β-cyclodextrin catalysis, too, if the correlation analysis involved phenyl acetates with such bulky substituents that they cannot be included within the β-cyclodextrin cavity.

A few examples have been reported in which no steric parameter is involved in the correlation analysis of cyclodextrin catalysis. Straub and Bender [108] showed that the maximal catalytic rate constant, k_2, for the β-cyclodextrin-catalyzed decarboxylation of substituted phenylcyanoacetic acid anions (*3*) is correlated simply by the Hammett σ parameter.

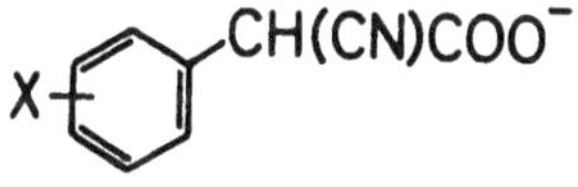

3

A similar correlation has been derived by Silipo and Hansch [77] with regard to the same reaction system (Eq. 41).

$$\log k_2 = \underset{(0.50)}{2.09\sigma} - \underset{(0.15)}{2.27} \tag{41}$$

$$n = 8, \quad r = 0.973, \quad s = 0.165$$

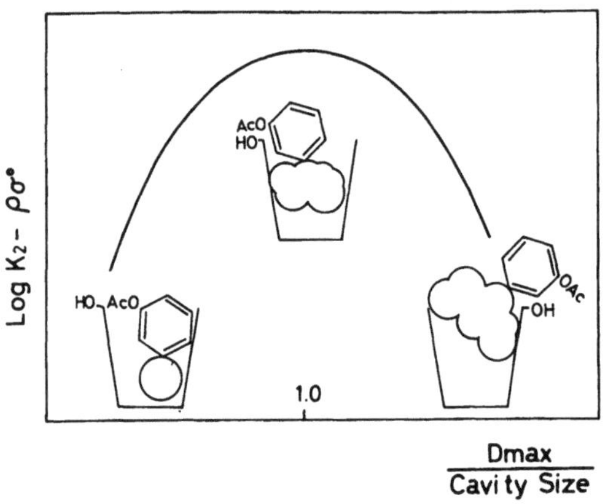

Fig. 8. Parabolic relationship between reactivity with cyclodextrin and substituent "diameter" of m-substituted phenyl acetates

The rate acceleration imposed by β-cyclodextrin was explained in terms of a microsolvent effect:[6] The inclusion of the substrate within the hydrophobic cavity of cyclodextrin simulates the changes in solvation which accompany the transfer of the substrate from water to an organic solvent. Uekama et al. [109] have analyzed the substituent effect on the alkaline hydrolysis of 7-substituted coumarins (*4*) in the

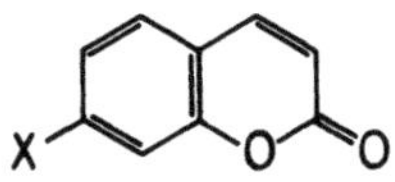

4

absence and in the presence of α-, β-, and γ-cyclodextrins by means of a modified Hammett equation (Eq. 42):

$$\log k_{obsd} = \varrho_m \cdot \sigma_m + \varrho_p \cdot \sigma_p + \log k_0 \tag{42}$$

where σ_m and σ_p refer to the ordinary meta and para substitution constants of the 7-substituent. This equation permits the estimation of the relative importance of two alternate paths in transmission of electronic effects of substituents to the reaction center. Good correlations were found, where both σ_m and σ_p were highly correlated to log k_{obsd}. The ϱ_m value was larger than the ϱ_p value for the uncatalyzed and γ-cyclodextrin-catalyzed hydrolyses, whereas the reverse was true for α- and β-cyclodextrin catalyzed hydrolyses. Again, these results were explained by invoking the microsolvent effect of cyclodextrin catalysis.

6 Concluding Remarks

As shown above, quantitative structure-reactivity analysis is very useful in elucidating the mechanisms of cyclodextrin complexation and cyclodextrin catalysis. This method enables us to separate several intermolecular interactions, acting simultaneously, from one another in terms of physicochemical parameters, to evaluate the extent to which each interaction contributes, and to predict thermodynamic stabilities and/or kinetic rate constants experimentally undetermined. Fortunately, the structures of cyclodextrin inclusion complexes as well as the binding and catalytic mechanisms of cyclodextrin have been widely and deeply investigated, so that it is possible to examine the validity of conclusions deduced from the correlation analysis. In most cases, the conclusions obtained are consistent with those deduced from experimental measurements. This fact provides a background for quantitative structure-reactivity analysis.

However, some problems remain to be solved. Changes of steric parameters used for the correlation analysis sometimes caused considerable modifications of the resulting regression equations, and, therefore, discussions on the relationships were modified to some extent. In order to overcome such discrepancies, it is necessary not only to increase the number of congeners examined but also to select an appropriate set of data for the correlation analysis. Furthermore, discussions on the magnitudes of regression coefficients were not always satisfactory, although the signs of the coefficients were always reasonable. In some cases, collinearities between such parameters as π, E_s, E_{min}, and σ° were so high that the "independent" variables for the correlation analysis were not always independent. Then, the appparent magnitudes of the regression coefficients do not correspond to the net contributions of intermolecular and/or intramolecular interactions. Again, the selection of appropriate substituents and an increase in the number of data for the analysis are essential to overcome these problems.

7 References

1. Hansch, C., Fujita, T.: J. Am. Chem. Soc. *86*, 1616 (1964)
2. Hansch, C.: Acc. Chem. Res. *2*, 232 (1969)
3. Hansch, C.: Drug Design, Vol. 1 (ed. Ariens, E. J.), p. 271, New York, Academic Press 1971
4. Fujita, T.: Adv. Chem. Ser. *114*, 1 (1972)
5. Hansch, C.: J. Med. Chem. *19*, 1 (1976)
6. Griffiths, D. W., Bender, M. L.: Adv. Catal. *23*, 209 (1973)
7. Bender, M. L., Komiyama, M.: Cyclodextrin Chemistry, Berlin, Springer-Verlag 1978
8. Saenger, W.: Environmental Effects on Molecular Structure (ed. Pullman, B.), p. 265, Dordrechit-Holland, D. Reidel 1976
9. Saenger, W.: Angew. Chem. Int. Ed. Engl. *19*, 344 (1980)
10. Tabushi, I.: Acc. Chem. Res. *15*, 66 (1982)
11. Breslow, R.: Adv. Chem. Ser. *100*, 21 (1971)
12. Breslow, R.: ibid. *191*, 1 (1980)
13. Siegel, B., Breslow, R.: J. Am. Chem. Soc. *97*, 6869 (1975)
14. Eftink, M. R., Harrison, J. C.: Bioorg. Chem. *10*, 388 (1981)
15. Gerasimovicz, W. V., Wojcik, J. F.: ibid. *11*, 420 (1982)
16. Cramer, F., Saenger, W., Spatz, H.-Ch.: J. Am. Chem. Soc. *89*, 14 (1967)
17. Turro, N. J., Okubo, T., Chung, C.-J.: ibid. *104*, 1789 (1982)
18. Jencks, W. P.: Catalysis in Chemistry and Enzymology, p. 393, New York, McGraw-Hill 1969
19. Lindner, K., Saenger, W.: Carbohydr. Res. *99*, 103 (1982)
20. Bergeron, R. J., Meely, M. P.: Bioorg. Chem. *5*, 197 (1976)
21. Bergeron, R. J., Channing, M. A., Gibeily, G. J., Pillor, D. M.: J. Am. Chem. Soc. *99*, 5146 (1977)
22. Harata, K.: Bioorg. Chem. *9*, 530 (1980)

23. Harata, K.: ibid. *10*, 255 (1981)
24. Matsui, Y., Mochida, K.: Bull. Chem. Soc. Jpn. *52*, 2808 (1979)
25. Komiyama, M., Bender, M. L.: J. Am. Chem. Soc. *100*, 2259 (1978)
26. Gelb, R. I., Schwartz, L. M., Laufer, D. A.: Bioorg. Chem. *9*, 450 (1980)
27. Uekama, K., Hirayama, F., Otagiri, M., Otagiri, Y., Ikeda, K.: Chem. Pharm. Bull. *26*, 1162 (1978)
28. Bergeron, R. J., Pillor, D. M., Gebeily, G. G., Roberts, W. P.: Bioorg. Chem. *7*, 263 (1978)
29. Gelb, R. I., Schwartz, L. M., Cardelino, B., Fuhrman, H. S., Johnson, R. F., Laufer, D. A.: J. Am. Chem. Soc. *103*, 1750 (1981)
30. Harata, K.: Bull. Chem. Soc. Jpn. *49*, 2066 (1976)
31. Harata, K., Uedaira, H., Tanaka, J.: ibid. *51*, 1627 (1978)
32. Tabushi, I., Kiyosuke, Y., Sugimoto, T., Yamamura, K.: J. Am. Chem. Soc. *100*, 916 (1978)
33. Harata, K.: Bull. Chem. Soc. Jpn. *48*, 2409 (1975)
34. Saenger, W., Beyer, K., Mannor, P. C.: Acta Crystallogr. *B32*, 120 (1976)
35. Harata, K.: Bull. Chem. Soc. Jpn. *50*, 1416 (1977)
36. Harata, K.: ibid. *52*, 2451 (1979)
37. Harata, K.: ibid. *53*, 2782 (1980)
38. Harata, K.: ibid. *54*, 1954 (1981)
39. Tokuoka, R., Abe, M., Matsumoto, K., Shirakawa, K., Fujiwara, T., Tomita, K.: Acta Crystallogr. *B37*, 445 (1981)
40. Harata, K.: Bull. Chem. Soc. Jpn. *55*, 1367 (1982)
41. Harding, M., Maclennan, J. M., Paton, R. M.: Nature *274*, 621 (1978)
42. Tokuoka, R., Fujiwara, T., Tomita, K.: Acta Crystallogr. *B37*, 1158 (1981)
43. Hamilton, J. A., Sabesan, M. N.: ibid. *B38*, 3063 (1982)
44. Matsui, Y.: Bull. Chem. Soc. Jpn. *55*, 1246 (1982)
45. Hill, T. L.: J. Chem. Phys. *16*, 399 (1948)
46. Saenger, W., McMullan, R. K., Fayos, J., Mootz, D.: Acta Crystallogr. *B30*, 2019 (1974)
47. Hingerty, B., Saenger, W.: J. Am. Chem. Soc. *98*, 3357 (1976)
48. Saenger, W., Noltemeyer, M.: ibid. *109*, 503 (1976)
49. Mannor, P. C., Saenger, W.: ibid. *96*, 3630 (1974)
50. Saenger, W., Noltemeyer, M., Mannor, P. C., Hingerty, B., Klar, B.: Bioorg. Chem. *5*, 187 (1976)
51. Klar, B., Hingerty, B., Saenger, W.: Acta Crystallogr. *B36*, 1154 (1980)
52. Chacko, K. K., Saenger, W.: J. Am. Chem. Soc. *103*, 1708 (1981)
53. Lindner, K., Saenger, W.: Acta Crystallogr. *B38*, 203 (1982)
54. Lindner, K., Saenger, W.: Angew. Chem. *90*, 738 (1978)
55. Maclennan, J. M., Stezowski, J. J.: Biochem. Biophys. Res. Commun. *92*, 926 (1980)
56. Lindner, K., Saenger, W.: ibid. *92*, 933 (1980)
57. Hamilton, J. A., Sabesan, M. K., Steinrauf, L. K.: Carbohydr. Res. *89*, 33 (1981)
58. Lindner, K., Saenger, W.: Carbohydr. Res. *107*, 7 (1982)
59. Harata, K., Uekama, K., Otagiri, M., Hirayama, F.: Bull. Chem. Soc. Jpn. *55*, 3904 (1982)
60. Cramer, F., Kampe, W.: J. Am. Chem. Soc. *87*, 1115 (1965)
61. Van Hooidonk, C., Breebaart-Hansen, J. C. A. E.: Recl. Trav. Chim. Pays-Bas. *90*, 680 (1971)
62. Hybl, A., Rundle, R. E., Williams, D. E.: J. Am. Chem. Soc. *87*, 2779 (1965)
63. Tokuoka, R., Abe, M., Fujiwara, T., Tomita, K., Saenger, W.: Chem. Lett. 491 (1980)
64. Harata, K.: Bull. Chem. Soc. Jpn. *49*, 1493 (1976)
65. Harata, K.: ibid. *50*, 1259 (1977)
66. Connors, K. A., Lin, S.-F., Wong, A. B.: J. Pharm. Sci. *71*, 217 (1982)
67. Leo, A., Hansch, C., Elkins, D.: Chem. Rev. *71*, 525 (1971)
68. Fujita, T., Iwamura, H.: Topics in Current Chemistry, Vol. 114, p. 119, Berlin, Heidelberg, Springer 1983
69. Fujita, T., Nishioka, T., Nakajima, M.: J. Med. Chem. *20*, 1071 (1977)
70. Casu, B., Rava, L.: Ric. Sci. *36*, 733 (1966)
71. Van Etten, R. L., Sebastian, J. F., Clowes, G. A., Bender, M. L.: J. Am. Chem. Soc. *89*, 3242 (1967)
72. Uekama, K., Hirayama, F., Nasu, S., Matsuo, N., Irie, T.: Chem. Pharm. Bull. *26*, 3477 (1978)
73. Rosanske, T. W., Connors, K. A.: J. Pharm. Sci. *69*, 564 (1980)

74. Harata, K.: Bull. Chem. Soc. Jpn. *52*, 1807 (1979)
75. Matsui, Y.: The 44th Ann. Meet. Chem. Soc. of Japan, Okayama, Oct. 1981, Abstr., p. 1186
76. Ozoe, Y., Mochida, K., Eto, M.: Agric. Biol. Chem. *45*, 2623 (1981)
77. Silipo, C., Hansch, C.: Bioorg. Chem. *8*, 237 (1979)
78. Nishioka, T., Fujita, T.: Symp. on Cyclodextrin, Kyoto, Dec. 1982, Abstr. p. 1
79. Fujita, T.: J. Pharm. Sci. *72*, 285 (1983)
80. Komiyama, M., Hirai, H.: Chem. Lett. 1467 (1980)
81. Komiyama, M., Hirai, H.: Bull. Chem. Soc. Jpn. *54*, 828 (1981)
82. Matsui, Y., Nishioka, T., Fujita, T.: Unpublished
83. Komiyama, M., Inoue, S.: Bull. Chem. Soc. Jpn. *53*, 3334 (1980)
84. Komiyama, M., Hirai, H.: Chem. Lett. 1251 (1980)
85. Van Etten, R. L., Clowes, G. A., Sebastian, J. F., Bender, M. L.: J. Am. Chem. Soc. *89*, 3253 (1967)
86. Gelb, R. I., Schwartz, L. M., Bradshaw, J. J., Laufer, D. A.: Bioorg. Chem. *9*, 299 (1980)
87. Gelb, R. I., Schwartz, L. M., Laufer, D. A.: ibid. *11*, 274 (1982)
88. Breslow, R., Czarniecki, M. F., Emert, J., Hamaguchi, H.: J. Am. Chem. Soc. *102*, 762 (1980)
89. Trainor, G. L., Breslow, R.: ibid. *103*, 154 (1981)
90. Ueno, A., Breslow, R.: Tetrahedron Lett. *23*, 3451 (1982)
91. Breslow, R., Czarnik, A. W.: J. Am. Chem. Soc. *105*, 1390 (1983)
92. Gillet, B., Nicole, D. J., Delpuech, J.-J.: Tetrahedron Lett. *23*, 65 (1982)
93. Bergeron, R. J., Burton, P. S.: J. Am. Chem. Soc. *104*, 3664 (1982)
94. Breslow, R., Trainor, G., Ueno, A.: ibid. *105*, 2739 (1983)
95. le Noble, W. J., Srivastava, S., Breslow, R., Trainor, G.: ibid. *105*, 2745 (1983)
96. Kurono, Y., Fukatsu, H., Yotsuyanagi, T., Ikeda, K.: Chem. Pharm. Bull. *27*, 963 (1979)
97. Komiyama, M., Bender, M. L.: Bull. Chem. Soc. Jpn. *53*, 1073 (1980)
98. Komiyama, M., Inoue, S.: ibid. *53*, 2330 (1980)
99. Komiyama, M., Inoue, S.: ibid. *53*, 3266 (1980)
100. Fujita, K., Shinoda, A., Imoto, T.: J. Am. Chem. Soc. *102*, 1161 (1980)
101. Fujita, K., Shinoda, A., Imoto, T.: Tetrahedron Lett. *21*, 1541 (1980)
102. Kitaura, Y., Bender, M. L.: Bioorg. Chem. *4*, 237 (1975)
103. Van Hooidonk, C., de Korte, D. C., Reuland-Meereboer, M. A. C.: Recl. Trav. Chim. Pays-Bas *96*, 25 (1977)
104. Matsui, Y., Okimoto, A.: Bull. Chem. Soc. Jpn. *51*, 3030 (1978)
105. Tabushi, I., Kuroda, Y., Sakata, Y.: Heterocycles *15*, 815 (1981)
106. Komiyama, M., Hirai, H.: Chem. Lett. 1471 (1980)
107. Matsui, Y., Nishioka, T., Fujita, T.: The 9th Meet. Enzyme-Simulated Organic Reactions, Tokyo, Mar. 1982, Abstr., p. 27
108. Straub, T. S., Bender, M. L.: J. Am. Chem. Soc. *94*, 8875 (1972)
109. Uekama, K., Lin, C.-L., Hirayama, F., Otagiri, M., Takadate, A., Goya, S.: Chem. Lett. 563 (1981)

The Role of Molecular Shape Similarity in Specific Molecular Recognition

Tadashi Endo

College of Science and Engineering, Aoyama Gakuin University, Chitosedai, Setagaya-ku, Tokyo 157, Japan

Table of Contents

1 Introduction

Molecules can in principle recognize one another. The recognition generally means that a given molecule binds to only a certain molecule out of a large number of structurally related molecules [1].

Amino acid side chains participate in the specific association of proteins with a variety of biomolecules, thus leading to enzyme-substrate [2], protein-protein [3], protein-nucleic acid [4], antibody-antigen [5], and hormone-receptor recognition [6]. For example, proteins such as restriction endonucleases and modification methylases can recognize the methyl group at the 5 and N6 positions of cytosine and adenine, respectively, lying within the recognition sequence in the DNA [7]; it is striking that quite a simple chemical modification, *methylation*, of these bases serves to protect the cell's own DNA from being degraded by its restriction endonucleases [8]. Furthermore, in the first step of the amino acid activation required for protein biosynthesis, each aminoacyl-tRNA synthetase must select only one amino acid from among the twenty kinds [2]. Indeed, the selection is extremely precise: in the case of the isoleucine-specific system, the dissociation constant for the enzyme-substrate complex indicates that L-isoleucine is preferred over L-valine by a factor of 100 [2], these two amino acids having the s-butyl and i-propyl groups, respectively, as their side chains.

As in the case of biological systems mentioned above, highly specific molecular recognition can also be achieved in chemical systems, if appropriate "structural devices" are embedded in molecules. The biological importance of nonpolar groups such as amino acid side chains (e.g., Me, i-Pr, i-Bu, and $PhCH_2$ groups) stimulated us to elucidate factors responsible for the specific recognition, especially of such nonpolar groups. One approach to this problem is to study the effect of the spatial relationship between the "interacting" groups in the respective molecules on the discrimination between them when they are held together.

The present review intends to be illustrative rather than comprehensive, and focuses on the results of this study leading to the hypothesis [9] — *the three-dimensional shape similarity between interacting groups in reacting molecules is responsible for more specific and precise molecular recognition than would otherwise be achieved* — and on the explanation of biological recognition on this basis.

2 Shape Similarity Effect in Chemical Systems

The problem of molecular recognition has attracted biologically oriented chemists since Emil Fischer's "lock-and-key" theory [10]. Within the last two decades, many model compounds have been developed: micelle-forming detergents [11], modified cyclodextrins [12], many kinds of crown-type compounds [13] including podands, coronands, cryptands, and spherands. Very extensive studies using these compounds have, however, not been made from a point of view of whether or not "shape similarity" affects the discrimination.

2.1 Specifically Associating Model Compound

It seems appropriate to utilize model compounds of relatively low molecular weight, in order for X-ray analysis to provide the highly accurate information on the spatial relationship between interacting groups in the respective molecules participating in the recognition. Our model compound comprises a pair of acylurea derivatives (Fig. 1), open-chain analogs of pyrimidine bases (e.g., uracil and thymine). Substituents A and D correspond to $HSCH_2$ and $HS(CH_2)_2$, respectively, when the model compound is used for oxidation with oxygen; for crystalline 1:1 complex formation as a model reaction, A and D are similarly shaped groups (p-$O_2NC_6H_4$ and p-$Me_2NC_6H_4$). As R and R′, various nonpolar groups were mainly used which included a series of straight-chain alkyl [$Me(CH_2)_n$], branched alkyl [$Me_2CH(CH_2)_n$], cyclohexylalkyl [$C_6H_{11}(CH_2)_n$], phenylalkyl [$Ph(CH_2)_n$], and *para*-substituted phenyl groups.

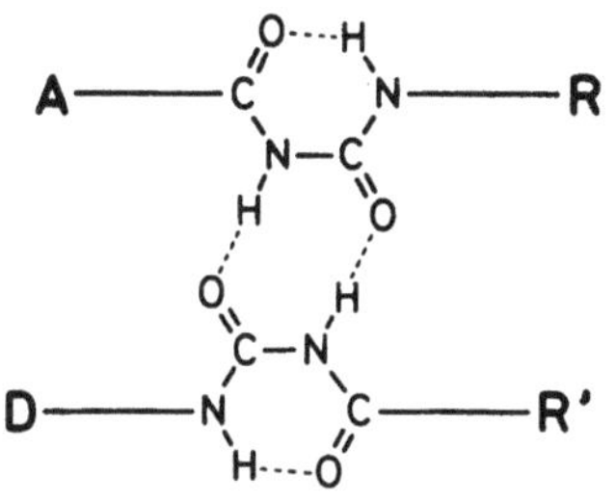

Fig. 1. Structure of the model compound. R and R′ represent the recognition sites. ----, Hydrogen bonding

Acylurea derivatives are readily prepared in one step by addition of amines to acyl isocyanates in high yields (Eqs. 1 and 2) [14].

$$A-\overset{\overset{O}{\|}}{C}-N=C=O + H_2N-R \longrightarrow A-\overset{\overset{O}{\|}}{C}\underset{H}{N}\underset{\underset{O}{\|}}{C}\overset{H}{N}-R \qquad (1)$$

$$D-NH_2 + O=C=N-\overset{\overset{O}{\|}}{C}-R' \longrightarrow D-\underset{H}{N}\overset{\overset{O}{\|}}{C}\overset{H}{N}\underset{\underset{O}{\|}}{C}-R' \qquad (2)$$

The chloroacetylurea derivative ($A = ClCH_2$) obtained from the above addition (Eq. 1), on treatment first with thioacetic acid and Et_3N and then with cysteamine, afforded the corresponding thiol ($A = HSCH_2$) (Eq. 3) [15].

$$ClCH_2\overset{\overset{O}{\|}}{C}\underset{H}{N}\underset{\underset{O}{\|}}{C}\overset{H}{N}-R \xrightarrow{HS\overset{\overset{O}{\|}}{C}CH_3 + Et_3N} CH_3\overset{\overset{O}{\|}}{C}SCH_2\overset{\overset{O}{\|}}{C}\underset{H}{N}\underset{\underset{O}{\|}}{C}\overset{H}{N}-R$$

$$\xrightarrow{H_2NCH_2CH_2SH} HSCH_2\overset{\overset{O}{\|}}{C}\underset{H}{N}\underset{\underset{O}{\|}}{C}\overset{H}{N}-R \qquad (3)$$

An interesting structural feature of the model compound (Fig. 1) is that two pairs of substituents (A — D and R — R') are each brought close together through specific association via two NH ··· O intermolecular hydrogen bonds between the inner $-\overset{\mathrm{H\ O}}{\mathrm{NC}}-$ units in the $-\underset{\mathrm{O}}{\overset{\mathrm{H\ O}}{\mathrm{CNCN}}}\underset{\mathrm{H}}{}-$ group[16] ("binding site"). Close proximity between a pair of substituents R and R′ ("*recognition site*") makes it possible to control the specificity of the model reactions. The pattern of hydrogen bonding shown in Fig. 1 is supported by X-ray diffraction (cf. Figs. 11 and 12).

2.2 Selectivity in Oxidation of Thiols in Solution

The correct pairing of half-cystine residues is shown to be dependent upon specific noncovalent bonds [17]. With this finding in mind, oxidation of a pair of associating thiols (*1* and *2*) was chosen as a model reaction. Thiol *1* has the same group as cysteine side chain ($HSCH_2$), *2* being a derivative of cysteamine.

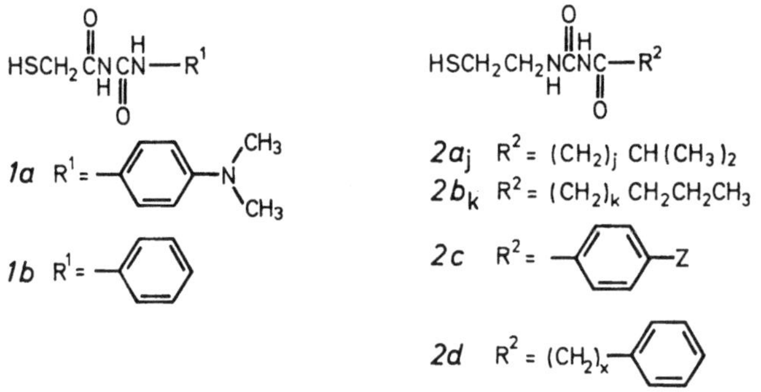

$$\mathrm{HS-X} + \mathrm{HS-Y} \xrightarrow{O_2} \begin{array}{c}\mathrm{S-X}\\ |\\ \mathrm{S-X}\end{array} + \begin{array}{c}\mathrm{S-X}\\ |\\ \mathrm{S-Y}\end{array} + \begin{array}{c}\mathrm{S-Y}\\ |\\ \mathrm{S-Y}\end{array} \quad (4)$$

1 *2* *3* *4* *5*

A 1:1 mixture of thiols (*1* and *2*), on treatment with oxygen in the presence of a catalytic amount of Et_3N, gives one unsymmetrical (*4*) and two symmetrical disulfides (*3* and *5*) (Eq. 4). As a measure of the degree of the recognition between *1* and *2* in the oxidation, the selectivity (r) is employed which is represented by the logarithmic ratio of the yield of *4* to twice that of *3* (Eq. 5). The r is so defined as to become zero when oxidation yields the three disulfides in a 1:2:1 ratio. In the present case, the recognition process is followed by covalent bond formation.

$$r = \ln([4]/(2[3])) \quad (5)$$

2.2.1 Structure Dependence of the Selectivity

2.2.1.1 Structures of the Recognition Sites

The oxidation described here was performed in 80% (v/v) acetonitrile—20% water (mole fraction of water = 0.42) at 35.0 °C. Figure 2 shows the selectivity as a function of the number of carbon atoms in R^2. In the case of oxidation of *1a* and *2a* (R^2 = branched alkyl groups), the selectivity reaches a sharp maximum ($r = 2.4$) at the isopentyl group (j = 2) [18]. For R^2 = straight-chain alkyl groups, alternation in the selectivity is clearly observed [18]. The difference between the *r* value for *1a* and $2a_2$ and that for *1a* and $2b_2$ reaches up to 3.7.

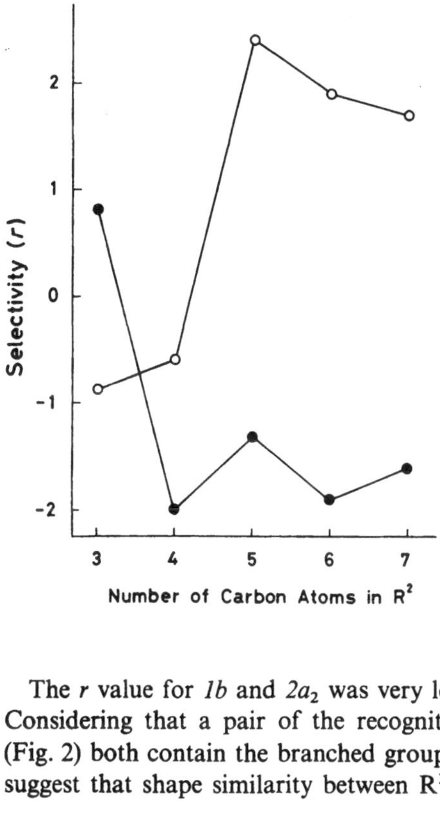

Fig. 2. Dependence of the selectivity (*r*) in oxidation of *1a* and *2* in 80% (v/v) MeCN — 20% H_2O at 35.0 °C on the number of carbon atoms in R^2 [18]. The selectivity was originally represented by the ratio of the yield of *4* to that of *3* [18]. The molar ratio *1a*:*2* is 1:1. ○, $R^2 = (CH_2)_jCHMe_2$; ●, $R^2 = (CH_2)_kCH_2CH_2Me$

The *r* value for *1b* and $2a_2$ was very low, compared with that for *1a* and $2a_2$ [19]. Considering that a pair of the recognition sites producing the highest selectivity (Fig. 2) both contain the branched groups (i.e., NMe_2 and i-Pr), these observations suggest that shape similarity between R^1 and R^2 increases the selectivity.

Table 1. The selectivity (*r*) in oxidation of *1a* and *2c* in 80% (v/v) MeCN — 20% H_2O at 35.0 °C for the *para*-substituent Z. [9] The molar ratio *1a*:*2c* is 1:1

Z	$CH(Me)_2$	NO_2	Me	Cl
r	1.2	1.2	— 0.86	— 1.0

This view is supported by further studies on the selectivity for *1a* and *2c* having *para*-substituted phenyl groups as R^2. The $CHMe_2$ and NO_2 groups as Z of *2c*, both with a branched structure similar to the NMe_2 group of *1a*, give higher selectivity than sphere-like groups (Me and Cl) (Table 1) [9].

Enzyme activity is reduced by binding of an inhibitor to the enzyme. Inhibitors have structures analogous to that of the specific substrate [20]. Thus, in a nonenzymatic associating system, the presence of a substance ("deceptor") similar in geometrical shape to the reacting molecule might be expected to affect chemical events. When a deceptor (*6* or *7*) — a derivative having the methyl group instead of the HS group of *1* or *2* — is added to a 1:1 mixture of *1* and *2*, thiol *1* (or *2*) could not differentiate

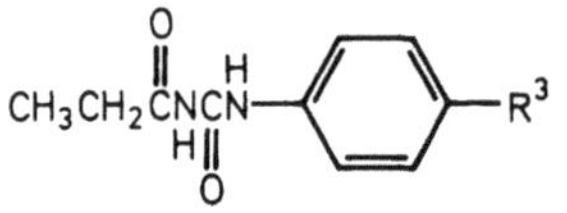

6

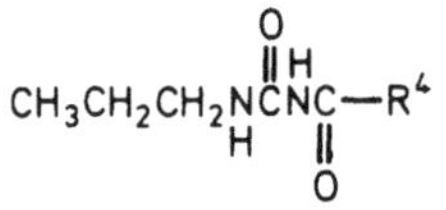

7

thiol *2* (or *1*) from deceptor *7* (or *6*), thereby resulting in changes in the selectivity. Indeed, for oxidation of *1a* and *2a₂* (R^2 = i-C_5H_{11}) in the presence of *6* at 35.0 °C in aqueous acetonitrile, the *r* value is reduced sharply from 2.4 (the value in the absence of *6*) to 0.79, [21] as shown in Fig. 3 where the selectivity is plotted against the A value [22] — the free energy difference ΔG (= $-RT \ln K$) involved in the axial-equatorial equilibrium of substituted cyclohexanes — for R^3. Of special interest is that (i) the *r* value becomes lowest at R^3 = NMe_2 (the same group as the *para*-substituent of *1a*) and (ii) the *r* value for R^3 = i-Pr is similar to that for R^3 = NMe_2. In addition, the selectivity decreases more markedly for R^3 = i-Pr (r = 0.97) than for R^3 = n-Pr (r = 1.8) [21].

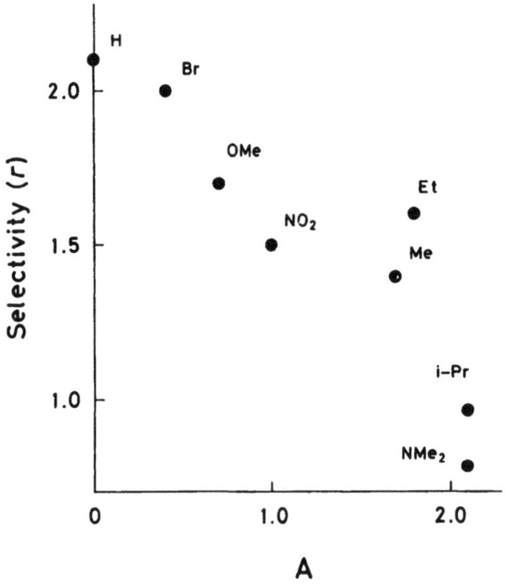

Fig. 3. Plot of the selectivity (*r*) in oxidation of *1a* and *2a₂* in the presence of *6* in 80% (v/v) MeCN — 20% H_2O at 35.0 °C against the A value for R^3 [21]. The molar ratio *1a*:*2a₂*:*6* is 1:1:1. The point for R^3 = n-Pr (r = 1.8) is not shown, since its A value is unavailable

The same phenomenon was observed for oxidation of *1a* and *2a*$_2$ in the presence of *7*. When a series of branched alkyl groups are used as R^4, the selectivity shows a minimum at the isopentyl group (Fig. 4) [21]. The *r* value of 1.4 for $R^4 = n\text{-}C_5H_{11}$ differs largely from that for $R^4 = i\text{-}C_5H_{11}$ ($r = 0.30$) [21]. These results demonstrate that the more closely the substituent R^4 of *7* resembles the isopentyl group of *2* in geometrical shape, the more remarkably the selectivity is lowered.

The selectivity in the presence of *7* also reached a minimum at the 2-phenylethyl group when a series of phenylalkyl groups were used as R^4 (Fig. 4) [21]. This finding, together with the selectivity data for oxidation of *1a* and *2a*$_2$ or *2d* (Fig. 4), suggests that the 2-phenylethyl group resembles the isopentyl group most closely among the phenylalkyl groups examined. The relationship between these two groups will be further discussed in Section 2.3.1.

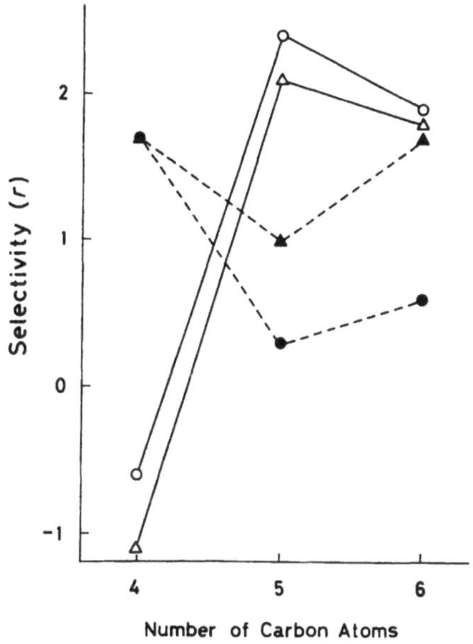

Fig. 4. Dependence of the selectivity (*r*) in oxidation of *1a* and *2a*$_2$ in the presence of *7* [$R^4 = (CH_2)_jCHMe_2$ (●) and $(CH_2)_xPh$ (▲)] and in oxidation of *1a* and *2* [$R^2 = (CH_2)_jCHMe_2$ (○) and $(CH_2)_xPh$ (△)] in 80% (v/v) MeCN — 20% H_2O at 35.0 °C on the number of carbon atoms in R^4 and R^2 [21]. With the $(CH_2)_xPh$ groups, the number of carbon atoms is counted as "x + 3". The molar ratio *1a*:*2a*$_2$:*7* is 1:1:1

2.2.1.2 Structures of the Binding Sites

Another example of the shape similarity effect on molecular recognition involves the similarity between the structures of the binding sites. Investigation was made for four reaction systems I–IV each consisting of a 1:1 mixture of thiols HS—X and HS—Y (Fig. 5), with the number (*m* or *n*) of amide unit $\left(\begin{smallmatrix} O \\ \| \; H \\ -C\!N- \end{smallmatrix}\right)$ being altered. The *m* or *n* of 2 corresponds to the acylurea bond. As is evident from Fig. 6, the *r* reaches a maximum when the numbers of the amide units in HS—X and HS—Y are equal to each other (i.e., when $m - n = 0$) in each of the four systems [23]. In system IV, the selectivity can be said to decrease largely on changing *m* from 2 to 3, though the *r* value could not be determined for technical reasons.

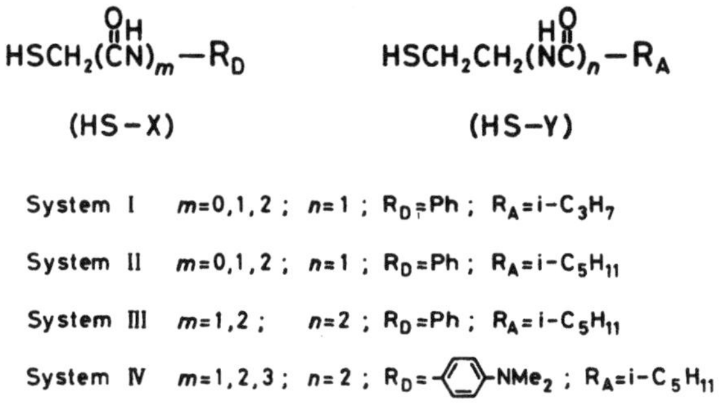

Fig. 5. Structure of reaction systems I—IV. Each system is composed of a pair of thiols (HS—X and HS—Y). The symbols *m* and *n* are the numbers of amide units in the thiols

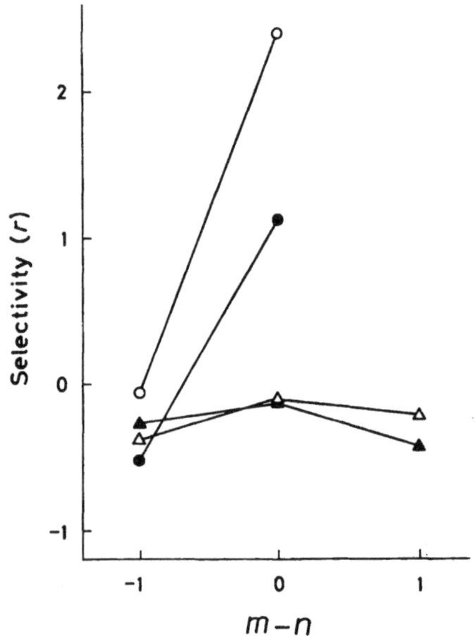

Fig. 6. Plot of the selectivity (*r*) in oxidation of systems I—IV in 80% (v/v) MeCN — 20% H_2O at 35.0 °C against "*m* — *n*" [23]. The symbols *m* and *n* are as in Fig. 5. The selectivity was originally represented by the ratio of the yield of *4* to that of *3* [23]. The molar ratio HS—X:HS—Y is 1:1. ▲, System I; △, system II; ●, system III; ○, system IV

2.2.2 Other Possible Factors Affecting the Selectivity

In this Section, possible factors influencing the selectivity other than shape similarity and shape-specific weak interactions (Sect. 2.4) are discussed. These mainly include intermolecular association, exchange reactions, and hydrophobic interaction. In connection with intermolecular association and crystalline 1:1 complex formation (Sect. 2.3), tetrameric intermediates are also discussed.

2.2.2.1 Intermolecular Association

The ability of a solute to associate with itself can be expressed by the degree of association (*f*). The *f* is obtained by dividing the stoichiometric mole fraction of the solute by the effective mole fraction of the solute. Assuming that a single multimer species in equilibrium with a monomer is a dimer, *f* values range from 1.0 to 2.0 [24].

At 36.0 °C and 0.01 M in benzene, *f* values for *8* and *9* are less than 1.05, those for *1* and *2* ranging from 1.3 to 1.5; [23] this suggests that *1* and *2* associate strongly with

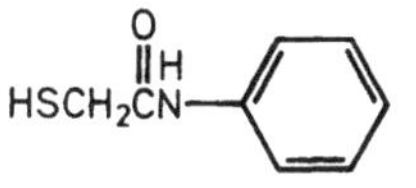

8

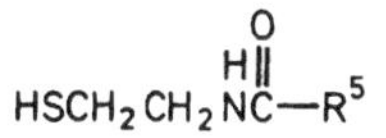

9a $R^5 = (CH_2)_jCH(CH_3)_2$
9b $R^5 = (CH_2)_xC_6H_5$

themselves, while *8* and *9* associate very weakly with themselves and that *8* associates very weakly with *9*, or even with *2a* because of the marked difference in *f* between *8* and *2a*. The selectivity in oxidation of *8* and *9* (j = 0 to 3 and x = 1 to 3) or of *8* and *2a* (j = 0 to 2) changed only slightly from 0 to — 0.2 or from — 0.3 to — 0.5, respectively [25]. This is also the case with oxidation of a pair of simple thiols, $HSCH_2CO_2Et$ and an alkanethiol [HS—R^6 (*10*), R^6 = $(CH_2)_jCHMe_2$ (j = 0 to 3)] [18]. Comparison of the structure dependence of the selectivity for the associating system (*1* and *2*) with that for the nonassociating system (e.g., *8* and *9*) leads to the conclusion that intermolecular association is the first requirement for molecular recognition.

Spectroscopic studies of *1* and *2*, made by using the dilution technique, confirmed their strong association in solution [16]. First, infrared spectra in $CDCl_3$ showed the NH stretching vibration bands due to intermolecular hydrogen bonds at 3266 cm^{-1} for *1a*, 3246 cm^{-1} for *2a₂*, and 3240 cm^{-1} for *2b₂* as well as intramolecular hydrogen bonds at 3355 cm^{-1} for *1a*, 3316 cm^{-1} for *2a₂*, and 3308 cm^{-1} for *2b₂*. Second, ^1H NMR spectra in $CDCl_3$ revealed that (i) on mixing of equimolar amounts of *1a* and *2a₂* (or *2b₂*), the NH^α and NH^γ protons were largely shifted downfield, while the NH^β and NH^δ protons underwent small shifts and (ii) the concentration dependence of the chemical shifts of the NH^α and NH^γ resonances was less sharp when *1a* and *2a₂* (or *2b₂*) were dissolved individually than when they were mixed together. These data indicate that *1a* and *2a₂* (or *2b₂*) form weak complexes with each other as well as with themselves.

If weak complexes present in solution consist of dimers alone (Fig. 7), association patterns of homodimers (*11* and *13*) are of the head-to-tail type, that of heterodimer (*12*) being of the head-to-head type, since the two acylurea bonds in *1* and *2* extend

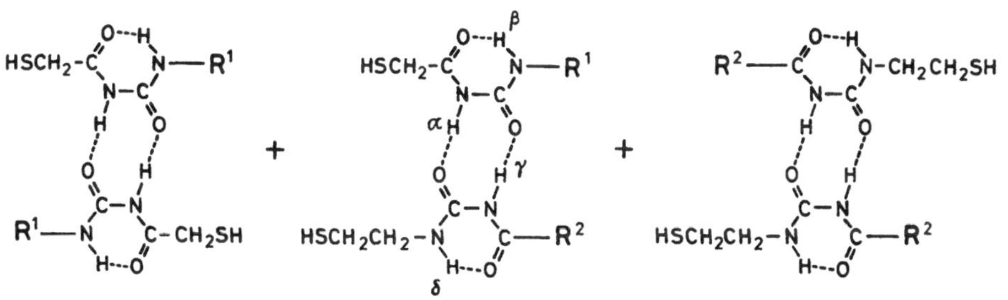

11 *12* *13*

Fig. 7. Association patterns of three dimers *11–13* [16]. ----, Hydrogen bonding

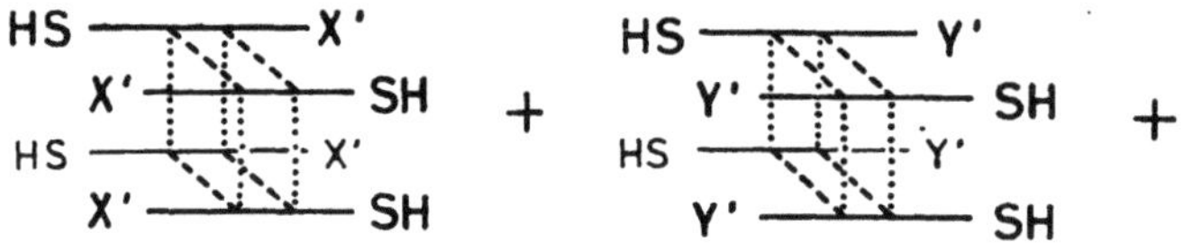

14 *15*

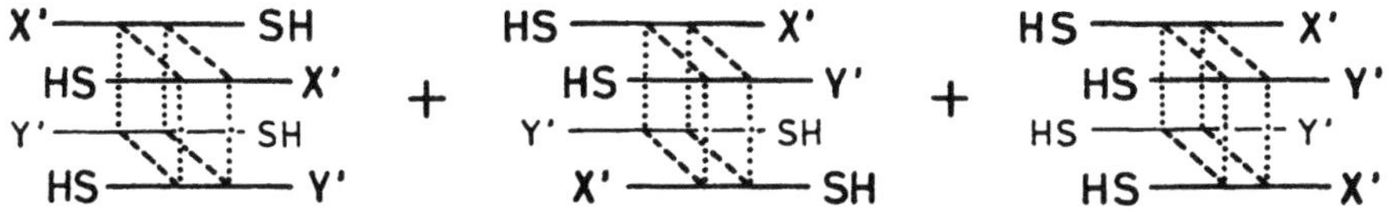

16 *17* *18*

Fig. 8. Association schemes of tetramers *14–18* formed by dimerization of dimers *11–13* [14b, 16, 26]. ----, Hydrogen bonding responsible for stabilization of dimers; ······, noncovalent weak interactions responsible for stabilization of tetramers. The symbols X′ and Y′ represent R^1 and R^2, respectively. In association schemes of tetramers *14*, *15*, and *17*, the configurations of the lower dimer units are so modified as to be consistent with the crystal structures (Figs. 13 and 15)

in the opposite direction to each other. However, homodimers of head-to-tail conformation cannot explain the selective formation of the symmetrical disulfide (i.e., $r < 0$) in some cases (Fig. 2). Furthermore, the f value of 1.95 for $2a_2$ at 0.06 M in CCl_4 suggests the presence of higher aggregates in addition to dimers. From these results, tetramers — two homotetramers (*14* and *15*) and three heterotetramers (*16–18*) — in equilibrium with dimers (*11–13*) and monomers (*1* and *2*) are suggested to be intermediates in the oxidation (Fig. 8) [16]. Among many possibilities, five tetramers are chosen according to the criteria that (i) the ratio of *1* to *2* in heterotetramers is 1:1 and (ii) the stabilization of tetramers may become strongest.

The finding that the oxidation of *1a* and *2c* (Z = H, Me, and Et) shows large negative activation entropies (— 51 to — 45 cal K^{-1} mol^{-1}) [26] can be understood by the tetrameric intermediates mentioned above.

There is considerable evidence for the presence of tetramers in chemical (e.g., cyclic tetramer formation of Et_2NH in cyclohexane [27] and of EtSH in argon matrices [28]) and biological systems (e.g., the hemoglobins of higher animals composed of two pairs of unlike polypeptide chains [29]).

2.2.2.2 Exchange Reactions

The thiol—disulfide exchange reaction between *2* and *3* was slow in the presence of a catalytic amount of Et_3N under argon [14b, 18]. Thus, the product ratio in the exchange reaction differed entirely from that in the oxidation. The disulfide—disulfide exchange reaction did not take place to any appreciable extent in the presence of a catalytic amount of both *1a* and Et_3N under argon [18]. These results, together with no appreciable changes in r values with the reaction time, indicate that the product ratio is kinetically controlled [18].

The selectivity is neither controlled by solubility differences between thiols [14b, 18] *1a* and *2* and between disulfides [18] *3–5* nor by whether the reaction is heterogeneous or homogeneous [18].

2.2.2.3 Hydrophobic Interaction

The solvent used in the oxidation described here is aqueous acetonitrile. Therefore, hydrophobic interaction [30] may operate between the recognition sites.

The strength of hydrophobic interaction (hydrophobicity) increases linearly with increasing surface area of the nonpolar group removed from contact with water [31]. The π value [32] (a parameter of hydrophobicity), defined as $\pi = \log P_X - \log P_H$ where P_H is the partition coefficient of a parent molecule between two solvents (1-octanol and water) and P_X is that for a derivative, increases in the order i-Pr < n-Pr < i-Bu < n-C_5H_{11} < n-C_6H_{13} < n-C_7H_{15}; [33] the order of π values for R^2 in *2a* and *2b* differs markedly from that of the corresponding *r* values (Fig. 2) [18]. Further, no relationship exists between the order of π values for Z (i.e., i-Pr > Cl > Me > NO_2) in *2c* and that of the corresponding *r* values (Table 1) [9].

Hydrophobicity represented by $\Delta G°$ for the transfer of solute from the pure liquid to aqueous solution increases progressively with increasing temperature [34]. There is, however, an extremum in the temperature—selectivity plot in some cases (e.g., R^2 = i-C_5H_{11}, Ph, and *p*-MeC_6H_4) [14b, 18]. It appears that the observed selectivity cannot be explained in terms of hydrophobic interaction.

2.3 Crystalline 1:1 Complex Formation

Weak complex formation between *1a* and *2* in solution has already been mentioned (Sect. 2.2.2.1). This suggests the possibility that crystalline complexes would be produced for a suitable substituent, R^2. When R^2 was various kinds of alkyl groups, however, attempts to form crystalline complexes failed. The $HSCH_2$ and $HSCH_2CH_2$ groups were then replaced by a pair of groups similar in shape to each other, *p*-$O_2NC_6H_4$ and *p*-$Me_2NC_6H_4$, with the expectation that these two groups would serve to stabilize weak complexes.

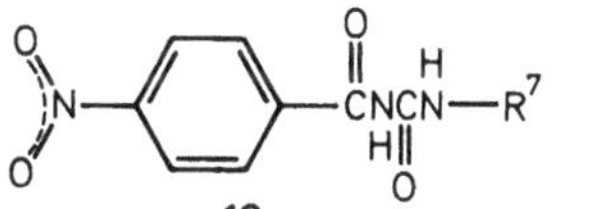

19

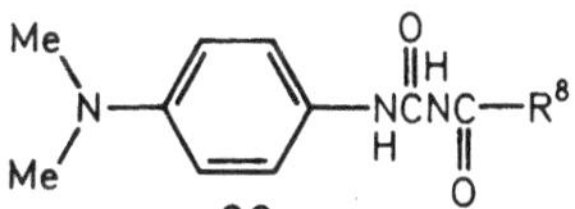

20

Compounds *19* and *20* gave the crystalline 1:1 molecular complex in some cases, when dissolved in acetonitrile or *N,N*-dimethylformamide followed by slow evaporation at room temperature [35]. The complex formation occured only for $n_7 + n_8 \geqq 5$ where n_7 and n_8 were the carbon numbers along the longest straight chains for R^7 and R^8, respectively. Branched alkyl and phenylalkyl groups tend to stabilize the crystalline complexes to a greater extent when paired with branched alkyl groups than when paired with straight-chain alkyl groups. Noteworthy is that (i) the color of the crystalline complexes varies markedly from pale yellow to reddish brown and to dark purple with the structures of R^7 and R^8 (cf. Fig. 9), in sharp contrast to their components (*19* and *20*) being generally colorless crystals and (ii) solutions of the crystalline complexes in acetonitrile are colorless in most cases [35].

2.3.1 Structure Dependence of the Complex Formation

Crystalline 1:1 complex formation can be regarded as molecular recognition in the process of crystallization. Since the formation of a new type of the crystalline 1:1 complex depends on the shapes of R^7 and R^8 [35], the influence of the spatial relationship between R^7 and R^8 on the complex formation was investigated.

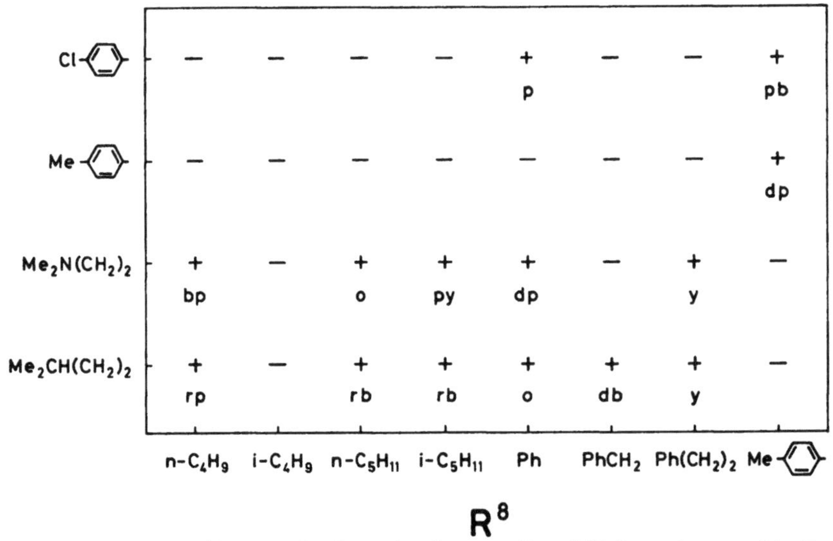

Fig. 9. Crystalline 1:1 molecular complex formation between *19* and *20* for various combinations of R^7 and R^8 [9]. +, Complex formation; —, no complex formation. The color of the complexes: bp, brownish purple; rp, reddish purple; o, orange; rb, reddish brown; py, pale yellow; p, purple; dp, dark purple; db, dark brown; y, yellow; pb, pale brown

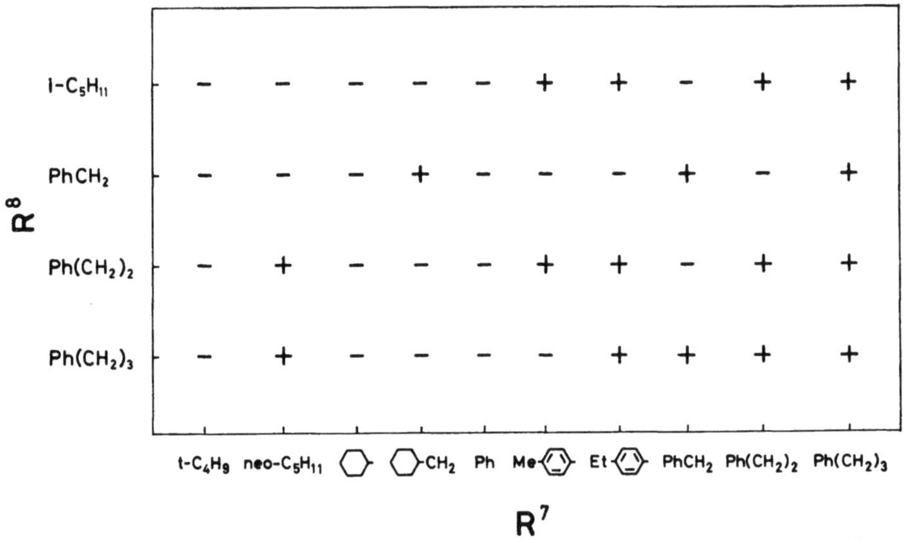

Fig. 10. Crystalline 1:1 complex formation between *19* and *20* for various combinations of R^7 and R^8 [21]. +, Complex formation; —, no complex formation

Among the nine possible combinations where R^7 is a phenylalkyl group [$Ph(CH_2)_x$, x = 0 to 2] and R^8 is a cyclohexylalkyl group [$C_6H_{11}(CH_2)_p$, p = 0 to 2], only two afford the complexes each having two spatially similar groups with the same number of carbon atoms: the pairs Ph—C_6H_{11} and $Ph(CH_2)_2$—$C_6H_{11}(CH_2)_2$ [9]. In this regard, it is interesting that the Ph and cyclohexyl groups are similar in three-dimensional shape to each other as indicated by CPK models. Figure 9 shows that (i) the groups $Me_2CH(CH_2)_2$ and $Me_2N(CH_2)_2$ as R^7, both with a branched structure, give the complexes with approximately the same probability when paired with a variety of groups as R^8 and (ii) the same is also true for the pair of *para*-substituted phenyl groups, *p*-MeC_6H_4 and *p*-ClC_6H_4 [9].

The idea of similarity between the isopentyl and 2-phenylethyl groups (Sect. 2.2.1.1) receives support from the results for the complex formation between *19* and *20* (Fig. 10). Comparison of the complex formation data for R^8 = i-C_5H_{11} with those for R^8 = $PhCH_2$, $Ph(CH_2)_2$, and $Ph(CH_2)_3$ indicates that, of the three phenylalkyl groups, the 2-phenylethyl group shows the closest similarity to the isopentyl group in the trend of the complex formation [21].

The need of the acylurea site participating in intermolecular hydrogen bonding (cf. Figs. 11 and 12) for the complex formation is exemplified by the fact that a 1:1 mixture of *N*-(*p*-dimethylaminophenyl)phenylacetamide (*21*) and *N*-isobutyl-*p*-nitrobenzamide (*22*) gives no crystalline complexes under the same conditions as with *19* and *20*. The trend of the complex formation often changes, when the combinations of R^7 and R^8 are reversed [35].

2.3.2 Crystal Structures of the Complexes

The pattern of hydrogen bonding between a pair of acylurea derivatives revealed by X-ray analysis was consistent with that predicted by spectroscopic studies. Typical examples are illustrated in Figs. 11 and 12; [35–37] two NH ··· O intermolecular hydrogen bonds connect the two molecules. This holds true for the other acylurea derivative *20* (R^8 = CH_2Ph) [38] and 1:1 complexes whose crystal structures were so far determined.

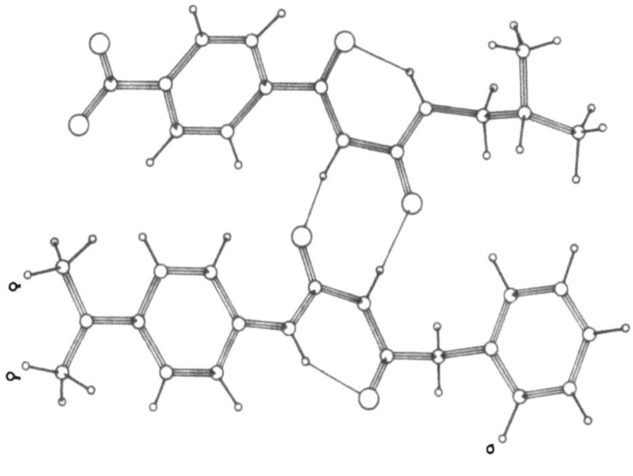

Fig. 11. Hydrogen bond pattern in the crystalline 1:1 complex (R^7 = i-Bu, R^8 = CH_2Ph) [35]. The thin lines indicate hydrogen bonds

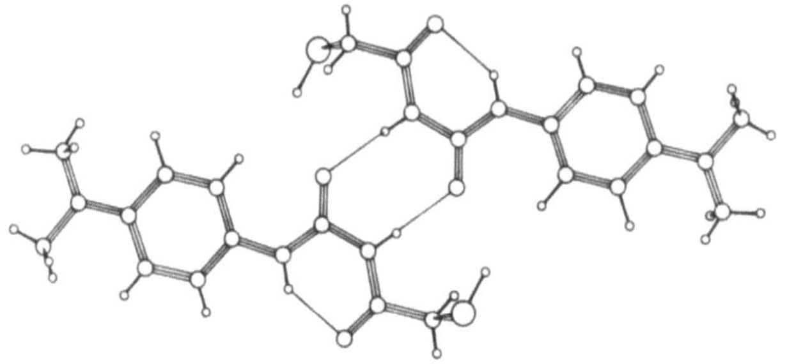

Fig. 12. Hydrogen bond pattern in the crystal of *1a* [37]. The thin lines indicate hydrogen bonds

The acylurea moiety, $-\overset{\mathrm{H}}{\mathrm{N}}\overset{\mathrm{O}}{\overset{\|}{\mathrm{C}}}\overset{\mathrm{H}}{\mathrm{N}}\overset{\mathrm{O}}{\overset{\|}{\mathrm{C}}}-$, of *1a* is planar within ± 0.017 Å, [37] that of *20* ($R^8 = CH_2Ph$) being less planar [38]. The mean plane of the acylurea moiety of *1a* makes an angle of about 4° with the benzene ring of the *p*-$Me_2NC_6H_4$ group, where the Me groups deviate significantly from the mean plane of the benzene ring [37]. The packing mode of the four molecules in the crystal of *1a* (Fig. 13) corresponds to the association scheme of tetramer *14* (Fig. 8). The acylurea moiety and the benzene ring of the $Me_2NC_6H_4$ group overlap each other [37].

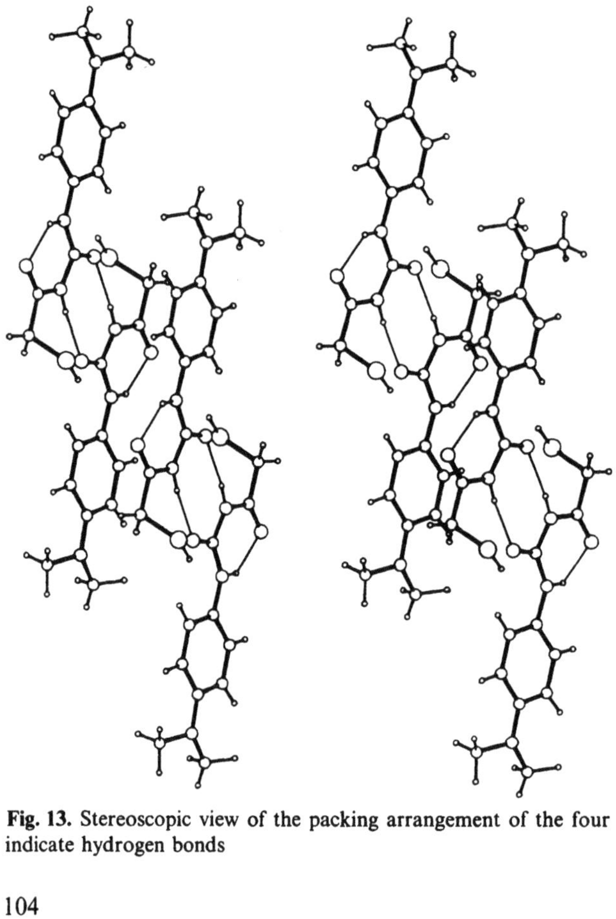

Fig. 13. Stereoscopic view of the packing arrangement of the four *1a* molecules [37]. The thin lines indicate hydrogen bonds

As can be seen from the crystal structure of the 1:1 complex for R^7 = i-Bu and R^8 = CH_2Ph (Fig. 14), the two Me groups of the *p*-$Me_2NC_6H_4$ group are capped by the CH_2Ph group in the other molecule of *20*.[36] Further, the two nonpolar groups — the phenylalanine and leucine side chains — are brought close together.

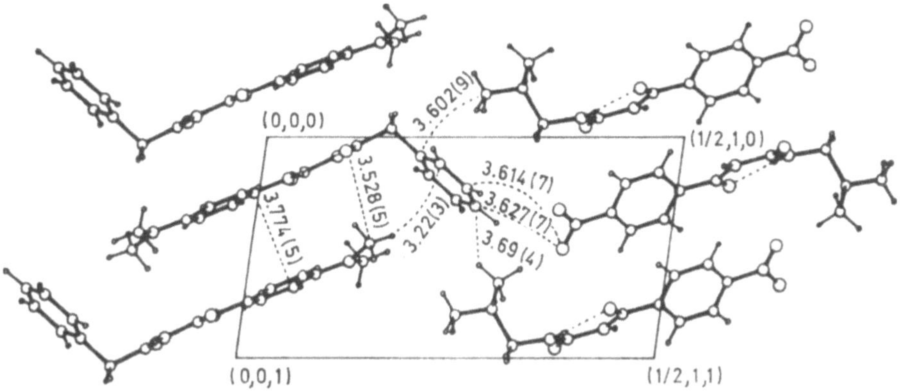

Fig. 14. Crystal structure of the 1:1 complex (R^7 = i-Bu, R^8 = CH_2Ph) showing the contacts of the CH_2Ph group with the i-Bu and Me groups[36]

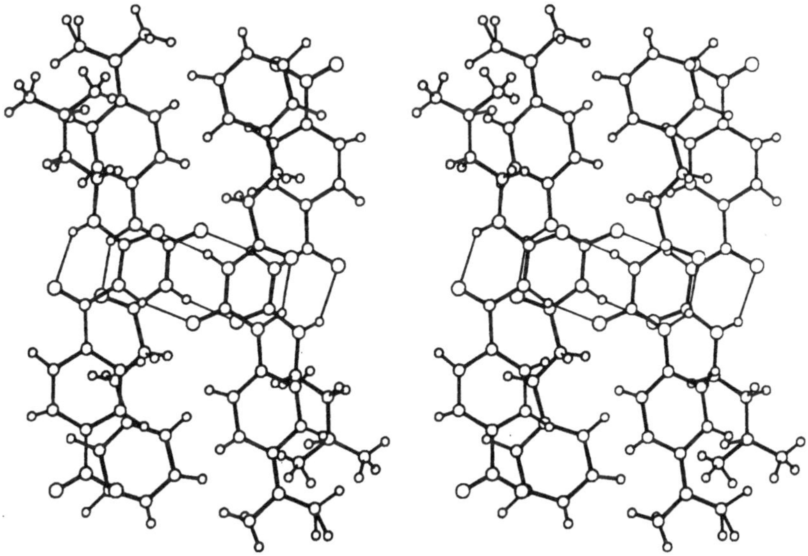

Fig. 15. Stereoscopic view of a tetramer composed of two hydrogen-bonded 1:1 complex units (R^7 = i-C_5H_{11}, R^8 = $(CH_2)_2Ph$)[9]. The thin lines indicate hydrogen bonds

Figure 15 shows a stereoscopic view of the crystalline 1:1 complex where R^7 = i-C_5H_{11} and R^8 = $(CH_2)_2Ph$[9]. The packing mode of the four molecules in the unit cell of this complex corresponds to the association scheme of tetramer *17* (Fig. 8). Of particular interest is that a pair of groups with similar geometrical shape, NMe_2 and $CHMe_2$ [part of $C_6H_4NMe_2$ and $(CH_2)_2CHMe_2$], are in close contact.

This provides substantial evidence for shape similarity effect on the specific recognition.

2.4 Shape-Specific Weak Interactions

The observed selectivity in the oxidation varies drastically with the shape of nonpolar groups R^2 (Sect. 2.2.1)[14b,18,25]. The crystalline 1:1 complex formation depends on the shape of nonpolar groups R^7 and R^8 (Sect. 2.3.1)[9,35]. These observations led us to assume the presence of specific weak interactions[19] related to geometrical shape of the recognition sites.

This type of interactions are considered to possess the following features[14b]:

(i) their energies are smaller than that for the hydrogen bond and similar in order of magnitude to that for the so-called van der Waals interactions;

(ii) they operate individually between groups;

(iii) they reflect the three-dimensional shape of interacting groups.

Recently, experimental evidence has been presented which indicates that the alkyl—phenyl interaction is specific for the three-dimensional shape of the alkyl group, by using gas—liquid partition chromatography[39]. The interaction of a phenyl group is more attractive with a cyclohexyl group, and probably with a branched alkyl group, than with the corresponding straight-chain alkyl group[39].

3 Shape Similarity Effect in Biological Systems

A receptor is a surface membrane component, usually a protein, which regulates some biological event in response to reversible binding of a relatively small molecule[40]. The precise three-dimensional structures of the binding sites of receptors still remain unknown today. Thus, this section mainly describes the correlation of shape similarity between the molecules which would bind to a given receptor with their biological activity.

Diphenylhydantoin (*23*), diazepam (*24*), phenacemide (*25*), and ethylphenacemide (*26*) are all potent anticonvulsants[41]. The first two drugs, though chemically unrelated, have similarities in their three-dimensional structures[42]. The other two open-chain acetylurea derivatives *25* and *26* in the crystalline state bear striking stereochemical

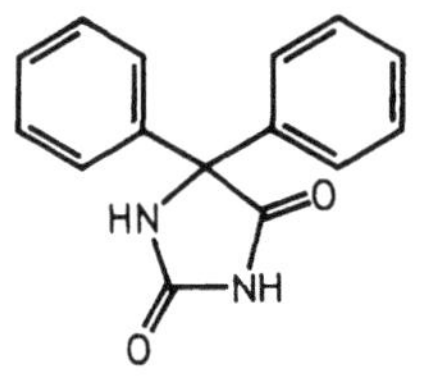

23

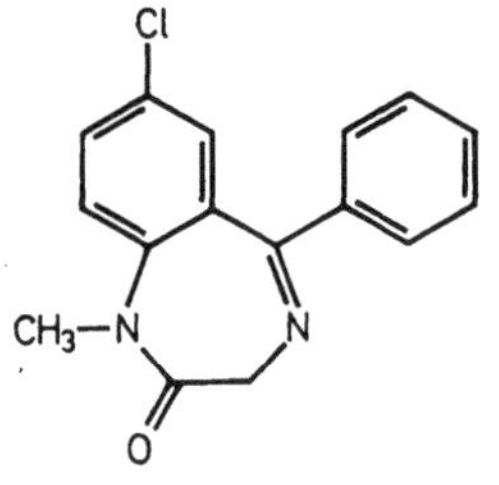

24

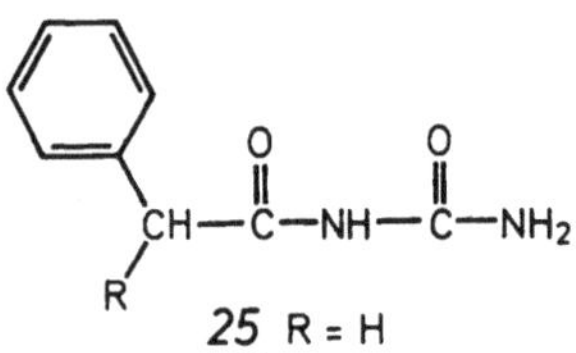

25 R = H
26 R = C_2H_5

resemblances to *23*, due to NH ··· O intramolecular hydrogen bond leading to a pseudocyclic conformation (Fig. 16)[41], as observed for *1a* (Fig. 12)[37].

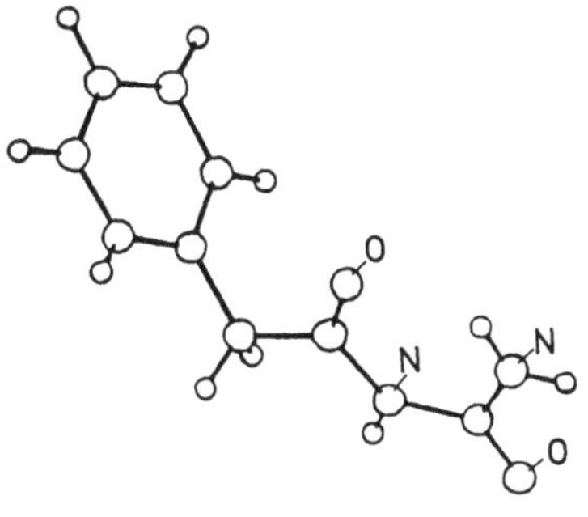

Fig. 16. Perspective drawing of the molecular conformation of *25*[41]

In anticholinergic agents (*27*) the Ph and cyclohexyl groups used as R are bioisosteric groups — chemical groups which, though not necessarily isosteric in the chemical sense, are considered to be equivalent and interchangeable as far as their contribution to the biological activity of drugs is concerned[43]. Moreover, aminopyrine (*28*) and its analog (*29*) each with the NMe_2 or i-Pr group possess antipyretic activity[44].

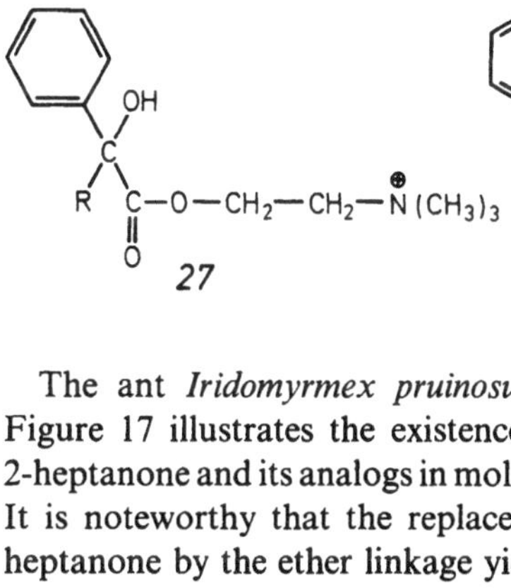

27

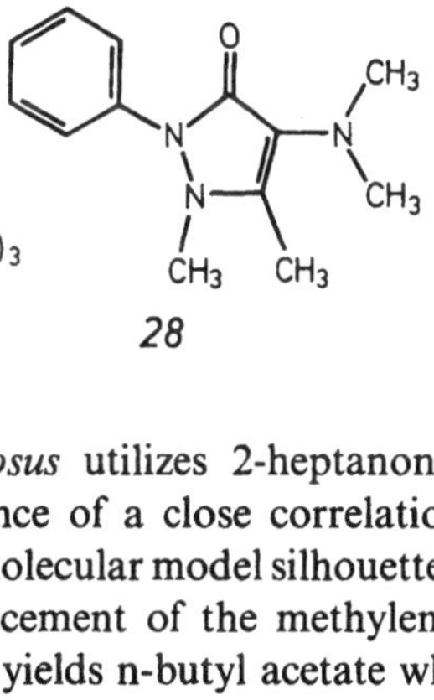

28

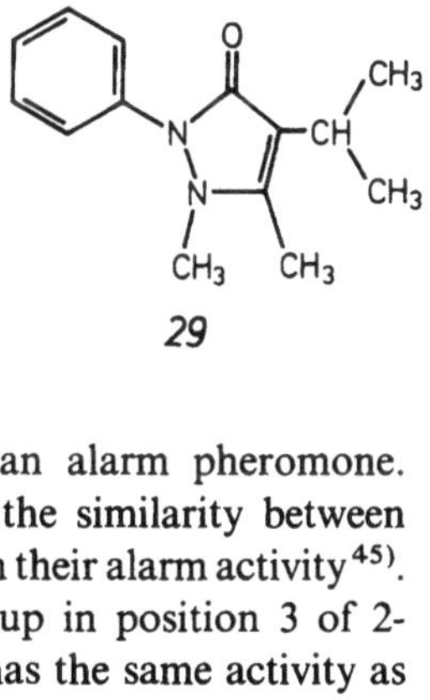

29

The ant *Iridomyrmex pruinosus* utilizes 2-heptanone as an alarm pheromone. Figure 17 illustrates the existence of a close correlation of the similarity between 2-heptanone and its analogs in molecular model silhouettes with their alarm activity[45]. It is noteworthy that the replacement of the methylene group in position 3 of 2-heptanone by the ether linkage yields n-butyl acetate which has the same activity as the natural pheromone. Considerable activity is still retained even when the carbonyl group is replaced by the hydroxyl group.

Olfactory and stereochemical similarities to benzaldehyde have been examined for a large number of homologs of benzaldehyde and nitrobenzene each with the Me, Et, i-Pr, or t-Bu group at the *ortho*, *meta*, or *para* position. There exists a strong correlation between the two: the more closely a compound resembles benzaldehyde in molecular shape, the more similar it is to benzaldehyde in almond odour[46].

Most enzymes consist of several identical or different subunits. It is known that subunits of similar activity but different origin, which therefore differ in size and amino acid composition and sequence, replace each other in oligomeric enzymes, leading to the formation of enzyme chimeras of catalytic activity[47]. The feasibility

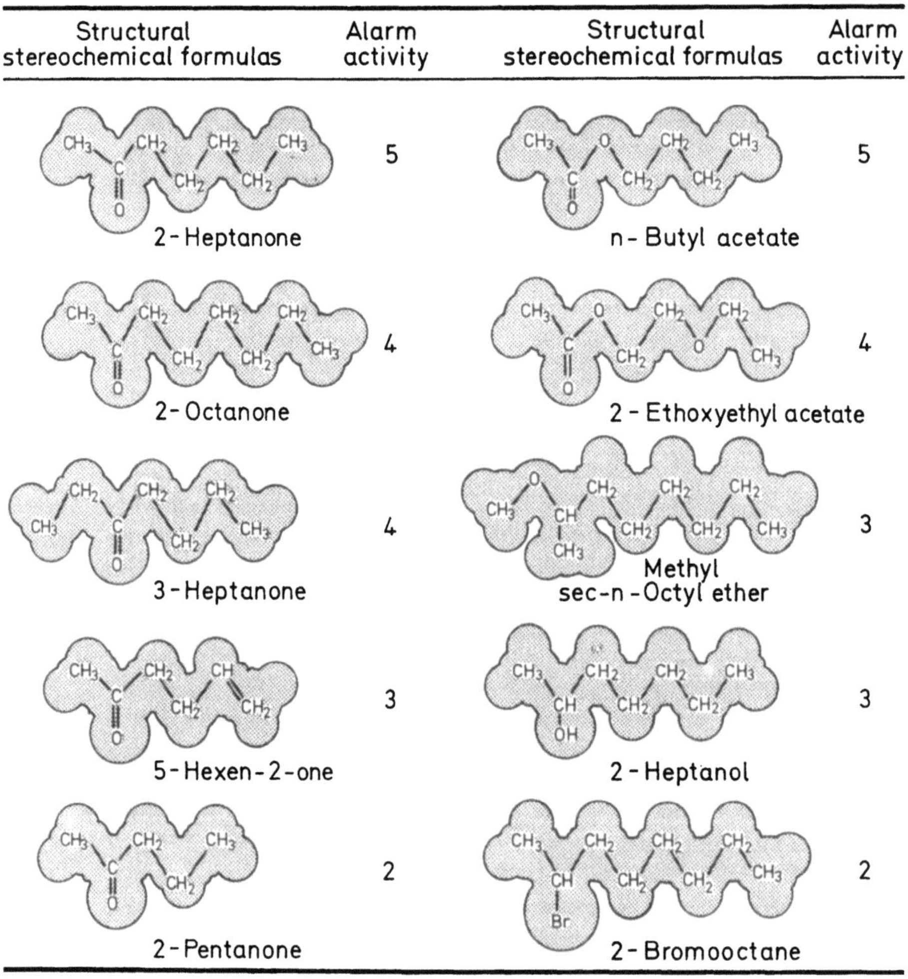

Fig. 17. Correlation of the molecular silhouettes of the natural alarm pheromone (2-heptanone) and its analogs with their alarm activity [45] (Copyright 1969 by the AAAS)

of exchange of subunits indicates that the shape required for specific association of the subunits is very similar [47]. In this connection, X-ray studies have revealed that whale myoglobin and the hemoglobin of an insect larva, while differing in more than 80% of their amino acid sequence, possess almost identical three-dimensional structures [47].

4 Concluding Remarks

Similarity effect on chemical events may be traced back to the old empirical rule that "like dissolves like" (*similia similibus solvuntur*). This rule, however, implies the similarity between solute and solvent molecules in polarity or functional group, but not in molecular shape.

The concept of "isosterism"[48] has been used in medicinal chemistry. Molecules or groups which possess physicochemical similarity (e.g., similarity in size or the number of valence electrons) are called isosteres. The classical isosteres include, for instance, the following two pairs of groups:

(i) Me and Cl;

(ii) $-CH_2-$ and $-O-$.

The selectivity in oxidation of *1a* and *2c* for Z = i-Pr and NO_2, one being an electron-donating group and the other an electron-withdrawing group, was exactly the same (Table 1). This observation cannot be interpreted by any existing substituent constants including the σ[49], π[50], A[22], and MR[50]. Furthermore, the selectivity for *1a* and *2a*$_2$ (R^2 = i-C_5H_{11}) differed sharply from that for *1a* and *2b*$_2$ (R^2 = n-C_5H_{11}) (Fig. 2), these two groups as R^2 being very similar in electronic effect; this would also not be understood by any existing substituent constants, the available constants for these two groups being only the υ[51] (0.68 for the two) and E_s[52] [− 0.40 and − 0.35 for n- and i-C_5H_{11} (cf. − 0.39 and − 0.93 for n- and i-Bu)]. These results would be better explained in terms of shape-specific weak interactions.

The "similarity recognition hypothesis" presented here would be applicable to the specific and precise discrimination in chemical and biological systems. It is hoped that this review will serve to stimulate further work on the physicochemical *origin* of the shape-similarity effect on specific molecular recognition, for example, work on weak interactions specific for the three-dimensional shape of interacting groups.

5 Acknowledgement

I wish to express my gratitude to Professor Teruaki Mukaiyama for his encouragement throughout this work. I would like to thank Professor Yoshio Sasada and Dr. Yuji Ohashi for reading the manuscript and supplying the unpublished figures, and for their valuable discussions. Appreciation is also expressed to my many co-workers who have actively contributed to the progress of the work; their names appear in the references listed. I gratefully thank Dr. Masato Ito for reading the manuscript and his stimulating discussions, and Mr. Shigeki Yagi and Mr. Koki Tajima for drawing several figures. This work was partly supported by a Grant-in-Aid for Scientific Research from the Ministry of Education, Science and Culture.

6 References

1. Simon, Z.: Angew. Chem. Int. Ed. Engl. *13*, 719 (1974)
2. Holler, E.: ibid. *17*, 648 (1978); see also Schulz, G. E., Schirmer, R. H.: Principles of Protein Structure, New York, Springer 1979
3. For example: (a) Jaenicke, R., Helmreich, E. (eds.): Protein-Protein Interactions, Berlin, Springer 1972,
   (b) Chothia, C., Janin, J.: Nature *256*, 705 (1975),
   (c) Frieden, C., Nichol, L. W. (eds.): Protein-Protein Interactions, New York, Wiley 1981
4. For example: (a) Seeman, N. C., Rosenberg, J. M., Rich, A.: Proc. Natl. Acad. Sci. U.S.A. *73*, 804 (1976),
   (b) Schimmel, P. R.: Acc. Chem. Res. *10*, 411 (1977),
   (c) Caruthers, M. H.: ibid. *13*, 155 (1980),

(d) Ohlendorf, D. H., Anderson, W. F., Fisher, R. G., Takeda, Y., Matthews, B. W.: Nature *298*, 718 (1982),
(e) Frederick, C. A., Grable, J., Melia, M., Samudzi, C., Jen-Jacobson, L., Wang, B.-C., Greene, P., Boyer, H. W., Rosenberg, J. M.: Nature *309*, 327 (1984)
5. For example: (a) Givol, D.: A Structural Basis for Molecular Recognition: the Antibody Case, in: Receptors and Recognition (Cuatrecasas, P., Greaves, M. F. eds.), Series A, Vol. 2, p. 1, London, Chapman and Hall 1976,
(b) Capra, J. D., Edmundson, A. B.: Sci. American *236*, 50 (1977)
6. For example: (a) Lübke, K., Schillinger, E., Töpert, M.: Angew. Chem. Int. Ed. Engl. *15*, 741 (1976),
(b) De Meyts, P., Van Obberghen, E., Roth, J., Wollmer, A., Brandenburg, D.: Nature *273*, 504 (1978)
7. Smith, H. O.: Science *205*, 455 (1979)
8. Arber, W.: Angew. Chem. Int. Ed. Engl. *17*, 73 (1978);
for other biological implications of DNA methylation, see, for example, (a) Razin, A., Riggs, A. D.: Science *210*, 604 (1980),
(b) Ehrlich, M., Wang, R. Y.-H.: ibid. *212*, 1350 (1981),
(c) Bird, A. P.: Nature *307*, 503 (1984)
9. Endo, T., Tasai, H., Miyazawa, K., Endo, M., Kato, K., Uchida, A., Ohashi, Y., Sasada, Y.: J. Chem. Soc., Chem. Commun. 636 (1983)
10. Fischer, E.: Chem. Ber. *27*, 2985 (1894)
11. For example: (a) Bunton, C. A.: Micellar Reactions, in: Applications of Biochemical Systems in Organic Chemistry (Jones, J. B., Sih, C. J., Perlman, D. eds.), Part 2, p. 731, New York, Wiley 1976,
(b) Moss, R. A., Lee, Y.-S., Alwis, K. W.: J. Am. Chem. Soc. *102*, 6646 (1980),
(c) Ohkubo, K., Sugahara, K., Yoshinaga, K., Ueoka, R.: J. Chem. Soc., Chem. Commun. 637 (1980)
12. For example: (a) Bender, M. L., Komiyama, M.: Cyclodextrin Chemistry, Berlin, Springer 1978,
(b) Saenger, W.: Angew. Chem. Int. Ed. Engl. *19*, 344 (1980),
(c) Tabushi, I.: Acc. Chem. Res. *15*, 66 (1982)
13. For example: (a) Cram, D. J., Cram, J. M.: Science *183*, 803 (1974),
(b) Stoddart, J. F.: Chem. Soc. Rev. *8*, 85 (1979),
(c) Lehn, J. M.: Acc. Chem. Res. *11*, 49 (1978),
(d) Vögtle, F. (ed.): Host Guest Complex Chemistry, Vol. I and II, Berlin, Springer 1981 and 1982,
(e) Cram, D. J.: Science *219*, 1177 (1983)
14. (a) Endo, T., Noguchi, S., Mukaiyama, T.: Bull. Chem. Soc. Jpn. *44*, 3424 (1971),
(b) Endo, T., Okubo, A., Kaneko, Y., Uehara, M., Tasai, H., Sato, A., Nikki, K., Nakagawa, N., Kamei, S.: ibid. *55*, 2224 (1982)
15. Endo, T., Oda, K., Mukaiyama, T.: Chem. Lett. 443 (1974)
16. Endo, T., Takeda, Y., Orii, T., Kaneko, Y., Kondo, M.: ibid. 1455 (1979)
17. Anfinsen, C. B., Sela, M., Cooke, J. P.: J. Biol. Chem. *237*, 1825 (1962)
18. Endo, T., Kuwahara, A., Tasai, H., Murata, T., Hashimoto, M., Ishigami, T.: Nature *268*, 74 (1977)
19. Endo, T., Takeda, Y., Orii, T., Murata, T., Sakai, M., Nakagawa, N., Nikki, K.: Chem. Lett. 1291 (1980)
20. For example: (a) Ondetti, M. A., Rubin, B., Cushman, D. W.: Science *196*, 441 (1977),
(b) Griffith, O. W., Meister, A.: Proc. Natl. Acad. Sci. U.S.A. *74*, 3330 (1977),
(c) Patchett, A. A., Harris, E., Tristram, E. W., Wyvratt, M. J., Wu, M. T., Taub, D., Peterson, E. R., Ikeler, T. J., ten Broeke, J., Payne, L. G., Ondeyka, D. L., Thorsett, E. D., Greenlee, W. J., Lohr, N. S., Hoffsommer, R. D., Joshua, H., Ruyle, W. V., Rothrock, J. W., Aster, S. D., Maycock, A. L., Robinson, F. M., Hirschmann, R., Sweet, C. S., Ulm, E. H., Gross, D. M., Vassil, T. C., Stone, C. A.: Nature *288*, 280 (1980)
21. Endo, T., Kaneko, Y., Mitsuhashi, M., Endo, M., Takeda, Y., Ito, M. M., Nikki, K., Nakagawa, N., Matsushita, K.: Chem. Lett. 551 (1984)
22. Eliel, E. L.: Angew. Chem. Int. Ed. Engl. *4*, 761 (1965)

23. Endo, T., Takeda, Y., Kamada, H., Kayama, S., Tasai, H.: Chem. Lett. 417 (1980)
24. Davies, M., Thomas, D. K.: J. Phys. Chem. *60*, 763 (1956)
25. Endo, T., Takeda, Y., Orii, T., Kuwahara, A., Ohta, M., Sakai, M., Okada, R., Hashimoto, M.: Bull. Chem. Soc. Jpn. *53*, 2687 (1980)
26. Endo, T., Hashimoto, M., Orii, T., Ito, M. M.: ibid. *57*, 1562 (1984)
27. Springer, C. S., Jr., Meek, D. W.: J. Phys. Chem. *70*, 481 (1966)
28. Barnes, A. J., Hallam, H. E., Howells, J. D. R.: J. Chem. Soc., Faraday Trans. II, *68*, 737 (1972)
29. Benesch, R., Benesch, R. E.: Science *185*, 905 (1974)
30. For example: (a) Tanford, C.: The Hydrophobic Effect: Formation of Micelles and Biological Membranes, New York, Wiley 1973,
(b) Franks, F.: The Hydrophobic Interaction, in: Water — a Comprehensive Treatise (Franks, F. ed.), Vol. 4, p. 1, New York, Plenum Press 1975,
(c) Ben-Naim, A.: Hydrophobic Interactions, New York, Plenum Press 1980
31. (a) Harris, M. J., Higuchi, T., Rytting, J. H.: J. Phys. Chem. *77*, 2694 (1973),
(b) Chothia, C.: Nature *248*, 338 (1974)
32. Fujita, T., Iwasa, J., Hansch, C.: J. Am. Chem. Soc. *86*, 5175 (1964)
33. Hansch, C., Steward, A. R., Iwasa, J., Deutsch, E. W.: Mol. Pharmacol. *1*, 205 (1965)
34. Némethy, G., Scheraga, H. A.: J. Chem. Phys. *36*, 3401 (1962)
35. Endo, T., Miyazawa, K., Endo, M., Uchida, A., Ohashi, Y., Sasada, Y.: Chem. Lett. 1989 (1982)
36. Uchida, A., Ohashi, Y., Sasada, Y., Moriya, M., Endo, T.: Acta Cryst. *C40*, 120 (1984)
37. Uchida, A., Ohashi, Y., Sasada, Y., Kaneko, Y., Endo, T.: ibid. *C40*, 115 (1984)
38. Ohashi, Y., Uchida, A., Sasada, Y., Kinoshita, K., Endo, T.: ibid. *C40*, 117 (1984)
39. Endo, T., Ito, M. M., Yamada, Y., Saito, H., Miyazawa, K., Nishio, M.: J. Chem. Soc., Chem. Commun. 1430 (1983)
40. Lindstrom, J.: Antibodies to Receptors for Acetylcholine and Other Hormones, in: Receptors and Recognition (Cuatrecasas, P., Greaves, M. F. eds.), Series A, Vol. 3, p. 1, London, Chapman and Hall 1977
41. Camerman, A., Camerman, N.: Proc. Natl. Acad. Sci. U.S.A. *74*, 1264 (1977)
42. Camerman, A., Camerman, N.: Science *168*, 1457 (1970)
43. Ariëns, E. J.: A General Introduction to the Field of Drug Design, in: Drug Design (Ariëns, E. J. ed.), Vol. I, p. 1, New York, Academic Press 1971
44. Korolkovas, A.: Essentials of Molecular Pharmacology — Background for Drug Design, Chapter 4, New York, Wiley 1970
45. Amoore, J. E., Palmieri, G., Wanke, E., Blum, M. S.: Science *165*, 1266 (1969)
46. Amoore, J. E.: Nature *233*, 270 (1971)
47. Hartmann, G. R.: Angew. Chem. Int. Ed. Engl. *15*, 181 (1976)
48. Thornber, C. W.: Chem. Soc. Rev. *8*, 563 (1979)
49. McDaniel, D. H., Brown, H. C.: J. Org. Chem. *23*, 420 (1958)
50. Hansch, C., Leo, A., Unger, S. H., Kim, K. H., Nikaitani, D., Lien, E. J.: J. Med. Chem. *16*, 1207 (1973)
51. Charton, M.: J. Am. Chem. Soc. *97*, 1552 (1975)
52. Taft, R. W., Jr.: Separation of Polar, Steric, and Resonance Effects in Reactivity, in: Steric Effects in Organic Chemistry (Newman, M. S. ed.), p. 556, New York, Wiley 1956

Macrocyclic Polyamines as Biological Cation and Anion Complexones — An Application to Calculi Dissolution

Eiichi Kimura

Department of Medicinal Chemistry, Hiroshima University School of Medicine, Kasumi, Hiroshima 734, Japan

Table of Contents

1 Introduction

Since first reported in 1967 by Pedersen [1)], synthetic macrocyclic polyethers ("crown ethers") have aroused great interest in almost every field of chemistry[1] [2–9)]. Considerable fundamental knowledge of the chemistry of these compounds has accumulated and now a great deal of effort is being directed toward their application. On the other hand, macrocyclic saturated polyamines having N donors in place of the O donors typical of crown ethers have, until recently, been viewed mostly as host ligands for transition metal and heavy metal ions. Studies of the macrocyclic polyamines have involved the thermodynamics and kinetics of complexation and the structures of their complexes [8, 10, 11, 12)].

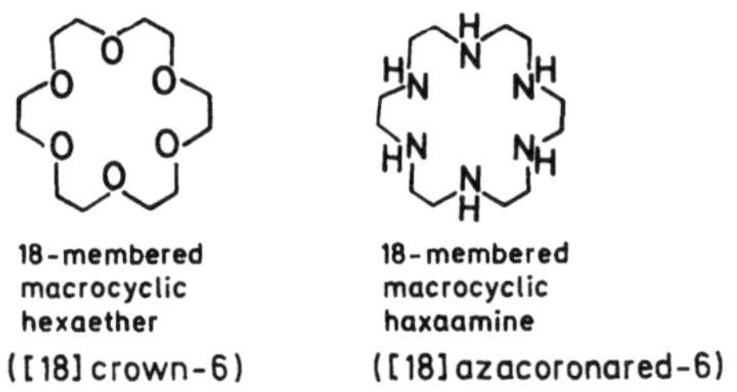

Interest in macrocyclic polyamines has been mostly limited to that shown by coordination chemists.

Macrocyclic polyamines may be viewed as an extended form of linear polyamines [13–16)] with one less degree of saturation. Polyamines with lower degrees of saturation, have important biological functions. Compounds such as macrocyclic polyimines [8, 17)] and porphyrines [18)], function as O_2 carriers and activators, promote photosynthesis, form the basic structure of vitamine B_{12}, etc., and for these reasons have been subjects of intense investigation.

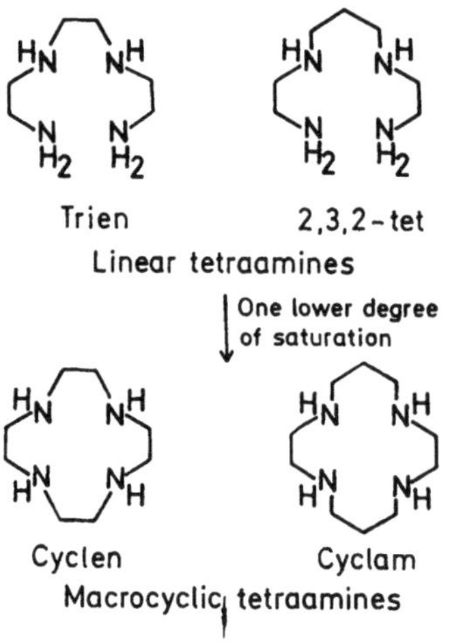

[1] Topics Curr. Chem. Vol. *8* (1981), *101* (1982), *121* (1984), Springer Verlag, Heidelberg/New York

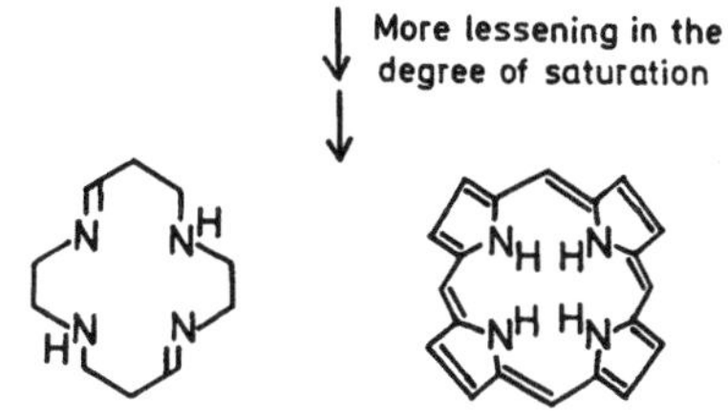

Macrocyclic polyimine   Porphyrin

In addition, the linear, biogenic polyamines, spermidine and spermine, originally isolated from human sperm [19], have been long known. Their structures are shown below.

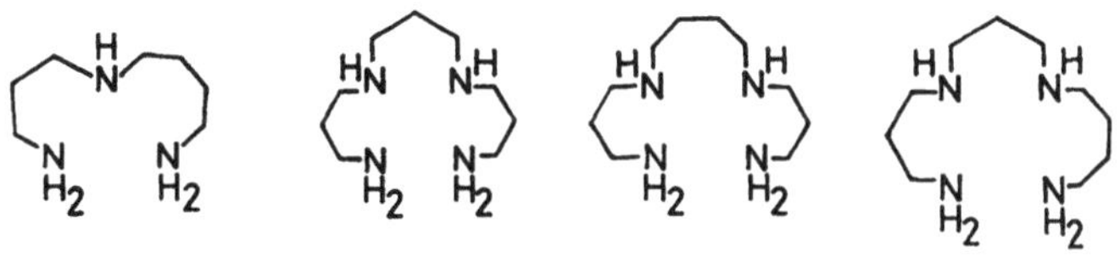

Spermidine   Thermine   Spermine   Thermospermine

Biogenic polyamines

Their biological and physiological significance have only recently been broadly recognized [20].

A correlation of enhanced synthesis of polyamines with rapid growth or cell proliferation has been observed [21]. From a physiological point of view, polyamines are implicated as regulators of cell proliferative activity [22]. It is well known that polyamines, as protonated polycations, can bind with nucleotide and nucleic acid anions [23,24] to affect biochemical reactivities and stabilize tertiary structures [25,26].

Recently considerable attention has been directed at anion binding ligands. Macrobicyclic [27–29] and macrotricyclic amines [30,31] were topologically designed to host anions such as spherical Cl^-, linear N_3^- [32]. These anion substrates are incorporated into macrocyclic cavities lined with appropriate anion-binding sites capable of forming hydrogen bonds like those of protonated amines (see *I*, below).

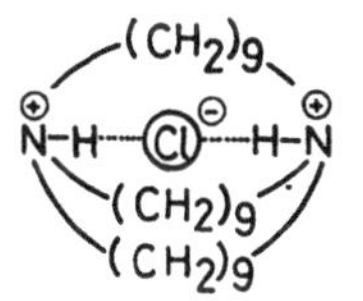

I

Other types of complexons for polyanions (e.g. polycarboxylates and phosphates) are linear ligands with polyguanidinium cations or polyammonium cations as functional groups [33].

Recently, systematic studies of macromonocyclic polyamines have revealed reactivities with biological cations and anions which may give significant insight

Chart 1. Structure and abbreviated nomenclature of macrocyclic saturated polyamines and their relevant compounds.

Triamines

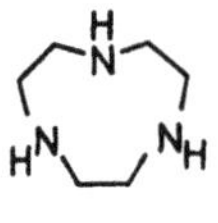
[9]ane N_3

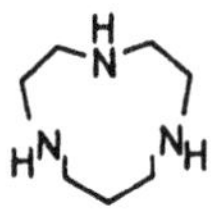
[10]ane N_3

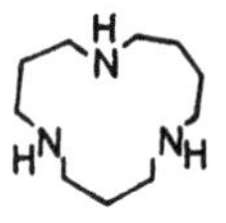
[14]ane N_3
(Cyclic spermidine)

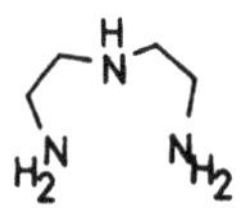
Dien

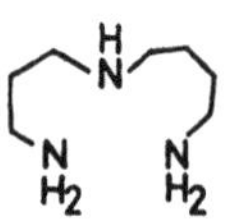
Spermidine

Tetraamines

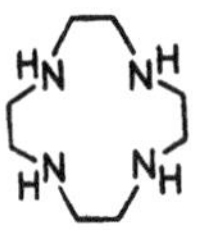
[12]ane N_4
(Cyclen)

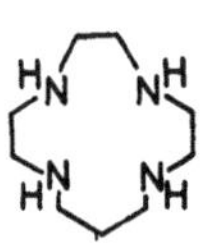
[13]ane N_4

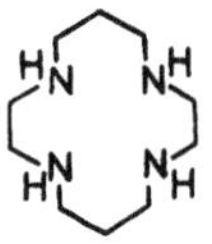
[14]ane N_4
(Cyclam)

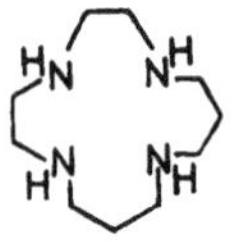
iso-[14]ane N_4

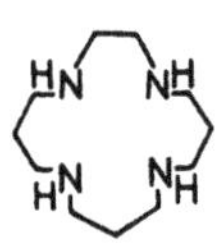
[15]ane N_4

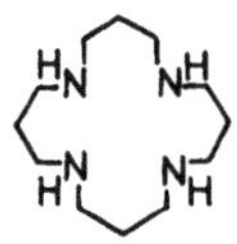
[16]ane N_4

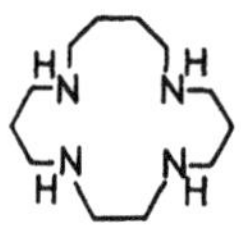
iso-[16]ane N_4
(Cyclic spermine)

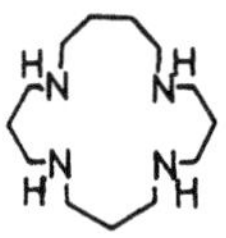
[17]ane N_4
(Cyclic spermine)

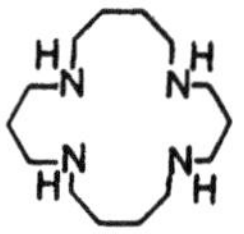
[18]ane N_4
(Cyclic spermine)

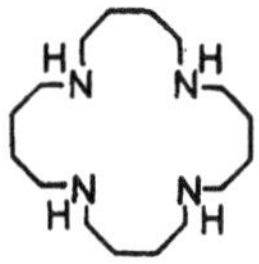
[20]ane N_4

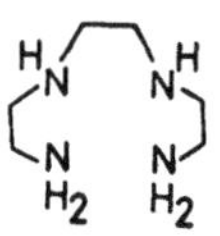
Trien

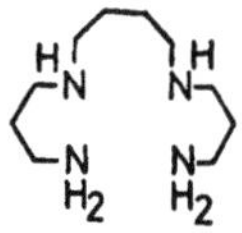
Spermine

Pentaamines

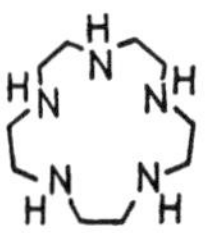
[15]ane N_5

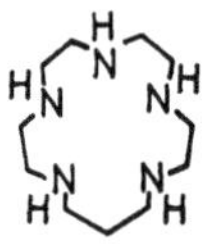
[16]ane N_5

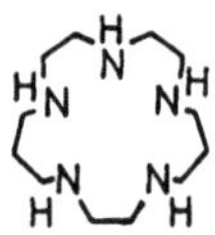
[17]ane N_5

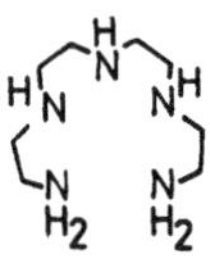
Tetren

Hexaamines

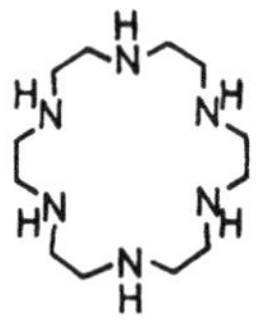
[18]ane N_6

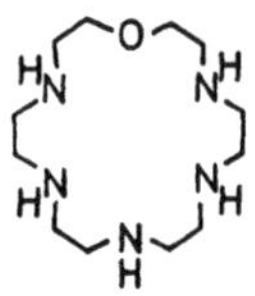
[18]ane N_5O

Hexaamines

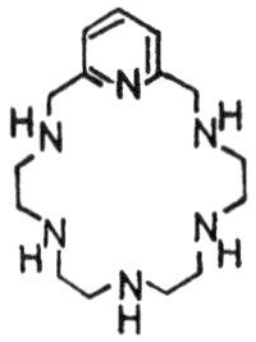
Py[18]ane N_6

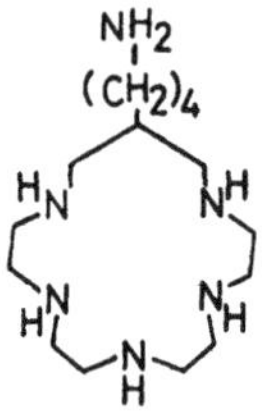
[16]ane N_5
$BuNH_2$-tailing

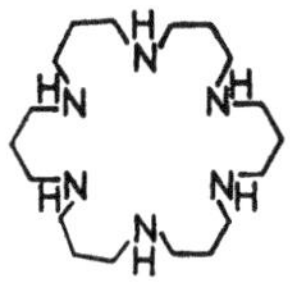
[24]ane N_6

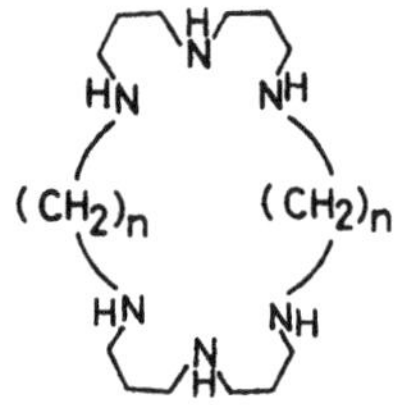
[32]ane N_6, n=7
[38]ane N_6, n=10

Octaamine

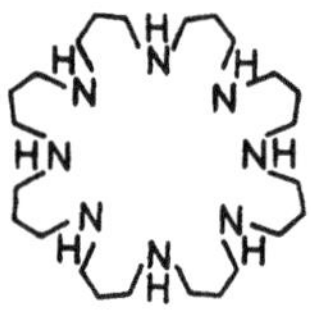
[32]ane N_8

into anion complex chemistry and implications for many biochemical, physiological, and pharmacological phenomena. As one potential application of macrocyclic polyamines, dissolution of urinary calculi has been tested [34,35].

2 Structures and Abbreviations

The macromonocyclic polyamines and the relevant compounds to be dealt with in this paper are shown in Chart I, along with abbreviation. For the synthetic procedures and physical properties of these compounds readers are requested to consult the references.

3 Proton Uptake Properties

The characteristic property distinguishing macrocyclic polyamines from their linear counterparts is seen in successive protonation. One is the higher N basicity to the first proton; and another is a sudden drop of N basicities in the later stages of protonation. Table 1 lists the protonation constants (Eq. 1) for the macrocyclic polyamines in comparison with the corresponding values for their linear homologues. When a linear triamine (*e.g.* dien) [36,37] is cyclized to, say, (9)aneN_3, the basicity of the first amine increases ($\log K_1 = 10.59$ *vs* 9.70), but the basicity of the second and especially the third amine diminish ($\log K_2 = 6.88$ *vs* 8.95, $\log K_3 < 1$ *vs* 4.25) [36]

$$H_{i-1}L^{(i-1)+} + H^+ \rightleftarrows H_iL^{i+}$$

$$K_i = \frac{[H_iL^{i+}]}{[H_{i-1}L^{(i-1)+}][H^+]} \tag{1}$$

By virtue of the conformational restraint imposed by the cyclic structure of (9)aneN_3, three of the N orbitals containing non-bonding electron pairs are directed toward the center of the cavity. The increased electron density in the macrocyclic cavity facilitates the first H^+ uptake, resulting in the higher K_1 value. With the first proton in place, the effect of the close proximity of the nitrogen bases retards the second protonation. The third protonation constant, $\log K_3$, of macrocyclic triamines is particularly small (less than 1), because the third proton is subject to steric and/or electrostatic repulsion from the two protons already coordinated and is probably inhibited from entering the nitrogen cavity.

A similar argument holds for tetraamines [16,36–40]. The basicities of macrocyclic systems for the first and second protons are higher and the subsequent basicities for the third and fourth protons are much lower than the corresponding basicities of linear tetraamines. According to the X-ray structural analysis of (14)ane$N_4 \cdot 2\,HClO_4$, the two protons are incorporated centrosymmetrically within the cavity [41]. Among the 12- to 15-membered tetraamines, (14)aneN_4 exhibits the highest values of the first and second protonation constants, indicating that it assumes the most stable conformation for enclosure of one and two protons. When the ring size expands and conformational restraint is removed, as in the case of cyclic spermines (e.g.

Table 1. Mixed protonation constants of polyamines. (25 °C and I = 0.2 M unless otherwise listed)

Polyamine	log K_1	log K_2	log K_3	log K_4	log K_5	log K_6	log K_7	log K_8
Triamines								
[9]aneN_3[a]	10.59	6.88	<1					
[10]aneN_3[b]	10.85	6.76	<1					
[13]aneN_3[c] (cyclic spermidine)	9.79	8.13	4.18					
dien[a]	9.90	9.05	4.02					
spermidine[c]	10.89	9.81	8.34					
Tetraamines								
[12]aneN_4[d]	10.70	9.70	1.73	<1				
[13]aneN_4e	11.10	10.10	1.7	1				
[14]aneN_4[a]	11.50	10.30	1.62	<1				
iso[14]aneN_4[f]	11.05	9.98	3.3	1				
[15]aneN_4[g]	11.20	10.10	<2	<1				
iso[16]aneN_4[c] (cyclic spermine)	10.04	9.69	6.80	3.54				
[17]aneN_4[c] (cyclic spermine)	10.23	9.66	7.40	5.31				
[18]aneN_4[h] (cyclic spermine)	10.36	9.97	7.00	6.70				
trien[d]	10.09	9.31	6.75	3.39				
spermine[c]	10.80	10.02	8.85	7.96				
Pentaamines[i]								
[15]aneN_5	10.85	9.65	6.00	1.74	1.2			
[16]aneN_5	10.64	9.49	7.28	1.7	1.5			
[17]aneN_5	10.32	9.62	7.36	4.10	2.38			
tetren	10.36	9.65	8.50	4.70	2.40			

Hexaamines								
[18]aneN_6[j]	10.19	9.23	8.73	4.09	2	1		
[24]aneN_6[k]	10.45	10.35	9.05	7.90	7.15	6.60		
[32]aneN_6[k]	10.85	10.60	9.80	9.05	7.40	6.65		
Octaamines[l]								
[32]aneN_8	10.70	10.45	9.65	9.00	8.05	7.50	6.95	6.45

[a] Ref. [36)], [b] Ref. [37)], [c] Ref. [42)], [d] Kodama, M., Kimura, E., *J. Chem. Soc., Chem. Commun.*, 326 (1975): *J. Chem. Soc., Dalton Trans.*, 116 (1976). [e] Kodama, M., Kimura, E., *J. Chem. Soc., Chem. Commun.*, 891 (1975): *J. Chem. Soc., Dalton Trans.*, 1720 (1976). [f] Kodama, M., Kimura, E., *J. Chem. Soc., Dalton Trams.*, 327 (1980). [g] Ref. [40)], [h] Ref. [43)], [i] Kodama, M., Kimura, E., *J. Chem. Soc., Dalton Trans.*, 104 (1978). [j] Ref. [45)], [k] $I = 0.1$ M (Ref. [46)]), [l] Ref. [51)].

(18)aneN$_4$)[42,43], the protonation trend characteristic of the smaller macrocyclic systems disappears and the protonations occur successively, as with linear tetra-amines.

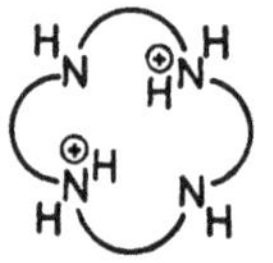

Diprotonated form of
[12]-[15]ane N_4

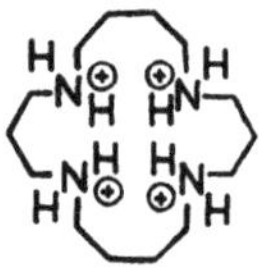

Tetraprotonated form of
[18]ane N_4
(a cyclic spermine)

Macrocyclic pentaamines follow similar amine basicity patterns[44]. Thus, as the ring size increases the values of K_4 and K_5 become higher, converging to the corresponding values for the linear pentaamine, tetren. With macrocyclic hexaamines, a similar protonation pattern persists. While the 18-membered hexaamine [18]ane N_6 is found to exhibit the protonation characteristics of conformationally restrained macrocyclic systems[45] the 24-membered hexaamine (24)aneN$_6$ is more like a linear system in its reaction with protons[46]. Thus, the latter can hold almost six protons to act as H_6L^{6+} and the former acts as H_3L^{3+} at neutral pH (Fig. 1).

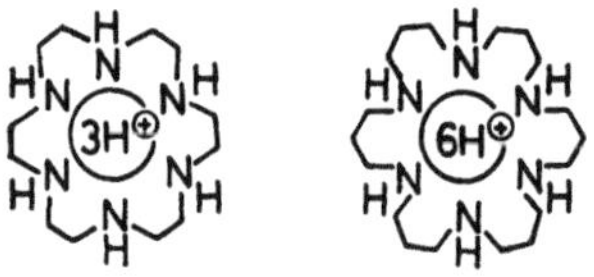

[18] ane N_6 [24] ane N_6 **Fig. 1.** The most predominant species at neutral pH

4 Uptake of Alkaline Metal Ions and Alkaline Earth Metal Ions. Crown Ether-like Functions

Since the N donor is softer than the O donor, macrocyclic polyamines tend to bind strongly with softer metal ions (*e.g.* transition metals) and were not considered to be better than macrocyclic polyethers in complexing hard metal ions such as alkali metals or alkaline earth metal ions[47]. Macrocyclic hexaamines turned out to be exceptional. (18)aneN$_6$[45] and its homologues[48] can form 1:1 complexes in aqueous alkaline solutions with K^+, Ca^{2+}, Sr^{2+} or La^{3+} (Table 2). For instance, Ca^{2+} is sequestered at pH values greater than 9. Considering the protonation constants, the reactive species are seen to be those having fewer than two protons[45]. It is an

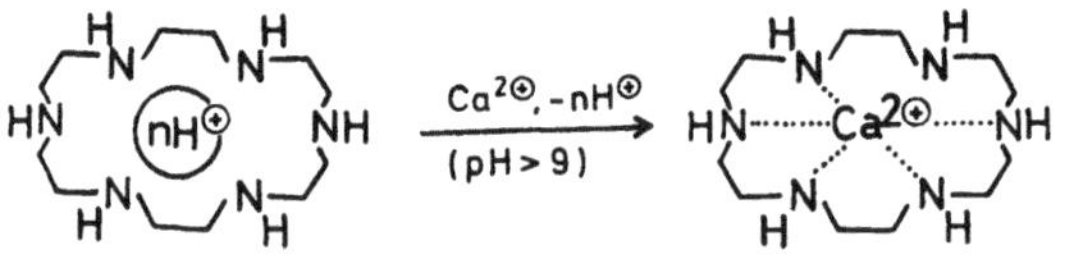

$\underline{n} \leq 2$

Table 2. 1:1 Complex formation constants log K_{ML} (25 °C, I 0.2 M)

Metal Ion	Ligand				
	[18]aneN_6[a]	[18]aneN_5O[b]	Py[18]aneN_6[b]	BuNH_2-[16]aneN_5[b]	18-Crown-6[c]
Na^+	Nil	Nil	Nil	Nil	0.8
K^+	0.8	Nil	Nil	Nil	2.03
Mg^{2+}	Nil	Nil	Nil	2.5	Nil
Ca^{2+}	2.5	2.30	2.70	Nil	0.5 >
Sr^{2+}	3.2				2.7
Pb^{2+}	14.1				4.3

[a]) Ref. [45], [b]) Ref. [48]), [c]) Izatt, R. M., Terry, R. E., Haymore, B. L., Hansen, L. D., Dalley, N. K., Avondet, A. G., Christensen, J. J., J. Am. Chem. Soc. *98*, 7620 (1976).

interesting comparison that $Ca(NH_3)_6^{2+}$ must be prepared under anhydrous conditions and is very unstable in the presence of water (immediately dissociating into Ca^{2+}(aq) and NH_3). The macrocyclic linkage of (18)aneN_6 establishes a high energy barrier to dissociation of the six amine donors from the central metal ion, thus making the 1:1 complexes relatively stable.

While [18]crown-6 has a higher affinity for K^+ than for Ca^{2+}, [18]azacoronand-6 displays the opposite preference. The strong coordinate bonding character of Ca^{2+}-(18)aneN_6 may account for its higher stability compared to Ca^{2+}-[18]crown-6 whose bonding nature is mostly electrostatic. This explanation als accounts for the higher stability of Sr^{2+}- and Pb^{2+}-(18)aneN_6 complexes [45]. Replacement of an amine donor of (18)aneN_6 by an O donor ((18)aneN_5O) or a pyridyl N donor (py (18)aneN_6) reduces the affinity for metal ions except for Ca^{2+}.

Neither of these 18-membered macrocyclic ligands can form complexes with Mg^{2+}. One probable reason is that the cavity size is too big for the smaller Mg^{2+} ion. In this respect, a macrocyclic pentaamine with an n-butyl amine tail (n-BuNH_2-(16)aneN_5) is very interesting in that it can bind with Mg^{2+}, but not with Ca^{2+}. Mg^{2+} would better fit the 16-membered ring cavity. However, the parent ligand, (16)aneN_5, does not show affinity for Mg^{2+}. The appended amine evidently serves to cap the Mg^{2+} ion and keep it within the macrocyclic cavity (see Fig. 2). A homologous n-BuNH_2-[14]aneN_4 does not complex with Mg^{2+} [48].

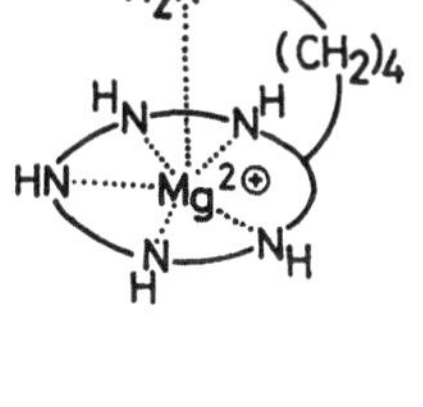

Fig. 2. A proposed structure for Mg^{2+}-n-BuNH_2-[16]aneN_5

5 Uptake of Biological Polyanions

Macrocyclic polyammonium cations containing more than one proton within the macrocyclic cavity have several unique features:

1) the cyclic conformations become more rigid due to intramolecular $N\text{–}H^+ \ldots N$ hydrogen bonding;

2) since a number of protons are condensed in a narrow space, they can behave as polycations;

3) unlike polycationic metal ions, they can hydrogen bond with oxyanions by providing their NH^+ protons, which assists association of ion pairs. Accordingly, the macrocyclic polyamines possessing greater numbers of protons can form stronger ion pairs with polyoxyanions.

5.1 Polycarboxylate Anions

The discovery of ion-pair formation with macromonocyclic polyamines stemmed from their electrophoretic analysis [49, 50]. Certain polycarboxylates used as buffers or electrolytes strongly influence the sequence and migration distances of some polyamines (Table 3). In monocarboxylate solutions such as acetate and lactate (at pH ~ 6), all the polyamines tested moved as anticipated. That is, they moved as protonated cations toward the cathode at more or less similar rates. On the other hand, in citrate buffers at the same pH, some of the polyamines showed unusually slow movement or migration toward the anode. The peculiarly moving polyamines were the larger macrocycles, namely, cyclic spermines ([16]aneN_4), [17]aneN_4, [15]–[17]aneN_5, and [18]aneN_6. By comparison, the smaller macrocyclic triamine such as the cyclic spermidine, and tetraamines, [12]–[14]aneN_4 were virtually unaffected. These findings led to the proposal of ion-pair formation between some macrocyclic polyamine polycations, H_3L^{3+} or H_4L^{4+}, and polycarboxylate anions [50]. A 1:1 complex for [18]aneN_6 and citrate is proposed as in Fig. 3.

At pH 7, [13]aneN_3 or [12]–[15]aneN_4 accommodate only two nitrogen-bound protons and these dipositive ammonium cations are apparently unable to provide sufficient electrostatic attraction to polycarboxylate anions for ion-pair formation. In contrast, the macrocyclic spermines, pentaamines and hexaamines accommodate more than three nitrogen-bound protons at pH 7 and for these ligands 1:1 associations

Table 3. Electrophoretic mobilities[a] of polyamines in various buffers[b]

0.1 M Buffer (pH)	Polyamine				
	[15]aneN_5	[16]aneN_5	[17]aneN_5	[18]aneN_6	tetren
acetate (5.8)	0.9	1.1	0.9	1.1	1.2
lactate (5.8)	1.0	1.0	0.9	0.9	1.1
malonate (5.9)	0.7	0.7	0.6	0.4	0.9
succinate (5.8)	0.7	0.7	0.7	0.5	0.8
malate (5.7)	0.7	0.8	0.7	0.4	0.8
fumarate (5.8)	0.9	0.9	1.0	0.8	1.0
o-phthalate (6.0)	0.6	0.9	0.5	0.3	0.5
m-phthalate (5.4)	0.8	0.9	0.8	0.7	1.0
citrate (6.0)	0.3	0	−0.2	−0.9	0

[a]) Relative values (at 25 °C I 0.2 M) with respect to [14]aneN_4.
[b]) Ref. [50].

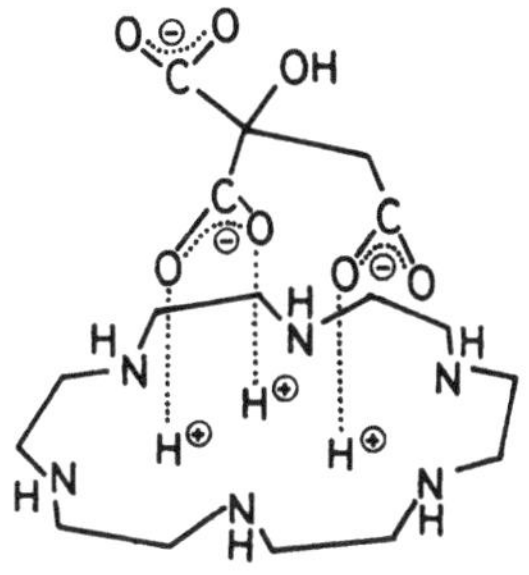

Fig. 3. A proposed model for 1:1 complex of [18]ane$N_6 \cdot 3\,H^+$ with citrate[49]

with di- and tricarboxylate anions have been confirmed by polarography with determination of the association constants (Table 4)[50]. The linear spermine, although acting as H_4L^{4+}, has only weak interaction with citrate[49]. This is due to the wide dispersion of the tetraammonium cation. The tricarboxylate anion, citrate, binds more strongly than any dicarboxylate anion, indicating that electrostatic interaction is fundamental to ion-pair formation in these systems. Furthermore, for a given receptor molecule, the anions most strongly complexed are usually the smallest and most highly charged ones. The stability sequence with [18]ane$N_6 \cdot 3\,H^+$ is oxalate > malonate > maleate > succinate > fumarate.

The linear homologue of pentaamine, tetren, has little interaction with polycarboxylates (except for citrate), which again indicates the operation of macrocyclic effects in stabilizing anion complexes. The higher stability with macrocyclic ligands is a result of the presence of the closed and relatively inflexible structure, which contributes a geometrically more advantageous positioning of the dense $^+$N–H binding sites, efficient hindrance to hydration of the $^+$N–H sites and resistance to deformation. The +3 cations most strongly bound to a given anion are usually those having the highest charge density. Thus, [16]ane$N_5 \cdot 3\,H^+$ is generally a stronger receptor to small anions than [17]ane$N_5 \cdot 3\,H^+$ or [18]ane$N_6 \cdot 3\,H^+$. The combination of electrostatic interaction and structural complementarity may account for the fact that [17]ane$N_5 \cdot 3\,H^+$ forms the most stable complex with citrate^{3-}.

If electrostatic interactions play a major role in the strength of anion binding, then it is natural to expect that the higher the polyammonium cation charge is the stronger the affinity for anions will be. This was found to be case with [24]aneN_6 and [32]aneN_8 which possess 6 and 8 protons, respectively, at pH $\sim$ 7 (see Table 4)[46].

Hexaammonium macrocycles [32]aneN_6 and [38]aneN_6 were designed as selective ditopic receptor molecules for dicarboxylates ^-O_2C—R—CO_2^- such as succinate, glutarate, or adipate[51]. Highest stability of the complex corresponds to the best fit between the substrate R length and the site separation of the receptor *II*.

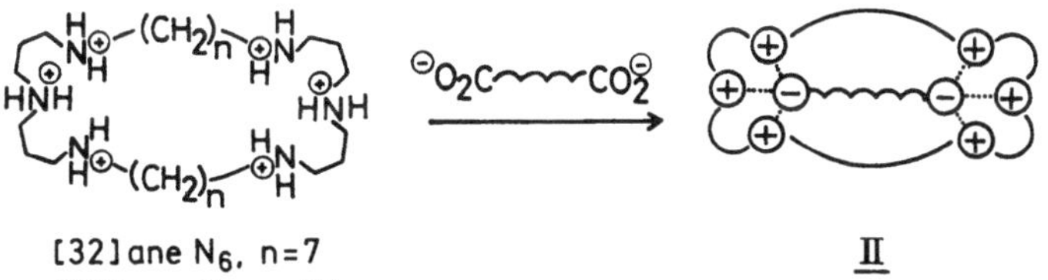

II

Table 4. 1:1 Association constants log β_L (M^{-1}) for polyamines with carboxylates at 25 °C and I = 0.2 M

	Ligand								
	[15]aneN_5[a]	[16]aneN_5[a]	[17]aneN_5[a]	[18]aneN_6[a]	tetren[a]	[24]aneN_6[b] (as H_6L^{6+})	[32]aneN_8[b] (as H_8L^{8+})	[32]aneN_6[c] (as H_6L^{6+})	[38]aneN_6[c] (as H_6L^{6+})
	(all as H_3L^{3+})								
citrate^{3-}	1.74	2.40	3.00	2.38	1.48	4.7	7.6		
succinate^{2-}	negligible	2.08	1.97	1.26	negligible	2.4	3.6	4.3	3.15
malonate^{2-}	negligible	1.82	1.40	1.52	negligible	3.3	3.9		
oxalate^{2-}		2.12		1.79		3.8	3.7		
malate^{2-}	slightly	1.70	1.42	1.18	negligible				
maleate^{2-}		1.88		1.46		3.7	4.1		
fumalate^{2-}		slightly		negligible		2.2	2.9		
glutarate^{2-}								4.4	3.3
adipate^{2-}								3.2	3.3
N-acetyl-(L) -aspartate^{2-}								4.1	3.35

[a] Ref. [50], [b] Ref. [46], at I = 0.1 M, [c] Ref. 51).

Substrates that are either too short or too long form less stable complexes.

Lipophilic N-tetrabenzylcylam, *III*, at pH 4–7 was found to be a good carrier for liquid membrane transport of dicarboxylate anions, such as *o*-phthalate, and of amino acid anions [52]. It was suggested that the diprotonated species, H_2L^{2+}, is an active carrier species for anions.

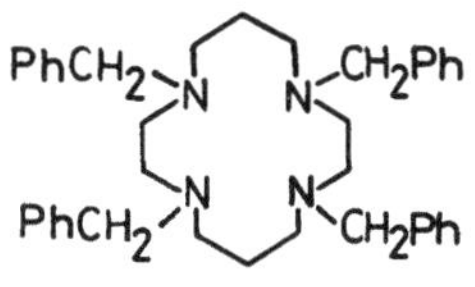

III

In connection with the anion binding ligands, it should be added that tetraprotonated $[18]aneN_6 \cdot 4H^+$ in aqueous solutions interacts with the monoanions Cl^- and NO_3^- having $\log K_{H_4L-X^-}$ of 1.8 and 2.3 [53]. Interestingly, the anion binding was found to be an entropy-driven reaction. Also the fully protonated tetraamines $[16]aneN_4$, $[18]aneN_4$ and $[20]aneN_4$ form 1:1 complexes with F^- having stability constants $\log K_{H_4L-F^-}$ of 1.9, 2.0, and 2.8, respectively [54].

5.2 Phosphate Anions

In biological anion transport systems, complexed species are often found to be phosphate anions rather than dicarboxylate anions [55, 56]. The ability of the polyoxyanions to form hydrogen bonds with N^+–H may operate in the biological recognition. The macrocyclic polyamines which act as polycarboxylate carriers indeed interact with phosphate anions to form 1:1 complexes [43, 46]. Macrocyclic polyamines having more than three protons form stable 1:1 complexes with inorganic phosphate, AMP, ADP and ATP at physiological pH (Table 5). A comparison of the values listed in Table 5 indicates that as in the polycarboxylate complexes, phosphate complex formation is governed by electrostatic forces. Thus, more negative nucleotides are more strongly bound to the protonated polyamine macrocycles. The stability order is $ATP^{4-} > ADP^{3-} > AMP^{2-}$ for all the polyamines. Despite having the same anionic charge as inorganic phosphate, AMP forms more stable complexes with $[18]aneN_6$ by a factor of 10 to 100 [43]. The extra stability for the AMP complex must come from an additional interaction of the adenine base. The active form of inorganic phosphate is HPO_4^{2-} (Fig. 4). Inorganic phosphate ions are bound to

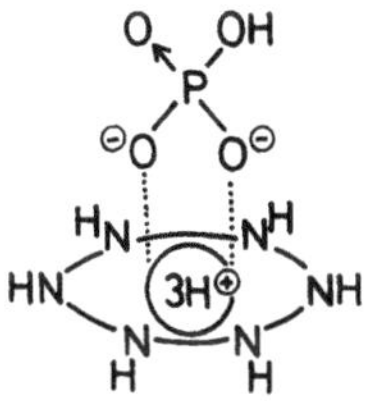

Fig. 4. A proposed $[18]aneN_6 \cdot 3H^+$-HPO_4^{2-} complex

Table 5. 1:1 Association constants log β_L for phosphate and nucleotides with polyamines or metal ions at 25 °C

Phosphate	Ligand or Metal								
	[18]aneN_4[a]	[15]aneN_5[a]	[16]aneN_5[a]	[17]aneN_5[a]	[18]aneN_6[a]	Mg^{2+}[a]	Ca^{2+}[a]	[24]aneN_6[b]	[32]aneN_8[b]
	(as H_4L^{4+})	(as H_3L^{3+})						(as H_6L^{6+})	(as H_8L^{8+})
HPO_4^{2-}			2.04		1.15				
AMP^{2-}	3.85	3.21	3.12	2.84	3.26	1.97	1.85	3.4	4.1
ADP^{3-}	4.48	3.90	3.18	3.0	5.65	3.34	2.86	6.5	7.5
ATP^{4-}	6.70	4.0	3.62	3.71	6.40	4.23	3.97	8.9	8.5

[a] Ref. [43]), [b] Ref. [46])

the polyamines with stability constants nearly comparable to those for dicarboxylate anions, implying a similar electrostatic force in phosphate complexes and dicarboxylate complexes. Here again, [12]–[15]aneN_4, which exist as diprotonated forms in neutral solutions, showed little complexing ability [43]. The phosphate binding with our model reminds us of a phosphate-transport protein in red cells that is supposed to contain a tripositive charge (by protonated amines) at the anion recognition site [57].

Of the equally +3 charged N_5 and N_6 macrocycles, [18]aneN_6 forms more stable complexes with ADP and ATP than [15]–[17]aneN_5 do. However, with inorganic phosphate and AMP, [15]–[17]aneN_5 form complexes with stabilities which are the same as (or larger than) those with [18]aneN_6. The outstanding stability of [18]aneN_6-ADP and [18]aneN_6-ATP probably results from the favorable structural effects between the large cation and the large nucleotide anions. It is postulated that ATP adopts a bent conformation to offer two donor sites, namely, adenine and phosphate (Fig. 5). The higher charged [18]ane$N_4 \cdot 4H^+$ binds more strongly with ATP^{4-} than [18]ane$N_6 \cdot 3H^+$.

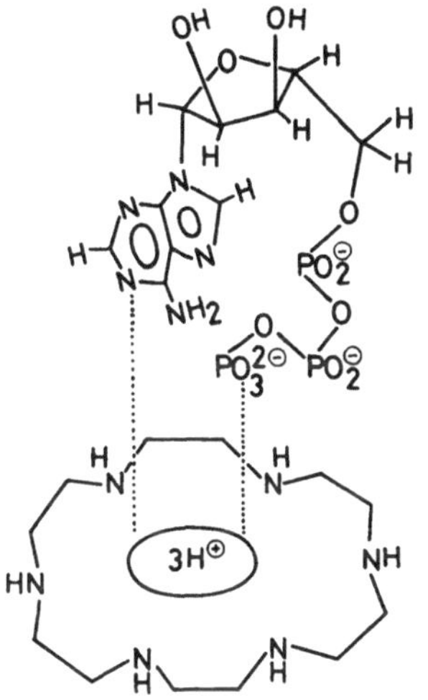

Fig. 5. A proposed structure for 1 : 1 complex of [18]ane$N_6 \cdot 3H^+$ with ATP at neutral pH [43]

The conditional association constant for [18]aneN_4 ("cyclic spermine") with ATP^{4-} at pH 7.5 is calculated from the β_L value (Table 5) and protonation constants (Table 1) to be 2.4×10^5 M^{-1}, which is larger than the association constants for the linear spermidine (9×10^2 M^{-1}) and spermine (9.5×10^3 M^{-1}) [23]. It is also of interest that cyclic spermine is selective for ATP over AMP (ratio association constants is 700), while linear spermine prefers ATP to AMP only by a ratio of 26 to 1 [43]. The selective complexation of biologically important anions is of particular interest, especially if the ligands are converted into selective anion carriers by attachment of lipophilic hydrocarbon chains.

The introduction of another organic cation function, guanidinium group, into macrocyclic structures such as (*IV*)–(*VI*) produces ligands which also display affinity for phosphate anions [58].

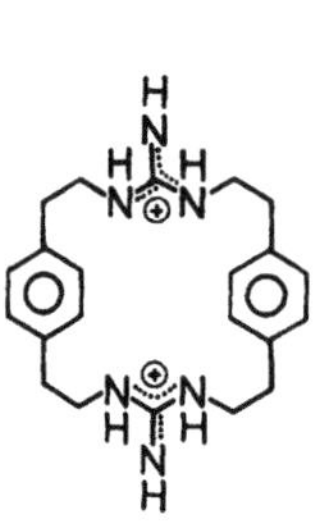

IV

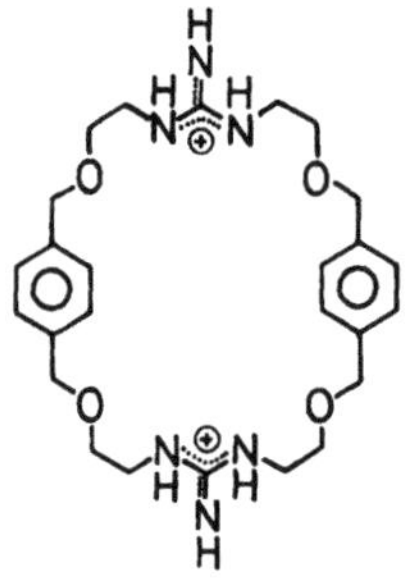

V

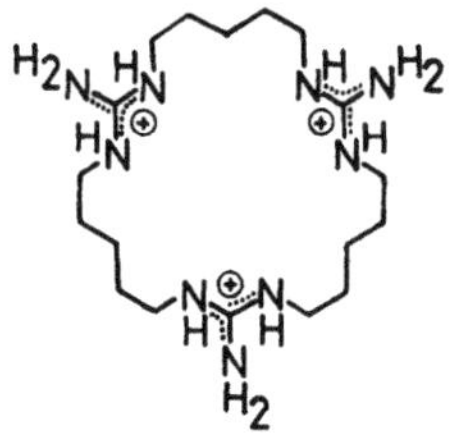

VI

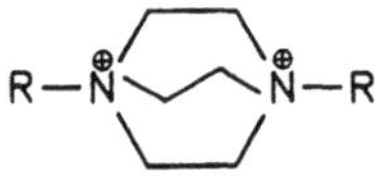

VII

They all form 1:1 complexes with PO_4^{3-} having stability constants log *K* of 1.7 (*IV*), 2.2 (*V*) and 2.4 (*VI*) in aqueous solutions. The macrocyclic effects are obvious when these values are compared with those for linear guanidines, which are 0.95 (N,N'-diethylguanidinium) and 0.7 (guanidinium). The guanidinium group has a very high pK_a (13.5) and therefore its cationic character is little affected by pH changes. Macrocyclic polyamines may be viewed as models for anion binding sites of carrier proteins involving lysine residues, while macrocyclic polyguanidinium ligands serve as models for anion binding sites composed of arginine residues of proteins.

By structural complementarity, dicationic 1,4-diazabicyclo[2.2.2]octane (*VII*) provides an appropriate recognition site for phosphate ions and two stearyl side chains attached to the amines add lipophilic properties [59,60]. Such a carrier model can selectively extract nucleotides from aqueous solution to chloroform solution *via* lipophilic salt formation. The order of nucleotide affinity is ATP > ADP > AMP. The selectivity ratios were 45 for ADP/AMP and 7500 for ATP/AMP at pH 3. The relative transport rate was ATP > ADP > AMP. The ratios were 60 for ATP/AMP and 51 for ADP/AMP. The modes of interaction of ADP and ATP are proposed to be as shown in Fig. 6.

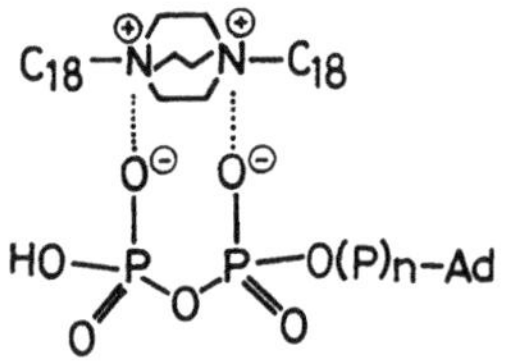

Fig. 6. A proposed structure for 1:1 (VII)-ADP (n = 0) and -ATP (n = 1) complexes [60]

Macrocyclic tetraammonium compounds *VIII* and *IX* [61] form stable 1:1 inclusion complexes with anionic molecules in aqueous solutions [62]. The anions are halides, carbonate, phosphate, AMP, ATP etc. The stability of the inclusion complexes depends on electrostatic as well as hydrophobic interactions. Whereas the complexes of *VIII* are dominated by the electrostatic component, the hydrophobic interaction plays the main part in complexes of *IX*.

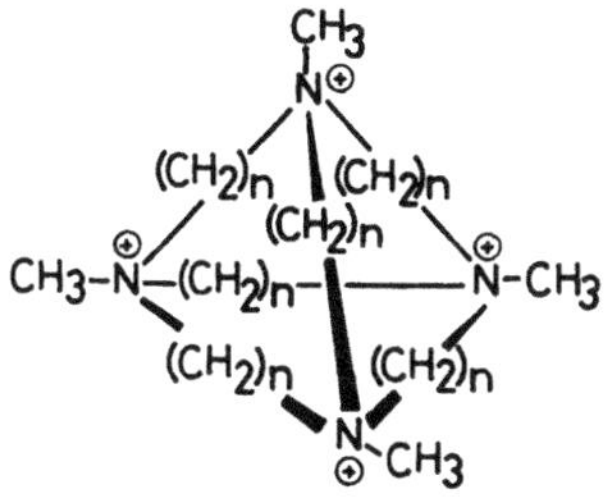

VIII : n = 6
IX : n = 8

Thus 1:1 association constants log *K* (M^{-1}) in aqueous solutions with hydrophilic HPO_4^- are 2.1 (*VIII*) *vs* 0.32 (*IV*), while with lipophilic *p*-nitrophenolate they are >0.7 (*VIII*) *vs* 2.25 (*IX*). With AMP^{2-} and ATP^{4-}, log *K* values are 1.99 and 2.46 (*VIII*), and 1.40 and 1.92 (*IX*). *IX* was shown to stabilize an anionic transition state and to accelerate the reaction for 2,4-dinitrofluorobenzene with N_3^-, [63].

5.3 Catechols

The macrocyclic hexamine [18]aneN_6 was further found to recognize catechol, catecholamines and biologically relevant compounds (see Chart II) [64]. It interacts with all of these donor compounds in neutral pH solutions to form 1:1 complexes, which were determined polarographically. The stability constants β_L are summarized in Table 6.

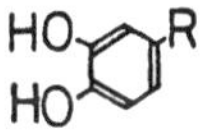

1. Catechol: R = H
2. Dopamine: R = $CH_2CH_2NH_2$
3. Dopa: R = $CH_2CH(CO_2H)NH_2$
4. Adrenaline: R = $CH(OH)CH_2NHCH_3$

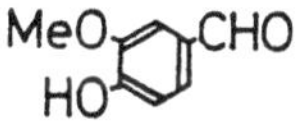

5. Vanilline

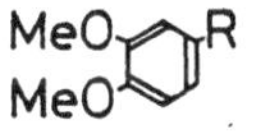

6. Veratraldehyde: R = CHO
7. Veratric acid: R = CO_2H

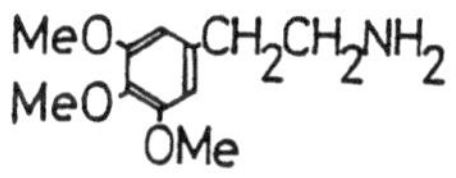

8. Mescaline

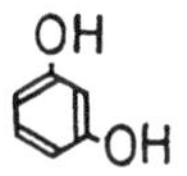

9. Resorcinol

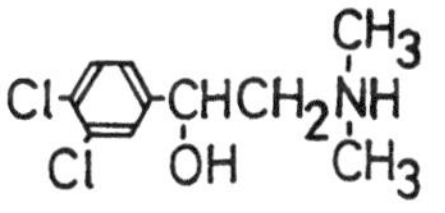

10. Dichloroisoproterenol

The catechol (*o*-dihydroxybenzene) function is essential for biological activity in a number of biogenic amines and their associated drugs. Pharmacological studies have proved the presence of the catechol recognition and binding sites in biological systems. However, chemical entities of the catechol receptors (like those of other receptors) remain almost unknown. Small molecular compounds having the efficient and selective catechol receptor functions are expected to be very useful not only in chemical elucidations but also in pharmacological and medicinal applications.

Table 6. 1:1 Association constants β_L for [18]aneN_6 (L) with catechols (A) at 25 °C, $I = 0.2$ M[a]

Catechol	Complex formula	β_L (M^{-1})
catechol *1*	H_3L^{3+} — $H_2A°$	1.6×10^2
dopamine *2*	H_3L^{3+} — H_3A^+	1.1×10^3
dopa *3*	H_3L^{3+} — $H_3A°$	3.7×10^3
adrenaline *4*	H_3L^{3+} — H_3A^+	1.0×10^3
vaniline *5*	H_3L^{3+} — A^-	4.9×10^2
veratraldehyde *6*	H_3L^{3+} — $A°$	2.5×10^2
veratric acid *7*	H_3L^{3+} — A^-	4.2×10^2
mescaline *8*	H_3L^{3+} — HA^+	5.8×10^2
resorcinol 9	H_3L^{3+} — $H_2A°$	1.3×10^3
dichloroisoproterenol *10*	H_3L^{3+} — $A°$	8.5×10^2

[a] Ref. [64].

As seen in the previous chapters, the triprotonated macrocyclic pentaamines are as good or in some cases better than hexaamine [18]aneN_6 as ligands for polycarboxylates or phosphates. For catechols, however, the only successful receptor molecule is [18]aneN_6. The narrower selection of the receptor molecules indicates a stricter geometric requirement for the catechol recognition. The pH-variation study established the involvement of five protons in the [18]aneN_6-catechol *1* association, permitting formulation of the complex as H_3L^{3+}–H_2A^0 (neutral catechol).

A monomethyl ether *5* and dimethyl ether *6* of catechol derivatives also form 1:1 complexes having the same order of β_L values, indicating that the methylated catechols are as good donors as the free catechol group. Catechol derivatives, regardless of the residual groups, bind with [18]ane$N_6 \cdot 3H^+$. The interaction would occur at the *o*-dihydroxy benzene part, which is not appreciably perturbed by the residual NH_3^+, CO_2^-, $^+NH_2CH_3$ groups. Thus, the structure of the catecholamine complex is postulated to be as shown in Fig. 8, which is compared with the oxygen donor counterpart of [18]aneN_6(*i.e.* [18]crown-6) recognizing dopamine by the ammonium cation. The reversed roles of donor-acceptor relations by the N- and O-coronand are clearly demonstrated (Fig. 7).

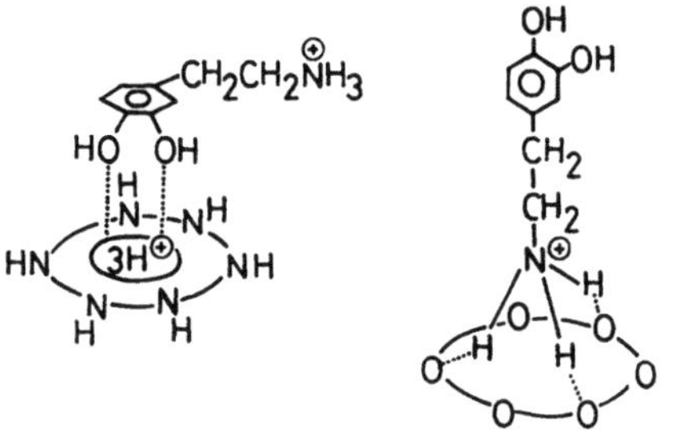

Fig. 7. Proposed structures for [18]aneN_6/*2* **vs**/[18]crown-6/*2* [64]

5.4 Competition of Phosphate Anions and Metal Cations for Macrocyclic Polyamines

As mentioned in Chapter IV and in section V.3, [18]aneN_6 has an appreciable affinity for both Ca^{2+} at alkaline pH and nucleotides at neutral pH. In addition, the interaction of Mg^{2+} or Ca^{2+} with nucleotides is well known [65]. Therefore, one naturally questions whether the macrocyclic polyamines (L) disrupt the metal nucleotide interactions in the manner expressed by Eq. (2) or Eq. (3)

$$M^{2+} - \text{nucleotide} + H_iL^{i+} \rightleftarrows M^{2+} - L + \text{nucleotide} + iH^{i+} \qquad (2)$$

or

$$M^{2+} - \text{nucleotide} + H_iL^{i+} \rightleftarrows H_iL^{i+} - \text{nucleotide} + M^{2+} \qquad (3)$$

The results with the [16]aneN_5-Mg^{2+} system indicate the formation of a ternary complex according to Eq. (4). The complex formation constant is $5.6 \times 10^4\ M^{-1}$ at 25 °C

$$M^{2+} - \text{nucleotide} + H_iL^{i+} \rightleftarrows M^{2+} - \text{nucleotide} - H_iL^{i+} \qquad (4)$$

The multiple cation binding sites on the nucleotides will permit the accommodation of both cations simultaneously. The fact that [16]aneN_5 in the triprotonated form binds to Mg^{2+}-ATP^{4-} complexes ($\beta_L = 5.6 \times 10^4\ M^{-1}$) more strongly than to ATP^{4-} itself ($\beta_L = 4.2 \times 10^3\ M^{-1}$) is a good indication that the binding of Mg^{2+} to ATP^{4-} does not conflict with the binding of [16]aneN_5. Hence, it is proposed that Mg^{2+} ion attaches to ATP^{4-} at a site (*i.e.* free oxyanions) different from that to which [16]aneN_5 binds.

The ternary Mg^{2+}-ATP-polyamine complex is more stable than the binary ATP-polyamine complex and may represent a model for the Mg^{2+} requirement in many enzymic reactions involving ATP. It is well known that most of the intracellular ATP is bound to divalent metal ions such as Mg^{2+}, and the resulting metal-ATP complexes are presumed to act as real substrates or cofactors in enzyme reactions. It has been suggested that metal-ATP complexes may be regulated by the natural polyamines by their association with the ATP. Hence, it will be of great interest to test whether macrocyclic polyamines perturb or promote the normal biochemical and biological functions of the natural polyamines.

6 Application of Cation and Anion Binding Properties of Macrocyclic Polyamines to Calculi Dissolution

Urinary lithiasis is a disease in which calculi form in the kidney and urinary tract. Roughly 5% of the human population suffers to some degree from urinary lithiasis. A number of severely afflicted patients (*e.g. ca.* 60,000 in West Germany and more than 100,000 in the U.S.) are hospitalized yearly for major surgical treatment. Obviously, nephrolithiasis is not only a common ailment but also an issue of great social and economic consequence.

In their composition human calculi are nearly 90% insoluble inorganic salts such as calcium phosphate, calcium oxalate, mixed calcium oxalate phosphate, or mixed calcium phosphate — magnesium phosphate. They can grow to a size of a few centimeters in diameter. The treatment of nephrolithiasis, especially of large calculi, is generally surgical. However, owing to the high percentage of patients who experience a recurrence of the disease and to physiological limitations, since the operation cannot be repeatedly performed on the same kidney, many endeavors have been undertaken to find nonsurgical methods for the removal of calculi. Recently developed physical techniques employing shock wave [66] or laser beams [67] may prove useful for disintegrating calculi in the kidney. However, the problem of elimination of elusive residual particles would remain. Hence, chemical dissolution by irrigation of the renal pelvis may still be a useful and complementary treatment.

6.1 Renal Stone Formation

The principal factors contributing to the formation of calculi are as follows:

1) The presence of sparingly soluble components in human urine, such as calcium oxalate, calcium phosphate, magnesium ammonium phosphate, uric acid and L-cystine. Kidney stones are composed mainly of these compounds.

2) The tendency of urinary components to crystalize. The "crystallizing propensity" (C.P.) of the urine has been expressed by formula (5) [68,69].

$$\text{C.P.} = \frac{(\text{Ca}^{2+})\,(\text{orthophosphate})\,(\text{oxalate})\,(\text{Mg}^{2+})\,(\text{uric acid})\,(\text{cystine})\,(\text{infection})}{(\text{citrate})\,(\text{Mg}^{2+})\,(\text{K}^{+})\,(\text{pyrophosplate})} \tag{5}$$

where (except for infection) the parentheses represent ionic concentrations. The components in the numerator favor crystallization. Components in the denominator tend to inhibit crystallization.

According to the authors, (Mg^{2+}) is included in the numerator as well as in the denominator because it reacts with oxalic acid to form $Mg(C_2O_4)$, which is more soluble than $Ca(C_2O_4)$ thus decreasing the C.P. But it also reacts with citrate to form magnesium citrate resulting in a reduction of (citrate).

3) Urinary pH. The solubility of phosphate salts increases at lower pH values, while pH scarcely affects the solubility of $Ca(C_2O_4)$ over a pH range of 5.7 to 7.5. Bacteria that metabolize urea contribute to an alkaline medium, thus decreasing the solubility of phosphates.

4) Metabolic factors. Hypercalciurea or cystinurea, for example, facilitate the formation of calculi.

6.2 Litholytic Agents

Litholytic agents in current use are classified as direct or indirect. Indirect type drugs decrease the C.P. of urine, thus inhibiting calculus formation. An example is citrate which helps prevent insoluble salts from crystallizing in the urinary tract. Potassium citrate is administered in pill form as a preventive drug. Direct type drugs dissolve renal calculi which have already formed.

Since Crowell first succeeded in dissolving calculi from the renal pelvis with an irrigating fluid [71], a number of attempts have been made to make this technique more practical. The direct application of drugs is usually performed with a double-channel catheter. The entrance channel permits the input of the solution containing dissolving agents, irrigation of kidney cavities, and contact of the dissolving agent with the calculus. The spent solution plus urine and calculus debris are withdrawn through the exit channel of the catheter. A major problem posed by chemical dissolution of renal calculi lies in the development of special and effective solutions which, by their physicochemical properties, will dissolve the calculi. These substances should combine low toxicity, high solvation capacity and lack of undesirable side effects. This method of administration has advantages and disadvantages.

Advantages
a. Direct contact of solubilizing agents with renal calculi.
b. Minimal dilution of the drugs.
c. Optimal adjustment of pH.
d. Constancy of dissolving solutions.
e. Freedom to administer the drug in any desired quantity.

Disadvantages
a. The treatment requires prolonged hospitalization.
b. There is danger of renal and other genitourinary tract infections via the catheter, although the infection can usually be kept under control through intraluminal application of antibiotics and by regulation of intrarenal pressure.

Current litholytic agents in clinical use are Renacidin (composition: glucuronic acid, citric acid, malic acid *etc.*) [71–74], Solution G (3.2% citric acid, 0.38% magnesium oxide, 0.33% Na_2CO_3) [75], and ethylenediamine tetraacetate, EDTA [76–81]. EDTA is currently the most efficient and most frequently used prescription for calculi which contain Ca^{2+} and Mg^{2+} because of its strong chelating action. EDTA solutions of up to 5.5% are normally used.

The following recipe was tested for *in vitro* dissolution of a calcium oxalate calculus [69,82].

Disodium EDTA	5.00 g
NaOH	0.55 g
KOH	0.02 g
Dist. water	100 ml
pH of the solution:	8.0

Using this preparation, a 500 mg oxalate calculus obtained surgically was almost completely dissolved within 60 hours. When the solution pH was lowered to 6.0, the dissolution efficiency dropped appreciably and 200 mg of the stone remained undissolved after 60 hours. In another experiment a 5% EDTA solution adjusted to pH 6.0 was applied *in vitro* to a phosphate kidney calculus (400 mg) which dissolved almost completely in 30 hours. However, clinical experience indicates that EDTA is still far from ideal for dissolving oxalate and phosphate calculi because of the very long time required for dissolution of the calculi.

6.3 Macrocyclic Polyamines as a New Type of Litholytic Agent

As seen in the preceding chapters, macrocyclic polyamines such as [18]aneN_6 can be cation chelators, anion chelators, or both depending on the conditions. It was expected that with these dual properties, the macrocyclic polyamines might serve as litholytic agents by removing Ca^{2+} and phosphate or oxalate anions from insoluble calculi.

In addition to [18]aneN_6, a newly synthesized [18]aneN_6-hexaacetic acid (*X*) has also been tested [34, 35]. (*X*) was originally designed to combine the anion-binding function of [18]aneN_6 with the cation-binding function of EDTA for more efficient dissolution of calculi.

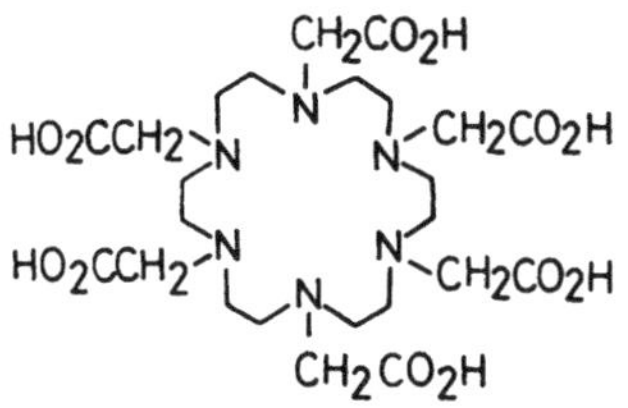

X

The *in vitro* dissolving efficiency of litholytic agents is usually evaluated by measuring the weight reduction of dissolving calculi [69] or by the progress of dissolution measured with a polarization microscope [80]. As a much simpler assay for preliminary tests of new candidate compounds, the following procedure was adopted [35]: A finely ground, accurately weighed human calculus or its model, $Ca_3(PO_4)_2$, $Ca(C_2O_4)$, $Mg_3(PO_4)_2$ and an excess of dissolving agent are placed in test tubes containing suitable buffer solutions, shaken in a water bath at 37 °C for one hour, filtered through filter paper and the appropriately diluted filtrates subjected to atomic absorption spectrophotometry for Mg^{2+} and Ca^{2+} analysis and to isotachophoretic analysis for determination of phosphate anion. It is to be noted that inorganic calculi may contain proteinaceous matrixes which can serve to retard the dissolution process by litholytic agents. Hence the above procedure may be suitable only for assaying the intrinsic solvability of inorganic components.

Table 7. Concentration of Ca^{2+} and PO_4^{3-} released from insoluble $Ca_3(PO_4)_2$ with ligands in pH 7.0 collidine buffer at 37 °C for one hr[a]

Ligand (10 mM)	free $[Ca^{2+}]$ in ppm	free $[PO_4^{3-}]$ in ppm	$[Ca^{2+}]/[PO_4^{3-}]$ ratio
None	21	16	
[18]aneN_6	157	92	3.6:2
	(136)	(76)	
X	833	556	3:2
	(811)	(540)	
EDTA	512	354	3:2
	(496)	(333)	

[a] Ref. [35].

Dissolution of Calculi Model. Dissolved Ions from $Ca_3(PO_4)_2$. Dissolution of calcium phosphate by macrocyclic polyamines proceeds at pH 7, which is established by measuring the freed cation concentration as well as the freed anion concentration with respect to the control values (Table 7). The molar ratio of $[Ca^{2+}]$ to $[PO_4^{3-}]$

dissolved with the aid of chelating agents, was, as expected, 3:2, justifying the now current practice of determining the extent of dissolution of inorganic salts and calculi by analyzing the cation concentration alone. The dissolution of $Ca_3(PO_4)_2$ by litholytic agents at pH 7 is virtually complete immediately. The extent of dissolution is almost entirely determined by thermodynamic factors. EDTA can dissolve up to an equivalent of calcium ion upon prolonged exposure, whereas the new macrocyclic ligand *X* dissolves more (up to two equivalents after 6 hr's treatment) (Table 8). The results listed in Table 9 show that [18]aneN_6 promotes dissolution of $Ca_3(PO_4)_2$ at acidic pH, while EDTA and the carboxylated ligands *X* are effective at neutral to alkaline pH. Solvation of $Ca(C_2O_4)$ by these ligands depends on pH in the same way. This opposite trend in pH effect on dissolution was rationalized by the reversed types of complexation as postulated in Fig. 8. The chelation of phosphate and oxalate with [18]aneN_6 is favorable in acidic solutions. Taking into consideration the protonation constants (log *K*) of 10.19, 9.23, 8.73, 4.09, ~2, ~1 for [18]aneN_6 and of 12.38, 7.20, 2.15 for PO_4^{3-}, the most stable complex form of [18]aneN_6 and phosphate at

Table 8. Time effect in dissolution of $Ca_3(PO_4)_2$ (20 mg) at pH 7 collidine buffer and 37 °C[a]

Ligand (3 mM)	Reaction time					
	20 min.	40 min.	1 hr.	2 hr.	4 hr.	6 hr.
	$[Ca^{2+}]$ in ppm[a]					
none	14	15	16	16	23	43
X	202	212	219	232	254	285
	(188)	(197)	(203)	(216)	(231)	(242)
EDTA	134	128	131	139	145	166
	(120)	(113)	(115)	(123)	(122)	(123)

[a] Ref. [35].
[b] Theoretical value is 134 ppm. provided that dissolution occurs by 1:1 complex.

Table 9. pH Effect on dissolution of $Ca_3(PO_4)_2$ (5 mg) at 37 °C and for 1 hr. (Ref. [35])

Ligand (10 mM)	pH				
	4.0[b]	5.0[b]	5.9[b]	7.0[c]	8.0[d]
	$[Ca^{2+}]$ in ppm[a]				
None	252	176	18	23	13
[18]aneN_6	437	212	99	113	72
	(185)	(36)	(81)	(90)	(59)
X	374	463	400	510	432
	(122)	(287)	(382)	(487)	(419)
EDTA	315	453	370	307	282
	(63)	(277)	(352)	(284)	(269)

[a] Theoretical $[Ca^{2+}]$ is 401 ppm, provided that dissolution occurs by 1:1 complexation.
[b] 0.2 M Acetate buffer (3 ml).
[c] 0.05 M Collidine buffer (3 ml).
[d] 0.05 M Tris buffer (3 ml).

acidic pH is H_4L^{4+} — $H_2PO_4^-$. Note that H_4L^{4+} was earlier found to associate with monoanions Cl^- and NO_3^- having log $K_{H_4L-X^-}$ of 1.8 and 2.3 [53]. To make the carboxylate ligands *X* and EDTA better agents for cation complexation, higher pH will be required.

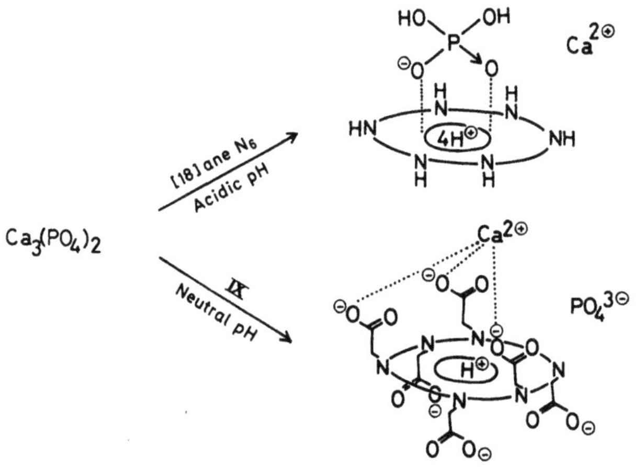

Fig. 8. A proposed dissolution mechanism of $Ca_3(PO_4)_2$ with [18]aneN_6 and **IX**

The new macrocyclic hexaamine ligand *X* has the mixed protonation constants, log K^i of 10.10, 10.01, 8.96 and 8.02 at 25 °C and I = 0.2 M for the four most basic amines. The values for other weaker bases, including the six carboxylates, are all less than 5. From a comparison with log K_i values of the parent macrocycle [18]aneN_6, it was deduced that the initial four protonations occur to the macrocyclic amine bases. Thus, the most abundant species of *X* at neutral pH is depicted as *XI*.

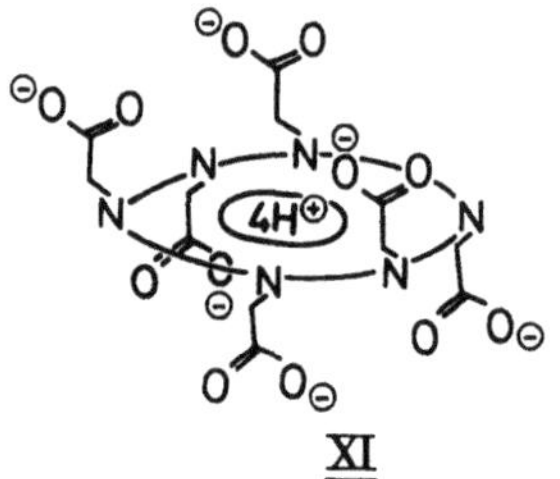

XI

The 1 : 1 complexation constants with Mg^{2+} and Ca^{2+} determined by potentiometric titration are 1.7×10^8 M^{-1} for Mg^{2+}–L and 8.7×10^7 M^{-1} for Ca^{2+}–HL (= [CaHL]/[Ca^{2+}] [HL]). These values may be compared with 4.9×10^8 M^{-1} for Mg^{2+}–$EDTA^{4-}$ and 5.0×10^{10} M^{-1} for Ca^{2+}–$EDTA^{4-}$ [83]. Utilizing the reported log K_i values of 10.34, 6.24, 2.75, 2.09 etc., and log K_{CaHY} (k_{CaHY} = [CaHY]/[CaY][H^+]) of 3.10 for EDTA [83] one can conclude that the total concentration of complexed Ca^{2+} is much larger with EDTA than with L. The conditional formation constants (= $[CaL]_{total}/[Ca^{2+}][L]_{uncomplexed}$) at pH 7 are calculated to be $10^{7.30}$ for EDTA and $10^{1.78}$ for *X*. However, the dissolution test indicates that *X* is more efficient than

EDTA, leading to a postulate that more than one equivalent of Ca^{2+} can be captured by *X* (*e.g.* one Ca^{2+} sequestered by the three amines and the three carboxylates and another Ca^{2+} by the remaining half the donor groups), as the Dreiding model suggests. The fact that there was no interaction at neutral pH of *X* with phosphate or oxalate anions was separately confirmed. Thus, the dissolution of $Ca_3(PO_4)_2$ and $Ca(C_2O_4)$ is entirely due to the cation complexation mechanism.

Table 10. Dissolution of insoluble $Ca_3(PO_4)_2$ (5 mg), $Ca(C_2O_4)$ (3 mg) and $Mg_3(PO_4)_2$ (7 mg) with ligands at pH 7.0 (Ref. [35])

Ligand (10 mM)	$Ca_3(PO_4)_2$	$Ca(C_2O_4)$	$Mg_3(PO_4)_2$
	$[Ca^{2+}]$ in ppm	$[Ca^{2+}]$ in ppm	$[Mg^{2+}]$ in ppm
None	23	17	126
[18]aneN_6	113	28	302
	(90)	(11)	(176)
X	510	222	316
	(487)	(205)	(190)
EDTA	307	309	303
	(284)	(292)	(177)

Comparison of $Ca_3(PO_4)_2$, $Ca(C_2O_4)$, and $Mg_3(PO_4)_2$. The effects of cation and anion composites were tested by comparing the dissolution of $Ca_3(PO_4)_2$, $Ca(C_2O_4)$, and $Mg_3(PO_4)_2$ at pH 7 (Table 10). The dissolution of $Ca_3(PO_4)_2$ is achieved more effectively with *X* than with EDTA. However, when the anion is oxalate, the dissolution of Ca^{2+} is drastically reduced with *X*. EDTA can dissolve $Ca(C_2O_4)$ to the same extent as $Ca_3(PO_4)_2$, *i.e.* the anion effect is insignificant. A better separation of Ca^{2+} from oxalate anion may be achieved by EDTA.

Dissolution of Human Urinary Calculi in Vitro. Five human urinary calculi containing various proportions of $Ca_3(PO_4)_2$, $Ca(C_2O_4)$, $CaCO_3$, and $MgNH_4(PO_4)$ were subjected to similar dissolution tests at pH 7 (Table 11). The same dissolution patterns as those of the model phosphate and oxalate calculi are found. That is, for phosphate calculi no. 1–4, *X* is more effective than [18]aneN_6 or EDTA and for oxalate calculus no. 5, EDTA is best.

The clinical use of EDTA to dissolve inorganic calculi is well established. Hence, comparison of dissolution *in vitro* constitutes a primary assay to find better litholytic agents than EDTA, for which the chemical principle of dissolution of calcium phosphate and calcium oxalate (major components of inorganic calculi) should be fully utilized. At present, the principle of cation chelation is fairly well exploited with linear polyaminocarboxylate ligands such as EDTA or CyDTA [69], but their clinical application is still limited due to insufficient solubility. For better reagents, cyclization to macrocyclic polyaminocarboxylates such as *X* is one approach. Another approach is anion chelation specific for oxalate or phosphate anions by macrocyclic polyamines, which was shown to be partially successful with [18]aneN_6 at acidic pH [35]. From those results it is predicted that the more suitable candidates will be those which are efficient chelating agents for oxalate and phosphates at neutral pH, for example

Table 11. Dissolved [Ca^{2+}] and [Mg^{2+}] from human urinary calculi with ligands at 37 °C for one hour in pH 7 collidine buffer (Ref. [35])

Ligand (5 mM)	Calculus-1	Calculus-2	Calculus-3	Calculus-4	Calculus-5
	Composition[a]				
	$Ca_3(PO_4)_2$ 78%	$Ca(C_2O_4)$ 69%	$MgNH_4PO_4$ 93%	$Ca_3(PO_4)_2$ 85%	$Ca(C_2O_4)$ >98%
	$CaCO_3$ 12%	$Ca_3(PO_4)_2$ 31%	$CaCO_3$ 7%	$CaCO_3$ 15%	
	[Ca^{2+}] in ppm				
None	22	21	18	19	26
[18]aneN_6	71	69	31	43	31
X	254	290	224	342	82
EDTA	143	168	89	140	173
	[Mg^{2+}] in ppm				
None	31	3	90	37	2
[18]aneN_6	43	4	103	51	2
X	54	8	149	73	2
EDTA	45	5	131	54	2

[a] Analyzed by IR spectral measurement (Japan Special Reference Lab., Inc.).

[24]aneN_6 [46]. It is also predicted that a new type of macrocycle which can complex both Ca^{2+} cation and oxalate anion simultaneously (*e.g.* compounds having both a crown ether and an azacrown site) may be another candidate. The successful ligands would, no doubt, find many other applications which are not yet visualized.

7 Concluding Remarks

The chemistry of anion chelation has just begun. Since many organic materials (especially in our body) take anionic forms, the potential of anion chelation is tremendous. It is certain that new strategies for the rational design of organic anion receptors and carriers with selectivities and multiple functions will appear. Future development will not be limited to those mentioned; biogenic polyamine (spermide, spermine etc.) functions, ATP/ADP specific transport, polycarboxylate and catechol membrane transport, catalytic functions or calculi dissolution. As bases for extensive applications, thorough theoretical studies on nature of anion chelation are urgently needed.

Very recently, a very bold chemical model for histamine H_2-receptors has been proposed using [18]ane$N_6 \cdot 3H^+$ [84], which can chemically recognize histamine, histamine H_2-agonists, and histamine H_2-antagonists such as cimetidine *XII* or ranitidine *XIII* that are currently in world-wide use for treatment of peptic ulcers [85].

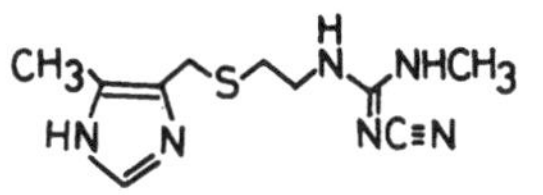

Cimetidine XII

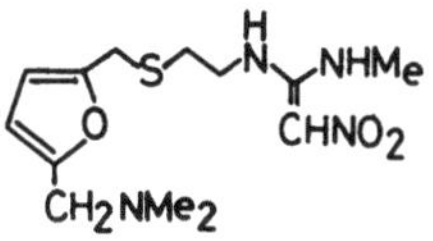

Ranitidine XIII

In this histamine H_2-receptor model, both histamine and its antagonists bind to [18]ane$N_6 \cdot 3H^+$ forming 1:1 complexes at neutral pH, and the histamine antagonists are more strongly bound to the receptor model (see Fig. 9). Although this picture

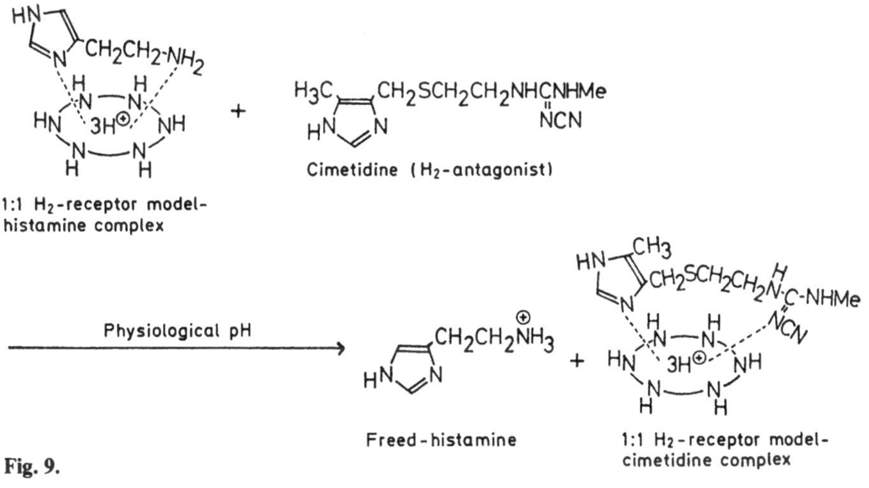

Fig. 9.

seems too simplified to explain the complex pharmacological facts, a significance lies in the translation of the biological concept of drug receptors into visual chemical terms (using simple compounds such as [18]aneN_6). Refined drug receptor models would eventually help chemically rationalize biological activity, as enzyme models do enzymic reactions. It would also assist in designing new drugs.

8 References

1. Pedersen, C. J.: J. Am. Chem. Soc. *89*, 2495, 7017 (1967)
2. Cram, D. J., Cram, J. M.: Science *183*, 803 (1974)
3. Izatt, R. M., Christensen, J. J. (ed.): Synthetic Multidentate Macrocyclic Compounds, Academic Press 1978
4. Hiraoka, M. (ed.): Crown Compounds — Properties and Application (in Japanese), Kodansha Scientific 1978
5. Ono, R., Shono, T., Tabushi, I. (ed.): Chemistry of Crown Ethers (in Japanese), Kagaku Zokan 74, Kagaku Dojin 1978
6. Kolthoff, I. M.: Anal. Chem. *51*, 1R (1979)
7. Takagi, M.: Kagaku no Ryoiki *94*, 1410 (1979)
8. Melson, G. A. (ed.): Coordination Chemistry of Macrocyclic Compounds, Plenum Press, New York and London 1979
9. Izatt, R. M., Christensen, J. J. (ed.): Progress in Macrocyclic Chemistry, Vol. 1 (1979) and 2 (1981), Wiley-Interscience, New York
10. Curtis, N. F.: Coord. Chem. Rev. *3*, 3 (1968)
11. Kimura, E., Kodama, M.: Yuki Gosei Kagaku (in Japanese) *35*, 632 (1977)
12. Busch, D. H.: Acc. Chem. Res. *11*, 392 (1978)
13. Collman, J. P., Shneider, P. W.: Inorg. Chem. *5*, 1380 (1966)
14. Iitaka, Y., Shina, M., Kimura, E.: ibid. *13*, 2886 (1974)
15. Hinz, F. P., Margerum, D. W.: J. Am. Chem. Soc. *96*, 4993 (1974)
16. Kodama, M., Kimura, E.: J. C. S. Chem. Comm., 326 (1975)
17. Busch, D. H., Farmery, K., Goedken, V., Kotavic, V., Melnyk, A. C., Sperati, C. R., Tokel, N.: Adv. Chem. Ser., *No. 100*, 44 (1971)
18. Dolphin, D. (ed.): The Porphyrins Vol. I–VII, Academic Press 1979
19. Cohen, S. S. (ed.): Introduction to the Polyamines, Prentice-Hall, Inc., New Jersey 1971
20. Campbell, R. A., et al. (ed.): Advances in Polyamine Research, Vol. 1–4, Raven Press, New York 1983
21. Tabor, C. W., Tabor, H.: Annu. Rev. Biochem. *45*, 285 (1976)
22. Bichrach, U. (ed.): Function of Naturally Occurring Polyamines, Academic Press, New York and London 1973
23. Nakai, C., Glinsman, W.: Biochemistry *16*, 5636 (1977)
24. Allison, S. A., Herr, J. C., Schurr, J. M.: Biopolymers *20*, 469 (1981)
25. Rich, A., Raj Bohandary, U. L.: Annu. Rev. Biochem. *45*, 805 (1976)
26. Vesna, N. L., Ivanc, W. D., Tomislav, Z., Zeljko, K.: Eur. J. Biochem. *117*, 263 (1981)
27. Park, C. H., Simmons, H. E.: J. Am. Chem. Soc. *90*, 2431 (1968)
28. Bell, R. A., Christoph, G. G., Fronzeck, F. R., Marsh, R. E.: Science *190*, 151 (1975)
29. Lehn, J. M., Sonveaux, E., Wielard, A. K.: J. Am. Chem. Soc. *100*, 4914 (1978)
30. Graf, E., Lehn, J. M.: ibid. *98*, 6403 (1976)
31. Schmitchen, F. P.: Angew. Chem. Int. Ed. Engl. *16*, 720 (1977)
32. Lehn, J. M.: Acc. Chem. Res. *11*, 49 (1978)
33. Dietrich, B., Fyles, D. L., Fyles, T. M., Lehn, J. M.: Helv. Chim. Acta *62*, 2763 (1979)
34. Kimura, E., Watanabe, A., Nihira, H.: Chem. Pharm. Bull. Jpn. *31*, 3264 (1983)
35. Kimura, E., Fujioka, H., Yatsunami, A., Nihira, H., Kodama, M.: ibid. in press
36. Kodama, M., Kimura, E.: J. Chem. Soc., Dalton Trans. 1473 (1977)
37. Kodama, M., Kimura, E.: ibid. 1081 (1978)
38. Kodama, K., Kimura, E.: J. Chem. Soc., Chem. Comm. 891 (1975)

39. Kodana, M., Kimura, E.: J. Chem. Soc., Dalton Trans. 2335 (1976)
40. Kodama, M., Kimura, E.: ibid. 2341 (1976)
41. Nave, C., Truter, M. R.: ibid. 2351 (1974)
42. Kimura, E., Yatsunami, T.: Chem. Pharm. Bull. Jpn. *28*, 994 (1980)
43. Kimura, E., Kodama, M., Yatsunami, T.: J. Ame. Chem. Soc. *104*, 3182 (1982)
44. Kodama, M., Kimura, E.: J. Chem. Soc. Dalton Trans. 104 (1978)
45. Kodama, M., Kimura, E., Yamaguchi, S.: ibid. 2536 (1980)
46. Dietrich, B., Hosseini, M. W., Lehn, J. M., Sessions, R. B.: J. Am. Chem. Soc. *103*, 1283 (1981)
47. Poonia, N. S., Bajaj, A. V.: Chem. Rev. *79*, 389 (1979)
48. Fujioka, H., Kimura, E., Kodama, M.: Chem. Lett. 737 (1982)
49. Yatsunami, T., Sakonaka, A., Kimura, E.: Anal. Chem. *53*, 477 (1981)
50. Kimura, E., Sakonaka, A., Yatsunami, T., Kodama, M.: J. Amer. Chem. Soc. *103*, 3041 (1981)
51. Hosseini, M. W., Lehn, J. M.: ibid. *104*, 3525 (1982)
52. Tsukube, H.: Tetrahedron Lett. *24*, 1519 (1983)
53. Cullinane, J., Gelb, R. I., Margulis, T. N., Zompa, L. J.: J. Amer. Chem. Soc. *104*, 3048 (1982)
54. Suet, E., Handel, H.: Tetrahedron Lett. *25*, 645 (1984)
55. Robinson, B. H., Williams, G. R.: Biochem. Biophys. Acta. *216*, 63 (1970)
56. Palmieri, F., Prexioso, G., Guagliariello, E., Klingenberg, M.: Eur. J. Biochem. *22*, 66 (1971)
57. Rothstein, A., Cabantchick, Z. I., Knauf, P.: Fed. Proc., Fed. Am. Soc. Exp. Biol. *35*, 3 (1976)
58. Dietrich, B., Fyles, T. M., Lehn, J. M., Pease, L. G., Fyles, D. L.: J. C. S. Chem. Comm. 934 (1978)
59. Tabushi, I., Imuta, J., Seko, N., Kobuke, Y.: J. Amer. Chem. Soc. *100*, 6287 (1978)
60. Tabushi, I., Kobuke, Y., Imuta, J.: ibid. *103*, 6152 (1981)
61. Schmidtchen, F. P.: Chem. Ber. *113*, 864 (1980)
62. Schmidtchen, F. P.: ibid. *114*, 597 (1981)
63. Schmidtchen, F. P.: Angew. Chem. Int. Ed. Engl. *20*, 466 (1981)
64. Kimura, E., Watanabe, A., Kodama, M.: J. Am. Chem. Soc. *105*, 2063 (1983)
65. Tu, A. J., Heller, M. J.: Structure and Stability of Metal-Nucleoside Phosphate Complexes, in: Metal Ions in Biological Systems Vol. 1 (ed. Sigel, H.), p. 1, Marcel Dekker, Inc. New York 1974
66. Chaussy, C., Schmiedt, E., Jocham, D., Brendle, W., Forssman, B., Walter, V.: J. Urol. *127*, 417 (1982)
67. Tanahashi, Y., Orikasa, S., Chiba, R., Tahira, K., Miyakawa, T.: Tohoku J. Exp. Med. *128*, 186 (1979)
68. Edwards, N., Russel, R., Hodgkinson, A.: Brit. J. Urol. *37*, 390 (1965)
69. Kallistratos, G.: Litholytic Agents: Preventive and Curative Drugs for Nephrolithiasis, in: Drug Design, Vol. IV, Academic Press, Inc., New York and London 1973
70. Crowell, A. J.: Surg. Hynec. Obst. *38*, 87 (1924)
71. Mulvaney, W. P.: J. Urol. *82*, 546 (1959)
72. Mulvaney, W. P.: ibid. *84*, 206 (1960)
73. Mulvaney, W. P.: ibid. *84*, 206 (1960)
73. Mulvaney, W. P.: ibid. *79*, 765 (1963)
74. Inada, T., Nihira, H., Kiriyama, T.: Acts. Urol. Jpn. *9*, 28 (1963)
75. Keyser, L. O., Scherer, D. C., Claffey, W.: J. Urol. *69*, 286 (1948)
76. Abeshouse, B. S., Weinberg, Y.: ibid. *65*, 316 (1951)
77. Gehres, R. F., Raymond, S.: ibid. *65*, 474 (1951)
78. Brozinski, M., Sengbusch, V., Timmermann, A.: Urol. Int. *10*, 307 (1960)
79. Timmermann, F., Kallistratos, G.: J. Urol. *95*, 469 (1966)
80. Yamada, S.: Jpn. J. Urol. *58*, 526 (1967)
81. Kuwahara, M., Kambe, K., Takahashi, K., Orikasa, S., Suzuki, M.: J. Urol. *128*, 1379 (1982)
82. Timmermann, A., Kallistratos, G., Fenner, O.: Int. Congr. Ther., 10th 1969, p. 161 (1969)
83. Schwarzenbach, G., Gut, R., Anderegg, G.: Helv. Chem. Acta *37*, 937 (1954)
84. Kimura, E., Koike, T., Kodama, M.: Chem. Pharm. Bull. Jpn. *32*, 3569 (1984)
85. The Chemical Regulation of Biological Mechanisms, (ed. Creighton, A. M., Turner, S.), The Royal Soc. of Chem. Special Publication No. 42, London 1982

Micellar Models of Zinc Enzymes

Waichiro Tagaki and Kenji Ogino

Department of Applied Chemistry, Faculty of Engineering, Osaka City University, Osaka 558, Japan

Table of Contents

1 Introduction

Uncovering of the three dimentional structure of catalytic groups at the active site of an enzyme allows to theorize the catalytic mechanism, and the theory accelerates the designing of model systems. Examples of such enzymes are zinc ion containing carboxypeptidase A [1-5] and carbonic anhydrase [6-11]. There are many other zinc enzymes with a variety of catalytic functions. For example, alcohol dehydrogenase is also a zinc enzyme and the subject of intensive model studies. However, the topics of this review will be confined to the model studies of the former hydrolytic metalloenzymes.

Enzyme model studies may be arbitrarily classified into the following two categories:

i) To account or to understand the mechanisms of the enormous rate acceleration or the strict specificities of enzymes.

ii) To design artificial enzymes.

Included in the first but not in the second category is that intramolecular catalysis, in which the reaction center and the catalytic groups are connected in the same molecule. In the first category is also included the chemistry of an individual catalytic group or a cofactor, such as imidazole base or zinc ion. These model studies provide basic informations for the design of artificial enzymes of the second category. Artificial enzymes may be studied from various viewpoints. Obviously, one important purpose is to simulate natural enzymes, in which one may be required to use the same catalytic groups as those at the active sites of enzymes. However, such restriction is unnecessary and any catalytic groups can be used for the design of active catalysts for other purposes. Substrate binding is important in designing artificial enzymes which is usually accomplished by the use of a surfactant micelle, a macrocyclic cavity, or a polymer network, etc.

In this review, we will describe the model studies of hydrolytic metalloenzymes of the second category, emphasizing those of micellar systems. Here it should be mentioned that such micellar models have been studied only little in the past so that the major part of this review bases on our own work.

2 Non-micellar Models

2.1 Divalent Metal Ion Complexes of Imidazoles and Pyridines Having Hydroxyl Groups

In 1965, Breslow and Chipman discovered that zinc or nickel ion complexes of (E)-2-pyridinecarbaldehyde oxime (*8*) are remarkably active catalyst for the hydrolysis of 8-acetoxyquinoline 5-sulfonate [12]. Some years later, Sigman and Jorgensen showed that the zinc ion complex of N-(2-hydroxyethyl)ethylenediamine (*3*) is very active in the transesterification from p-nitrophenyl picolinate (*1*) [13]. In the latter case, noteworthy is a change of the reaction mode at the aminolysis in the absence of zinc ion to the alcoholysis in the presence of zinc ion. Thus, the zinc ion in the complex greatly enhances the nucleophilic activity of the hydroxy group of *3*. In search for more powerful complexes for the release of p-nitrophenol from *1*, we examined the activities of the metal ion complexes of ligand *2*–*12* [14,15].

2.1.1 Ligand Hydroxyl Group in Catalysis

Observed first-order rate constants (k_{obsd}) under the conditions of excess ligand over the substrate are shown in Table 1. The uncatalyzed buffer rate constant is $k_0 = 1.67 \times 10^{-5}\ s^{-1}$ at pH 7.05 and 25 °C. In the presence of 1.33×10^{-3} M Zn^{2+} ion, the

Table 1. Pseudo-first-order rate constants ($k_{obsd} \times 10^5\ s^{-1}$) for p-nitrophenol release from p-nitrophenyl picolinate *1* in the presence of metal ion and ligand under the condition of [ligand] ≫ [*1*][a]

L	None	M^{2+}				
		Ni^{2+}	Zn^{2+}	Co^{2+}	Ca^{2+}	Mg^{2+}
None	1.67	68.5	47.5	38.7	28.5	4.33
2	19.3	—	82	—	—	—
3	15.2	223	227	85.5	22	15.8
4a	—	—	635[b]	—	—	—
4b	—	—	1680[b]	—	—	—
4c	5.3	2050	3530	447	12.2	22.8
5	—	—	91.7[b]	—	—	—
6	—	—	403	—	—	—
7	95.7	—	2270	—	126	103
8	56.8	4320	2620	—	56	57.8
9	23.3	875	4220	—	27.7	26.5
10	41.7	1200	39200	19500	55	68
11	30.5	553	24000	8120	34.8	29.7
12	14.7	132	40.7	33.8	15	14.8

[a] Observed at 25 °C, pH 7.05 (0.1 M 2,6-lutidine-HNO_3 buffer), $\mu = 0.2$ (KNO_3), $[L]_T = [M]_T = 1.33 \times 10^{-3}$ M, and [*1*] = 1×10^{-4} M

[b] $[L]_T = 1 \times 10^{-3}$ and $[M]_T = 1.33 \times 10^{-3}$ M.

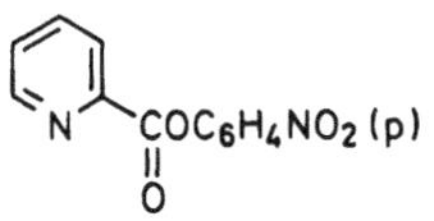

1

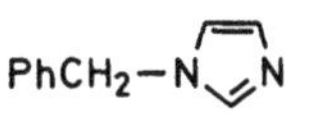

2

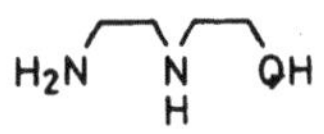

3

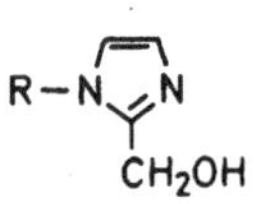

4a R=H
4b R=CH_3
4c R=$PhCH_2$

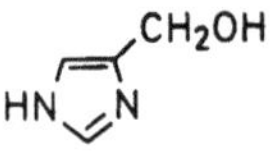

5

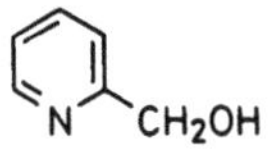

6

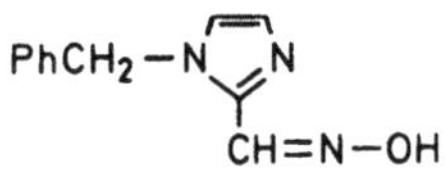

7

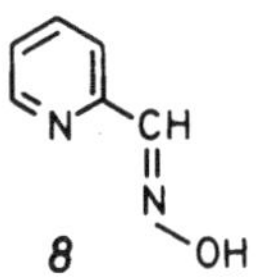

8

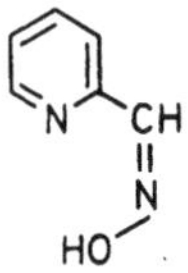

9

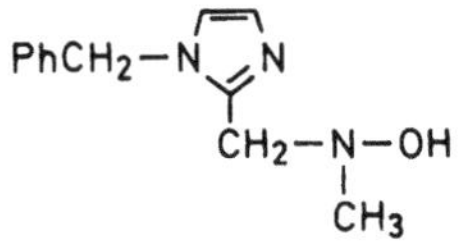

10

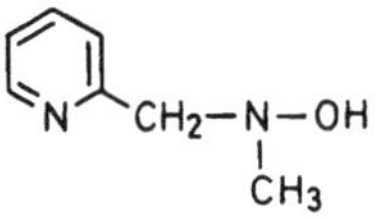

11

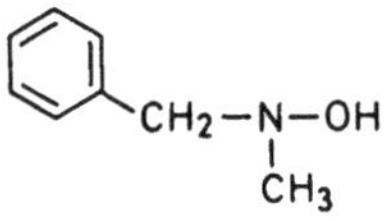

12

rate increased about 30-fold. This rate increase is little affected by addition of such a ligand as *2* having no hydroxyl group or *12* which is a poor ligand toward the Zn^{2+} ion. In the absence of metal ion, the rate enhancement by ligand alone is relatively small. A large rate enhancement can be obtained when ligands having both hydroxyl and neighboring nitrogen groups such as *3–11* are used in the presence of either Ni^{2+}, Zn^{2+}, or Co^{2+} ions.

Pseudo-first-order kinetics are also observed under the condition of five molar excess of substrate over ligand for all ligands except for *3*, indicating the reaction to occur catalytically. In the case of (Z)-2-pyridinecarbaldehyde oxime (*9*), the reaction proceeds in a biphasic manner in which the release of p-nitrophenol is initially rapid but then slows down. Such a biphasic behavior is an evidence for a two-step process involving, in this case, the acylation of the hydroxyl group of catalyst followed by rate determining deacylation to regenerate the catalyst [16,17]. In the case of *3*, as already demonstrated by Sigman and Jorgensen [13], the second deacylation step is so slow that the reaction ceased in effect at the first acylation step. The acylated intermediates are likely involved in the catalysis of other ligands than *3* and *9* having hydroxyl groups, although direct evidences are lacking except their enhanced reactivities.

2.1.2 Kinetic Equations

The rates of reaction of *1* are dependent on both metal ion and ligand concentrations

$$k = k_0 + k_L[L]_T + k_M[M]_T + k_{ML}[L]_T[M]_T \tag{1}$$

In the case of *2* having no hydroxyl group, the k_{obsd} values are given by Eq. 1, where k_0, k_L, k_M, and k_{ML} are the rate constants due to buffer, ligand, metal ion, and co-catalysis of metal ion and ligand, respectively, and $[L]_T$ and $[M]_T$ are the total concentration of ligand and metal ion, respectively. For other ligands having a hydroxyl group, saturation kinetics are observed consistent with the reaction scheme (1) and rate Eq. 2–4, where L, M, k_0, k_L and k_M are defined in Eq. 1, S refers to the substrate, K is the association constant for the complexation between metal ion and ligand, and k_c is the catalytic second-order rate constants for the reaction of the complex with the substrate *1*. Complexation between metal ion and substrate, or the formation of a ternary complex between the three components (M, L, S) appears to be un-

$$M + L \overset{K}{\rightleftarrows} \text{complex}$$

$$S + \text{complex} \xrightarrow{k_c} \text{products}$$

$$S \xrightarrow{k_0'} \text{products}$$

$$k_{obsd} = k_0' + \frac{k_c K[L]_T[M]_T}{1 + K([L]_T + [M]_T)} \tag{2}$$

$$k_0' = k_0 + k_L[L]_T + k_M[M]_T \tag{3}$$

$$\frac{1}{(k_{obsd} - k_0')} = \frac{1}{k_c[L]_T} + \frac{1 + K[L]_T}{k_c K[L]_T} \cdot \frac{1}{[M]_T} \tag{4}$$

Scheme 1.

Table 2. Association constants (K) and second-order rate constants (k_L and k_c) for p-nitrophenol release from p-nitrophenyl picolinate *1* in the presence of ligand (L) and metal ion (M)[a, b]

L-M	k_L $M^{-1}\,s^{-1}$	k_c $M^{-1}\,s^{-1}$	K M^{-1}	k_cK or k_{ML} $M^{-2}\,s^{-1}$
2-Zn^{2+}	0.133	—	—	1.62×10^2
5a-Zn^{2+}	—	42.7	142	60.6×10^2
4b-Zn^{2+}	—	136	104	141×10^2
4c-Zn^{2+}	0.028	167	162	271×10^2
4c-Ni^{2+}	0.028	138	110	152×10^2
5-Zn^{2+}	—	5.17	95	4.91×10^2
6-Zn^{2+}	—	21	178	37.4×10^2
8-Zn^{2+}	0.70	66.7	559	373×10^2
8-Ni^{2+}	—	105	1120	1180×10^2
9-Ni^{2+}	0.20	667	44	293×10^2
10-Zn^{2+}	1.94	463	563	2610×10^2
11-Zn^{2+}	0.20	208	1260	2620×10^2

[a] Observed at 25 °C, pH 7.05 (0.1 M 2,6-lutidine-HNO_3 buffer), $\mu = 0.2$ (KNO_3), $[1] = 1\times10^{-4}$ M.
[b] $k_0 = 1.67\times10^{-5}\,s^{-1}$, $k_M(Zn^{2+}) = 0.255\,M^{-1}\,s^{-1}$.

important, since the rates are linear to the metal ion concentration in the absence of ligands without showing any rate saturation by increasing metal ion concentration. The 1:1 stoichiometry in the complexation of M and L (Scheme 1) can be confirmed by examining the kinetic version of Job plots, i.e. the plots of k_{obsd} values as a function of the mole fraction of the ligand gives a maximum at the mole fraction of 0.5 corresponding to the 1:1 stoichiometry for the kinetically active species. The kinetic results based on Eq. 4 are shown in Table 2.

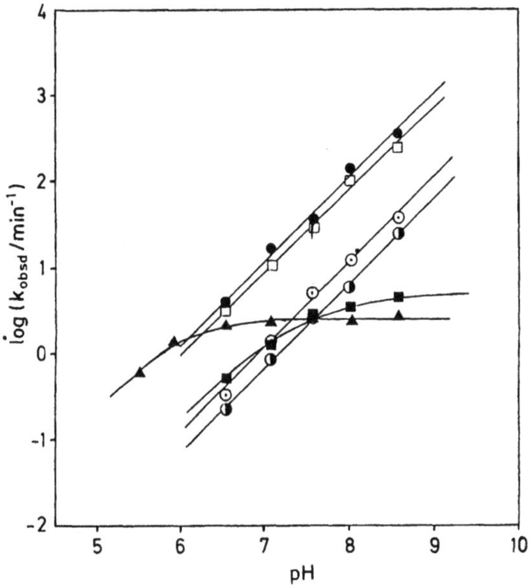

Fig. 1. Plots of k_{obsd} vs. pH for the release of p-nitrophenol from *1* in the presence of 1.33×10^{-3} M ligand and 1.33×10^{-3} M metal ion; see Table 1 for other reaction conditions. ◑: *4c*-Ni^{2+}, ⊙: *4c*-Zn^{2+}, ■: *8*-Zn^{2+}, ▲: *8*-Ni^{2+}, ●: *10*-Zn^{2+}, □: *11*-Zn^{2+}

2.1.3 pH-Rate Profile

Figure 1 shows the pH-rate profiles of some active complexes. Both Ni^{2+} and Zn^{2+} ion complexes of *8* afford saturation curves with inflection at around pH's 6 and 8, respectively, which represent, most likely, the ionization of the hydroxyl group complexed with a Ni^{2+} or a Zn^{2+} ion. The $pK_a = 8.6$ was assigned for the ionization of the hydroxyl group of the latter complex [12]. The lower pH for the ionization of the Ni^{2+} ion complex in respect to that of the Zn^{2+} ion complex indicates that the ligand *8* coordinates to Ni^{2+} ion more tightly than to Zn^{2+} ion, which is in conformity with a larger K value (1120 M) for the Ni^{2+} ion than for the Zn^{2+} ion complex (559 M) at pH 7.05 (Table 2).

Other complexes of *4c*, *10*, and *11* show linear plots with slopes close to unity in the a pH range 6.5–8.5. These linear plots also seem to represent the ionization of hydroxyl groups, but their pK_a values must be higher than 8.5. Unfortunately, it is difficult to examine higher pH's due to precipitation of the metal hydroxide.

2.1.4 Effects of Metal and Ligand Structure on the Mechanism of Catalysis

The activities of metal ions in the absence of ligands are in the order $Ni^{2+} > Zn^{2+} > Co^{2+} > Ca^{2+} > Mg^{2+}$ (Table 1), as it may be predicted from the ligand affinity of a pyridine base like *1* toward these metal ions. In the absence of metal ions, the activities of ligands are in the order *7* > *8* > *10* > *11* > *3* > *4c*, as anticipated from the nucleophilic reactivities of the hydroxyl groups of these ligands. These activities are enhanced 10^2–10^3 times in the presence of Ni^{2+}, Zn^{2+}, or Co^{2+} ions although Ca^{2+} and Mg^{2+} ions seem to have no activating effect. It may be of interest to note in Table 1 that the activating effect of Ni^{2+} ions is somewhat different from those of Zn^{2+} and Co^{2+} ions, i.e. the activity order is *8* > *4c* > *10* > *11* > *3* in the presence of Ni^{2+} ions, while *10* > *11* > *4c* > *8* > *3* in the presence of Zn^{2+} or Co^{2+} ions. The activity order is also pH-dependent, i.e. the complexes of *4c*, *10*, and *11* become much more active than that of *8* at higher pH's (Fig. 1).

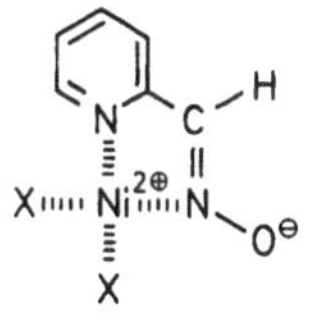

13

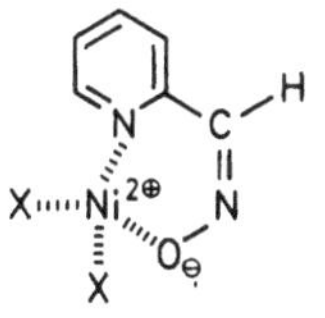

14

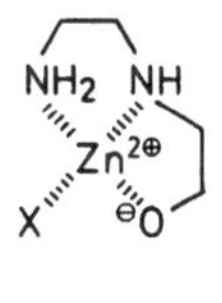

15

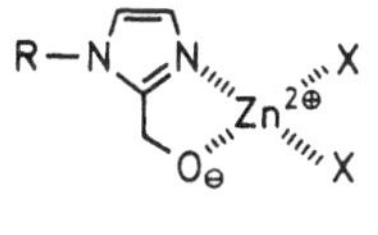

16

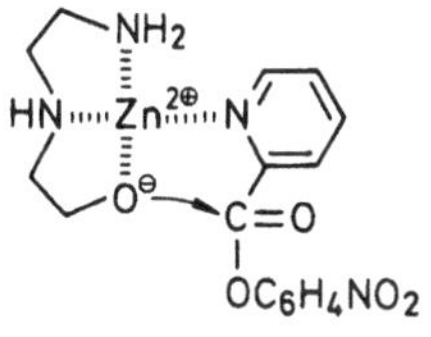

17

It is most likely that a metal ion activates a ligand hydroxyl group by coordination. Comparison of (E)- and (Z)-oxime *8* and *9* is of interest. The association constants in Table 2 indicate that *8* forms a much more stable complex with a Ni^{2+} ion than *9*, presumably because in *8* the two nitrogen atoms are in a coplanar position which is favorable for chelation with a metal ion, in contrast to *9* (compare the structures *13* and *14*). The activity of the complex itself (k_c value), however, is much larger for *9* than for *8*, presumably because the hydroxyl group is bonded to the metal ion much more tightly in *9*(*14*) than in *8*(*13*). Similarly the diamine *3* must form a much more stable complex with Zn^{2+} ion than *4c* (*15* vs. 16), but the resulting activation of the hydroxyl group in the former complex is not so large as in the latter complex. Such activation may also be influenced by the nature of the nitrogen ligand, such as whether it is aliphatic or aromatic. It is interesting to note with Table 2 that in each pair of *4c*-Zn^{2+} and *6*-Zn^{2+} and *10*-Zn^{2+} and *11*-Zn^{2+}, the imidazole ligands give larger k_c values, whereas the pyridine ligands give larger K values than those of the other. It is also important to note that the 2-(hydroxymethyl)imidazole-*4a*-Zn^{2+} ion complex is more active than the 4-(hydroxymethyl)imidazole-*5*-Zn^{2+} ion complex. At any rate, it should be pointed out that the activity of *10*-Zn^{2+} ion complex seems to be really remarkable as compared to those of known examples in the literature.

Sigman and Jorgensen [13] proposed the formation of a reactive ternary complex *17* composed of Zn^{2+} ion, *3* and *1* in the transesterification reaction between *1* and *3*. In this ternary complex the Zn^{2+} ion perturbs the pK_a (estimated to be 8.4) of the hydroxyethyl group of *3* to provide a high concentration of the effective nucleophile, which in turn acts as a template to orient *1* for the intracomplex nucleophilic attack of the ionized hydroxyethyl group on the carbonyl group of *1*. The importance of the formation of the intracomplex for the catalysis was shown by the inability of the *3*-Zn^{2+} ion complex to catalyze the transesterification of p-nitrophenyl acetate or p-nitrophenyl nicotinate which are unable to form any intracomplex. We also observed similar substrate specificity in the present systems of *4c*-Zn^{2+} and *10*-Zn^{2+} ion complexes. Thus the formation of the ternary complex *18b*, analogous to *17*, is also conceivable for the catalysis by the *10*-Zn^{2+} ion complex as illustrated in Scheme 2. However, as mentioned earlier, such a complex as *18b* should be very low in concentration as compared to the 1:1 complex of *18a*. It is rather more reasonable to consider that the Zn^{2+} ion stabilizes the tetrahedral addition intermediate *18c* through which the acylated intermediate *18d* is formed. As already mentioned, the Zn^{2+} ion must also catalyze the deacylation of *18d*. The complex *18e* may explain how the hydroxide ion bound on the Zn^{2+} ion facilitates the deacylation through the formation of another tetrahedral intermediate *18f*. Alternatively, the Zn^{2+} ion may act as an electrophile to assist the carbonyl sttack by the hydroxide ion as illustrated in *18g*, although it is difficult to distinguish between the two mechanisms (*18e* and *18g*) kinetically. Whichever mechanism is true, the breakdown of the tetrahedral intermediate *18f* into products should also be fast for an effective catalysis to occur.

We have disclosed that the ligands *4c*, *10*, and *11*, when complexed with a metal ion such as Zn^{2+}, Ni^{2+}, or Co^{2+}, become highly active toward the hydrolysis of p-nitrophenyl picolinate (*1*). The catalysis is most likely to occur through formation of a ternary complex in the transition state or in reactive intermediates. The metal ion in such a complex serves to activate the ligand hydroxyl group for nucleophilic attack and to orient the substrate into a favorable position to undergo the reaction.

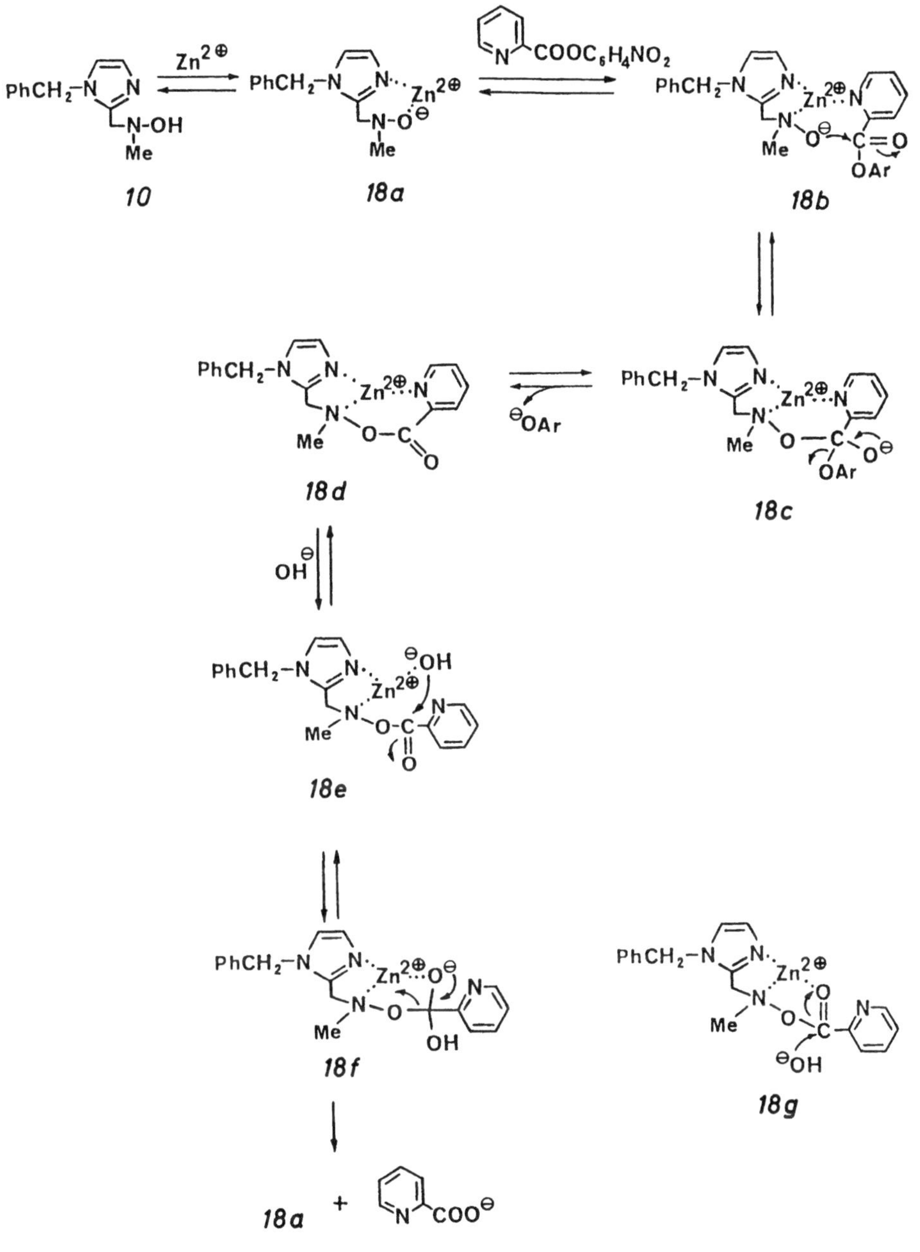

Scheme 2. Mechanism of the hydrolysis of *1* catalyzed by the *10*-Zn^{2+} complex.

2.2 Carbonic Anhydrase Models

In carbonic anhydrase [6], the Zn^{2+} ion is bound to three imidazole ligands, and in carboxypeptidase [2] the Zn^{2+} ion is bound to two imidazole ligands together with a carboxylate anion. In other metalloenzymes, imidazole rings of histidine are also frequently the metal-binding site. Therefore, it is important to develope general synthetic methods of polyimidazole ligands to mimic such metalloenzymes. The initiation was first made by Breslow et al. who reported the syntheses of various bis- and tris-imidazole ligands [18, 19]. Few examples are shown in *19–22*. Among them, the ligand *22* is particularly interesting and appears to be a good enzyme model, although the catalytic activity has not yet been reported. The ligand *22* forms a stable 1:1 complex with Zn^{2+} ions without forming octahedrally coordinated 2:1 complexes by dimerization or with added other ligands, or a hydroxide-bridged dimer, presumably due to a cage effect by the bulky substituents. Brown et al. also reported several tris-imidazole derivatives, such as *23–25* [20, 21]. These ligands show some carbonic anhydrase activity, and a moderate catalytic activity in the hydrolysis of *1* was also observed for the *25*-Co^{2+} complex [22]. A problem for these complexes (*23–25*) is

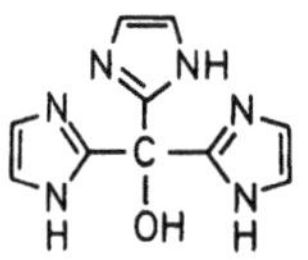

19

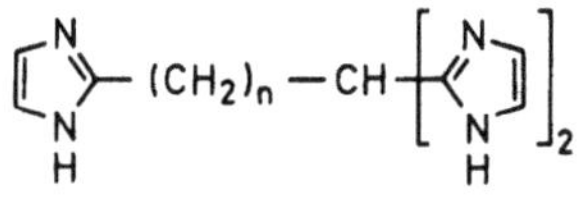

n = 1, 3

20

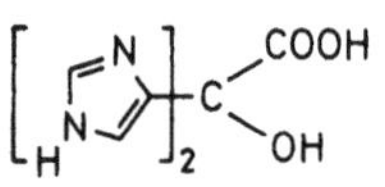

21

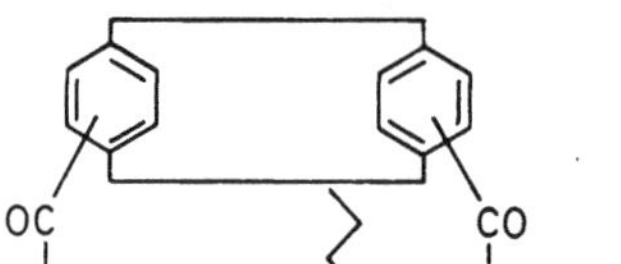

22

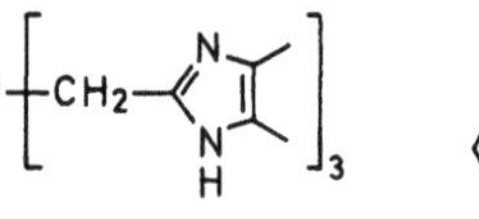

23

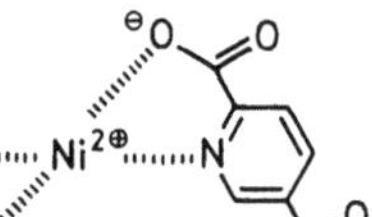

24

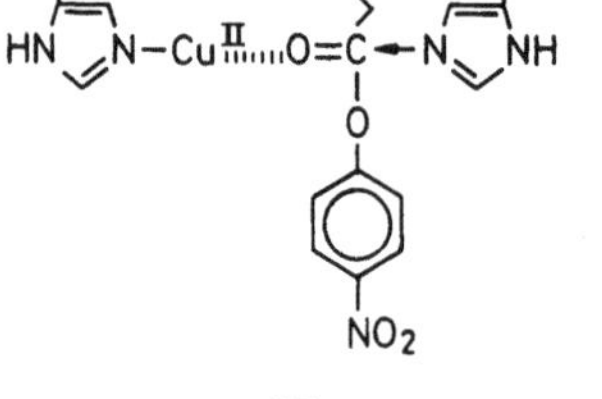

27

25

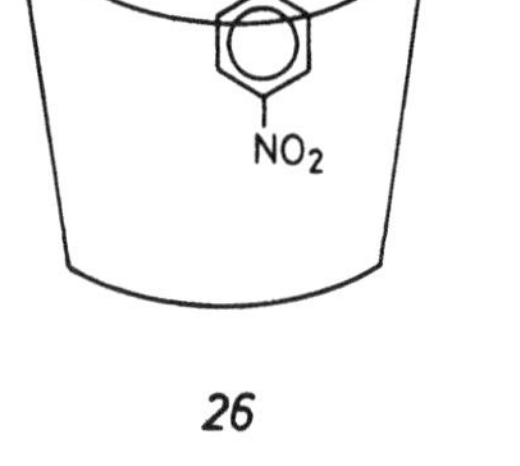

26

their insolubility in water so that the reactions were examined in heavily organic mixtures (80% ethanol-water). Tabushi et al. reported that the rates of hydration of carbon dioxide is strongly accelerated by a zinc ion complex of cyclodextrin bis-(histamine) [23]. Yet the above examples indicate that the model study of carbonic anhydrase still remains at an infant stage.

2.3 Models Having a Substrate Binding Cavity

An artificial metalloenzyme (*26*) was designed by Breslow et al. [24]. It was the first example of a complete artificial enzyme, having a substrate binding cyclodextrin cavity and a Ni^{2+} ion-chelated nucleophilic group for catalysis. Metalloenzyme (*26*) behaves a real catalyst, exhibiting turnover, and enhances the rate of hydrolysis of p-nitrophenyl acetate more than 10^3 fold. The catalytic group of *26* is a *8*-Ni^{2+} complex which itself is active toward the substrate *1*, but not toward such a substrate having no metal ion affinity at a low catalyst concentration. It is appearent that the metal ion in *26* activates the oximate anion by chelation, but not the substrate directly as believed in carboxypeptidase.

Murakami et al. reported that a cyclophane *27* having two imidazole groups is activated by Cu^{2+} ions in the hydrolysis of p-nitrophenyl dodecanoate [25,26], although the activation seemed to be small.

The above two models together with Tabushi's cyclodextrin bis(histamine) [23] are really elabolate ones, each having a substrate binding cavity, but their catalytic activities are yet far behind of those of natural enzymes. They suggest the difficulties associated with the design of a metal ion center inside of a cavity which activates both substrate and catalytic groups.

3 Micellar Models

Micelles in water are described as spherical aggregates of a surfactant monomer [27–30]. They somewhat resemble to enzyme proteins in structures and functions, although the details are yet the subjects of recent controversies [29,30]. There are numerous studies of micellar models of enzymes [28], but the examples of those of metalloenzymes are very few [31–37]. In particular, there are no examples of micellar models of carboxypeptidase or carbonic anhydrase except ours [36,37].

3.1 Basic Ideas in Designing Micellar Metalloenzymes

There are cationic, anionic, and non-ionic micelles. Divalent metal ions having positive charges are highly hydrophilic and cannot be incorporated into cationic micelles. Anionic micelles tend to form water insoluble salts with divalent metal ions. Interactions of non-ionic micelles with divalent metal ions appear to be small. Thus incorporation of a divalent metal ion into a micelle to form a catalytic center

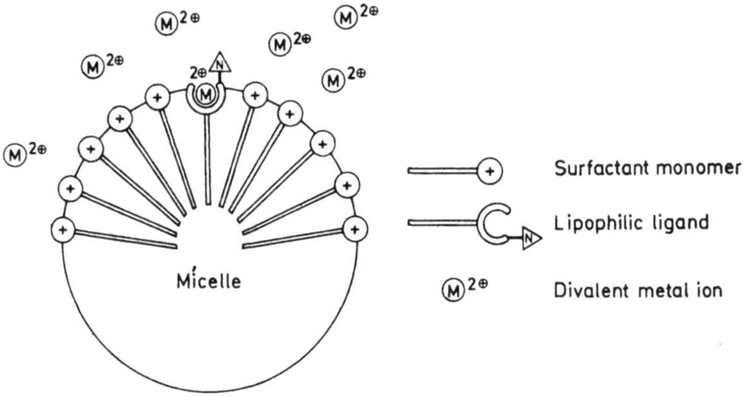

Fig. 2. Schematic illustration of a micellar model of a metalloenzyme

is not a straightforward process without some devices. A simple device is to use a mixed micellar system which is composed of chemically inert surfactant molecules and lipophilic ligands having a strong chelating ability towards metal ions together with a nucleophilic group for catalysis, such as illustrated schematically in Fig. 2. In more elaborate models, the surfactant molecule itself would have both metal ion binding and catalytic sites, although such molecules have not yet appeared. Both cationic and non-ionic surfactants may be used, but anionic ones may not be adequate because of the formation of insoluble salts. Stereo-selectivity may also be examined by introducing stereo-controlling elements in the ligand structures.

3.2 Lipophilic Monoimidazole Ligands

Non-micellar reactions (Table 1, 2) reveal that imidazole ligands are more active than the corresponding pyridine ligands, in particular when used as the Zn^{2+} ion complexes. Consequently ligand screening for higher activity was first carried out for lipophilic monoimidazole ligands, such as *28–37*. These ligands are soluble in chloroform, but virtually insoluble in water. Their chelating ability can be readily visualized by extracting Cu^{2+} or Co^{2+} ion color from an aqueous phase into an organic phase. Surfactants employed were cationic cetyltrimethylammonium bromide (CTAB) and non-ionic Triton X-100.

3.2.1 Kinetic Equations and Ligand Activities under [ligand] $\gg$ [substrate]

Pseudo-first-order rate constants (k_{obsd}) for the reaction of *1* are shown in Table 3. The rate increases slightly by the addition of CTAB of more than the CMC concentration (10^{-3} M). The Zn^{2+} ion is more activating, but further addition of CTAB has no effect. A large rate enhancement is only observed when the three components, a ligand, Zn^{2+} ion and CTAB are mixed together. The presence of CTAB is essential, although CTAB itself has almost no rate enhancing effect without the other com-

Table 3. Kinetic parameters for the release of p-nitrophenol from p-nitrophenyl picolinate *1* under micellar conditions of CTAB

catalyst	$k_{obsd} \times 10^3\ s^{-1}$	$k_c\ (M^{-1}\ s^{-1})$	$K\ (M^{-1}$ or $M^{-2})$[c]
none	0.013	—	—
CTAB	0.020	—	—
Zn^{2+}	0.25	—	—
Zn^{2+} +CTAB	0.28	—	—
28-Zn^{2+} +CTAB	2.05[a]	—	—
29-Zn^{2+} +CTAB	166[a]	3.7×10^3	$1.7 \times 10^5\ (M^{-2})$
29-Zn^{2+} +CTAB	92.2[b]	—	—
30-Zn^{2+} +CTAB	42.2[a]	2.7×10^2	$8.57 \times 10^5\ (M^{-2})$
31-Zn^{2+} +CTAB	1.18[c]	—	—
32-Zn^{2+} +CTAB	114[a]	5.3×10^2	$6.10 \times 10^2\ (M^{-1})$
33-Zn^{2+} +CTAB	64.7[a]	1.8×10^2	$4.95 \times 10^6\ (M^{-2})$
34-Zn^{2+} +CTAB	111[a]	1.4×10^2	$6.04 \times 10^6\ (M^{-2})$
35-Zn^{2+} +CTAB	83.7[a]	1.55×10^2	$6.22 \times 10^6\ (M^{-2})$
36-Zn^{2+} +CTAB	11.6[b]	7.14×10^3	$1.26 \times 10^5\ (M^{-2})$
37-Zn^{2+} +CTAB	31.6[b]	2.67×10^3	$0.34 \times 10^5\ (M^{-2})$

[a] pH 7.05 (2,6-lutidine buffer 0.1 M, $\mu = 0.2\,(KNO_3)$), 25 °C, $[1] = 1 \times 10^{-4}$ M, [ligand] $= 5 \times 10^{-4}$ M, $[Zn^{2+}] = 1 \times 10^{-3}$ M, [CTAB] $= 5 \times 10^{-3}$ M.
[b] [CTAB] $= 1 \times 10^{-2}$ M; other conditions are the same as in (a).
[c] $[1] = 1 \times 10^{-4}$ M, [ligand] $= 1 - 5 \times 10^{-4}$ M, $[Zn^{2+}] = 0 - 1 \times 10^{-2}$ M, [CTAB] $= 1 \times 10^{-2}$ M; other conditions are the same as in (a).

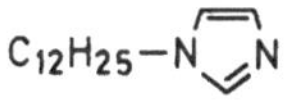

28

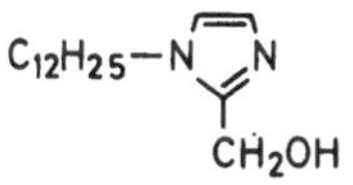

29

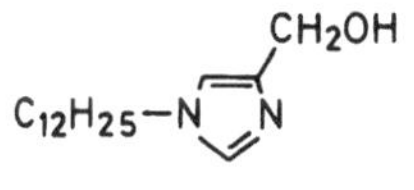

30

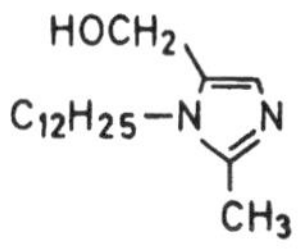

31

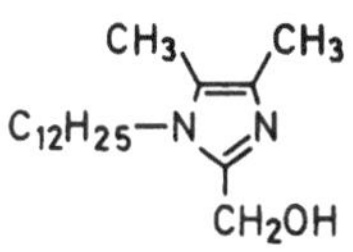

32

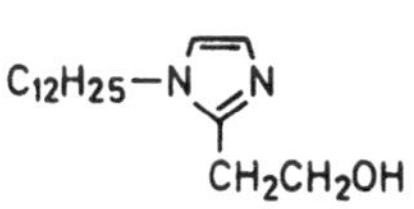

33

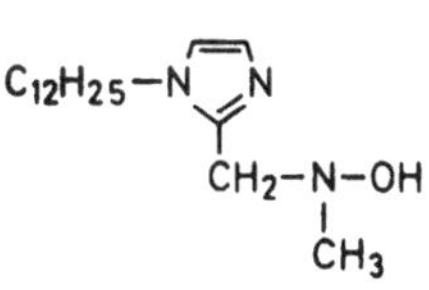

34

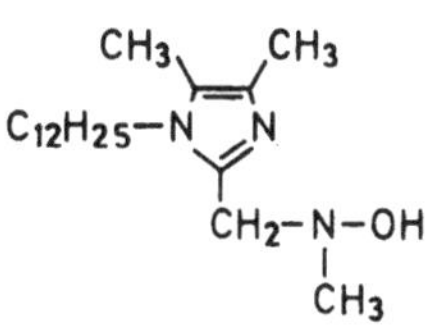

35

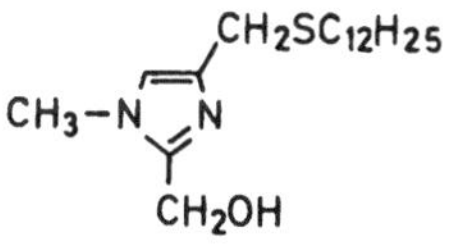

36

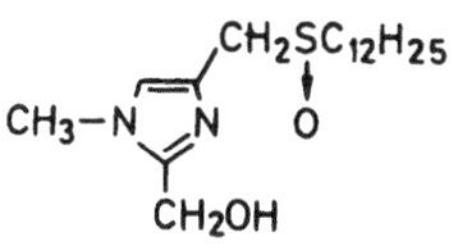

37

ponents. CTAB micelles solubilize a ligand in water and greatly activate the ligand-Zn^{2+} ion complex as mentioned later.

The rates of this micellar reaction are also dependent on both metal ion and ligand concentration (Scheme 1 and Eq. 1–4) [36].

However, there is an important difference between these two systems in the ligand-metal ion ratio in complexation. Namely, micellar reactions require a more generalized reaction Scheme 3, where the molarity of ligand n is either 1 or 2 depending upon the structure of the ligands. This scheme gives rates: Eq. 2–4 for n = 1 and Eq. 3, 5, 6 for n = 2. The results of the kinetic analysis are shown in Table 3.

$$M + nL \overset{K}{\rightleftarrows} \text{complex}$$

$$S + \text{complex} \xrightarrow{k_c} \text{products}$$

$$S \xrightarrow{k_0} \text{products}$$

$$n = 2, \qquad k_{obsd} = k_0' + \frac{k_c K[L]_T^2[M]_T}{1 + K[L]_T^2 + 4K[L]_T[M]_T} \tag{5}$$

$$\frac{1}{(k_{obsd} - k_0')} = \frac{4}{k_c[L]_T} + \left(\frac{1}{k_c} + \frac{1}{k_c K[L]_T^2}\right) \cdot \frac{1}{[M]_T} \tag{6}$$

Scheme 3.

Table 3 indicates that five ligands form 2:1 complexes (n = 2) and one ligand forms a 1:1 complex (n = 1). As illustrated in Fig. 3, it is possible to build a molecular model of the 2:1 complex of *29*-Zn^{2+}, but it is difficult to do so in the case of the *32*-Zn^{2+} complex because of steric repulsion between 4(5)-methyl groups. The ligand *35* has also 4(5)-methyl groups, but it forms a 2:1 complex. Presumably, coordination by the hydroxyamino group to the metal ion in *35* widen the space between the 4(5)-carbons of the imidazole rings, reducing the repulsion between the 4(5)-methyl groups. Thus, the formation of a 2:1 or a 1:1 complex seems to depend on a delicate balance between the structural elements of the complexes. Presumably, the hydration of the metal ion to stabilize a 1:1 complex under non-micellar conditions becomes weak in a micellar phase so that the coordination of the second ligand occurs to form the 2:1 complex.

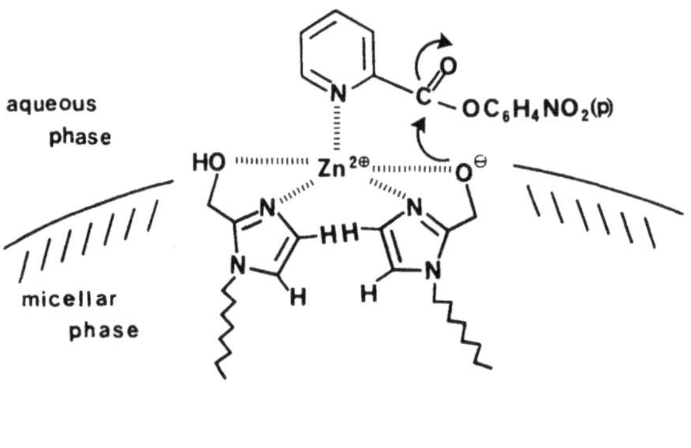

Fig. 3. Schematic illustration of the *29*-Zn^{2+} ion complex embedded near the micellar surface

Comparison of the *28* complex with those of *29* and *30* (Table 3) indicates that the hydroxyl group is essential for a high activity as in the non-micellar reaction (Table 1, 2). Comparison of the *29* and *30* complexes reveals the 2-hydroxymethyl group being more active than the 4-hydroxymethyl group. The k_c value of *29*-Zn^{2+} (Table 3) is 27 fold larger than that of *4b*-Zn^{2+} (Table 2), indicating an extent of micellar activation of the complex. However, the k_c values of the *34* and *35* complexes (Table 3) are smaller than that of the *10*-complex (Table 2); although the K values of these complexes are much larger than that of the *29*-Zn^{2+} complex. Thus micellar effects appear to depend on the nature of the complex.

3.2.2 Effects of Micelle and Salt Concentration

Figure 4 indicates the effect of the CTAB concentration on the rate constant of the complexes of *29* and *32*. The CMC of CTAB is near 1×10^{-3} M. Below CMC, the rates cannot be measured because of insolubility of the ligands. Although unmeasured, the rates of the *29* and *32* complexes must be greatly enhanced in the presence of CTAB micelles up to CMC, but further increase of the micelle concentration above CMC cause a rate decrease. This type of micellar effect can be seen in many micellar reactions [27]. Hence, it should be noted that the rate constants in Table 3 would be several times larger if they are measured by using a lower concentration of CTAB than 5×10^{-3} M.

The effect of salt (KNO_3) concentration is small, i.e. less than a two fold rate increase by increasing the salt concentration up to 0.2 M.

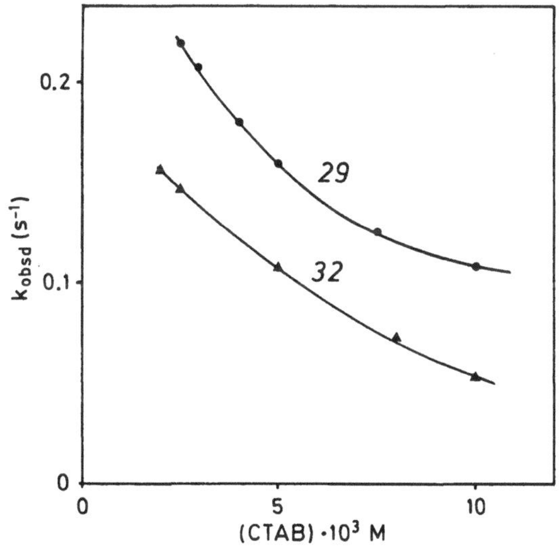

Fig. 4. Effect of CTAB concentration on the pseudo-first-order rate constants at pH 7.00 at 25 °C; [*I*] = 1×10^{-4} M, [ligand] = 5×10^{-4} M, [Zn^{2+}] = 1×10^{-3} M

3.3 Lipophilic Bisimidazole Ligands

The next subject is to examine bisimidazole ligands in which two imidazole rings are connected by a covalent bond. Thus, we have prepared a number of bisimidazole derivatives as listed in *38–44* according to the synthetic methods of Breslow [18]

Among them, *38a* and *38b* are water soluble and they can be considered as the non-micellar counterparts. All other ligands are water insoluble but can be solubilized in CTAB or Triton X-100 micelles.

3.3.1 Effect of CTAB Micelle Concentration and Ligand Lipophilicity

Figure 5a indicates the effect of the CTAB concentration on the rate constants of the complexes of *38b* and *38c*. In the case of the water soluble *38b* ligand, the rate increases with increasing CTAB concentration up to a saturation level. This type of saturation kinetics is usually interpreted to show the incorporation of a ligand-metal ion complex into a micellar phase from a bulk aqueous phase, and the catalytic activity of the complex is higher in the micellar phase than in the aqueous phase. In the case of lipophilic *38c*, a very similar curve as in Fig. 4 is obtained. At a first glance, there appears to be a big difference between these two curves. However, they are rather common in micellar reactions and obey the same reaction mechanism [27].

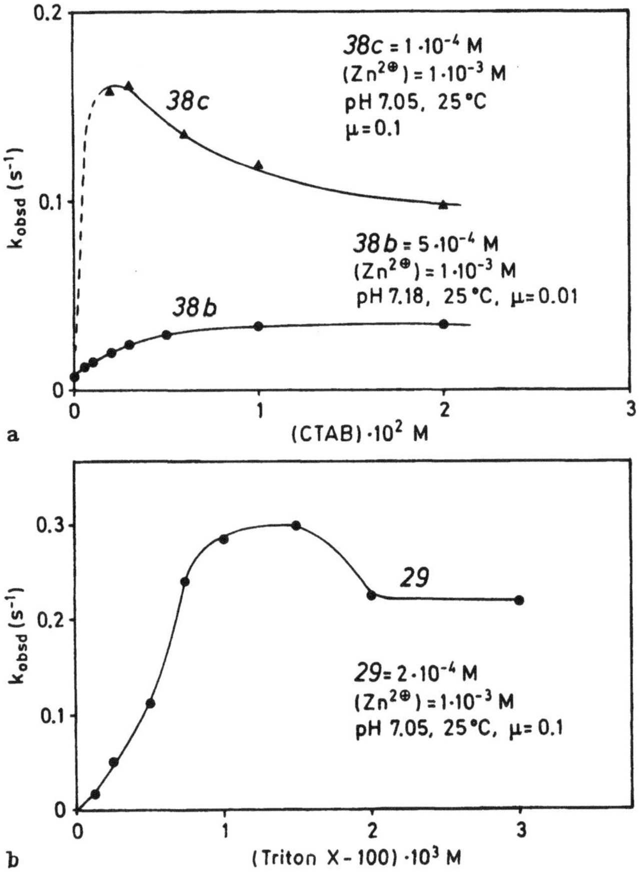

Fig. 5a and b. a Plots of k_{obsd} vs. concentration of CTAB for the release of p-nitrophenol from *1*. **b** Plots of k_{obsd} vs. concentration of Triton X-100 for the release of p-nitrophenol from *1*

The apparent difference seems to be due to the difference in the binding constants of the complexes to micelles which is much larger in the lipophilic *38c* than in the hydrophilic *38b* complex [27]. A somewhat different, but not an unusual micellar effect is observed in the case of the non-ionic surfactant Triton X-100 as shown in Fig. 5b.

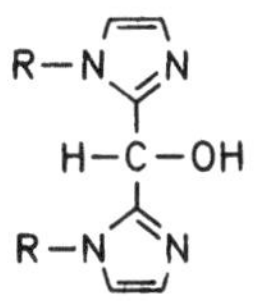

38a R = CH_3
38b R = C_4H_9
38c R = C_8H_{17}
38d R = $C_{12}H_{25}$

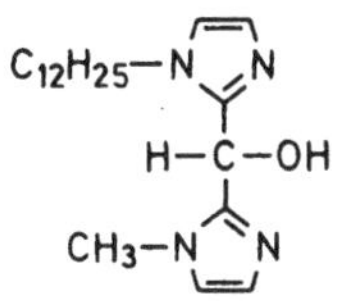

39

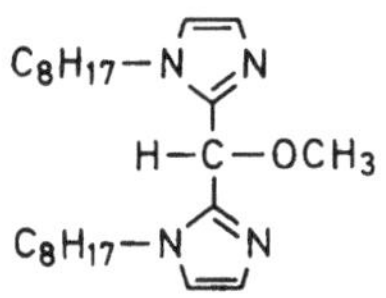

40

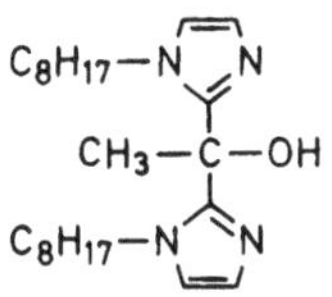

41

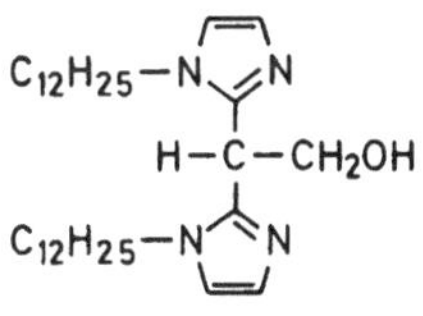

42

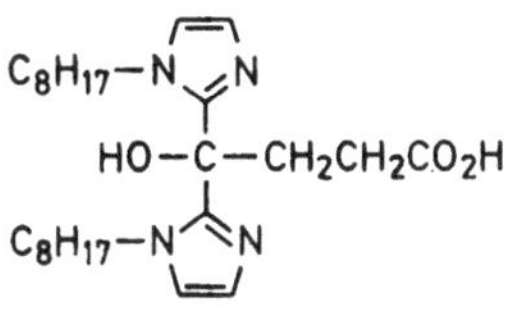

43

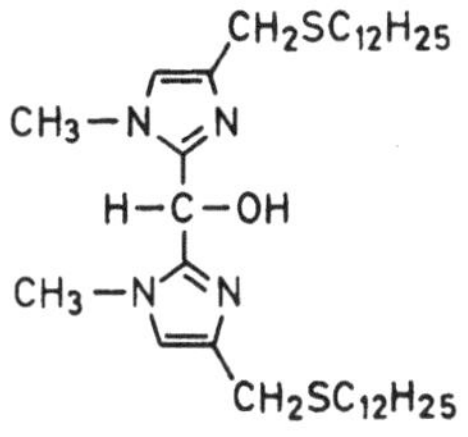

44

Another features of the ligand lipophilicity and the stability of the complex on the rates are shown in Fig. 6: Rate saturation corresponds to the formation of a 1:1 or 2:1 ligand-metal ion complex. Non-micellar reactions of curves b and c indicate that the N-butyl ligand *38b* forms a more active complex than N-methyl ligand *38a* does. It may be interesting to note that in the micellar reaction of *38b*, a flat

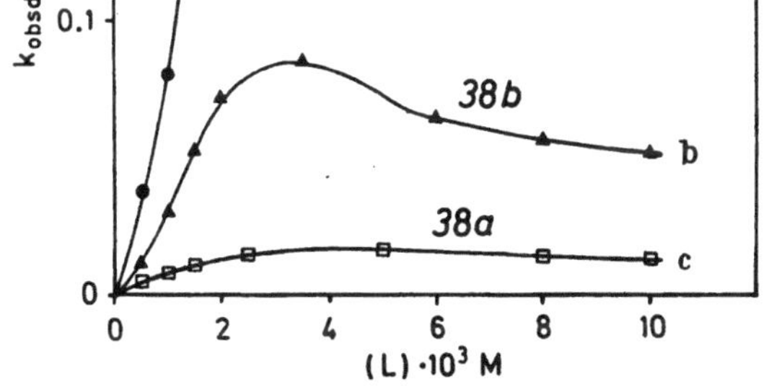

Fig. 6. Effects of CTAB and ligand concentrations on the hydrolysis of p-nitrophenyl picolinate(*1*): 25 °C, $[Zn^{2+}] = 1 \times 10^{-3}$ M; a: [CTAB] $= 1 \times 10^{-2}$ M, pH 7.05; b: [CTAB] = 0, pH 7.18; c: [CTAB] = 0, pH 7.18

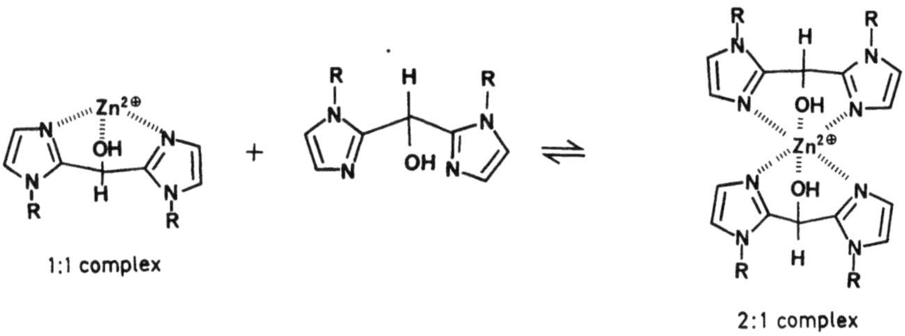

Fig. 7. Formation of 1:1 and 2:1 complexes in the bis-imidazole/Zn^{2+} ion equilibrium

saturation level is obtained, while in the non-micellar reaction, the rates go down after a maximum. A flat saturation curve is also observed in the reaction of the more lipophilic ligands *38c*. These results suggest that there is an equilibrium between 1:1 and 2:1 complexes as illustrated in Fig. 7, and the formation of the 2:1 complex which is less active than the 1:1 complex is hindered in the micellar phase.

3.3.2 Activity Profiles under [ligand] ≫ [substrate]

The results are shown in Table 4. In CTAB micelles, the complexes of the lipophilic bisimidazole ligands, *38c*, *38d*, *39*, and *44* are much more reactive than those of the lipophilic monoimidazole ligands, *29* and *36*, and the rate enhancement by the *38c*-Zn^{2+} ion is 21500 fold as compared to the rate in CTAB alone.

Table 4. Pseudo-first-order rate constants (k_{obsd}) for p-nitrophenol release from p-nitrophenyl picolinate *1*[a]

ligand and Zn^{2+}	$k_{obsd} \times 10^2$ s^{-1}		
	CTAB	Triton X-100[b]	PEG-$C_{12}H_{25}$
none	0.00193	0.00148	0.00201
29	0.192	0.742	—
36	0.088	—	—
38c	0.307	0.414	—
29-Zn^{2+}	3.02	27.5	32.3
36-Zn^{2+}	3.15	24.6	—
38b-Zn^{2+}	0.987	—	—
38c-Zn^{2+}	41.5	21.5	18.9
38d-Zn^{2+}	30.9	—	—
39-Zn^{2+}	22.1	—	—
40-Zn^{2+}	0.0457	—	—
41-Zn^{2+}	0.0183	—	—
42-Zn^{2+}	0.418	—	—
44-Zn^{2+}	10.8	—	—

[a] 25 °C, pH 7.05 (0.1 M 2,6-lutidine-HNO_3 buffer), μ = 0.1 (KNO_3), [*1*] = 5×10^{-5} M, [ligand] = 2×10^{-4} M, [Zn^{2+}] = 1×10^{-3} M, [CTAB] = 1×10^{-2} M, [Triton X-100] = 2×10^{-3} M, [PEG-$C_{12}H_{25}$] = 2×10^{-3} M.

[b] Triton X-100 = $HO(CH_2CH_2O)_{9.7}C_6H_4C_8H_{17}$

[c] PEG-$C_{12}H_{25}$ = $HO(CH_2CH_2O)_{25}C_{12}H_{25}$

Comparable activity of the *39*-Zn^{2+} ion having unsymmetrical N-dodecyl and N-methyl groups with those of *38d*-Zn^{2+} ions having symmetrical N-alkyl groups indicates that lipophilicity is more important than symmetry of dialkyl groups. The importance of the hydroxyl group for the activity is indicated in a low activity of the methylated *40*-Zn^{2+} ion complex. At a first glance, it seems strange that the *41*-Zn^{2+} ion having a free hydroxyl group shows even lower activity than the *40*-Zn^{2+} ion. However, the CPK molecular model indicates that the methyl and the two octyl groups of *41* are too much repulsive when placed at the same side so that the hydroxyl group has to take the side with the two non-alkylated imidazole nitrogen atoms to coordinate to the metal ion (Fig. 8).

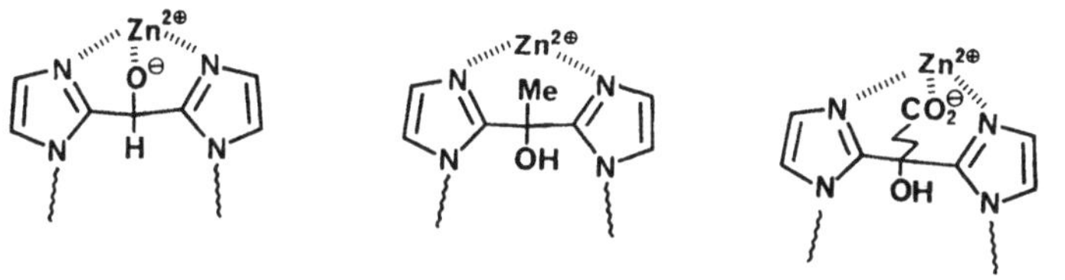

38-$Zn^{2\oplus}$ *41*-$Zn^{2\oplus}$ *43*-$Zn^{2\oplus}$

Fig. 8. Possible structures of 1:1 complexes of bis-imidazole-Zn^{2+} ions

Ligand *43* forms a very strong 1:1 complex with the Zn^{2+} ion in a CTAB micelle, but the catalytic activity is only three fold larger than that of the Zn^{2+} ion. The *43*-Zn^{2+} ion complex formally resembles the Zn^{2+} ion center of carboxypeptidase A having two imidazole and one carboxylate groups, but lacks a nucleophilic group (i.e. the hydroxyl group is forced to be away from the Zn^{2+} ion) (Fig. 8). On the other hand, it is difficult to explain the low activity of the *42*-Zn^{2+} ion complex in view of a relatively high activity of the corresponding monoimidazole complex (*33*-Zn^{2+} ion). It should be noted that the complex reactivities are dependent upon the types of surfactants, i.e. non-ionic surfactants enhance the activities of mono-imidazole complexes much more than those of bisimidazole complexes.

3.3.3 Mechanism of the Catalysis of the Bis(N-octyl-2-imidazolyl)-carbinol(*38c*)-Zn^{2+} ion and of Related Complexes

3.3.3.1 K and k_c values obtained under [ligand] ≫ [substrate]

The activity of the *38c*-Zn^{2+} ion complex is remarkable as compared to those of the *29*-Zn^{2+} and *38b*-Zn^{2+} ion complexes, as illustrated in Fig. 9. Here, as described already, the reaction of the *29*-Zn^{2+} ion can be represented by Scheme 2 (n = 2).

On the other hand, the reaction of the *38c*-Zn^{2+} ion can be represented by Scheme 1 (or Scheme 2, n = 1) and by Eq. 7, which is equal to Eq. 2 when the second term $\left(\frac{K}{k_c}(k_{obsd} - k'_0)\right)$ is neglible as compared to the first term in the denominator.

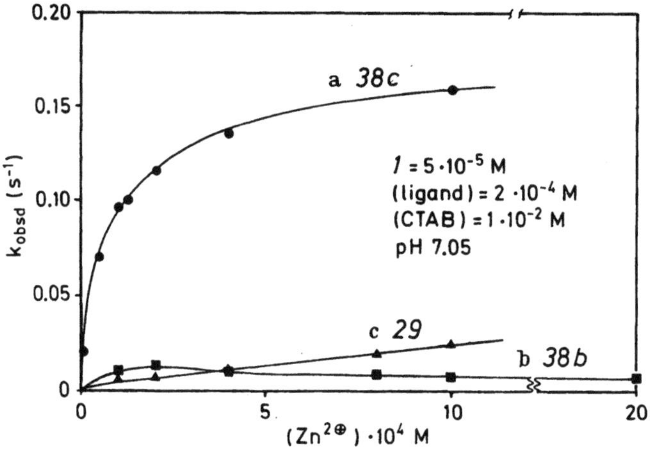

Fig. 9. Plots of k_{obsd} vs. concentration of Zn^{2+} ion for the release of p-nitrophenol from *1* under fixed concentration of [ligand] = 2×10^{-4} M and [CTAB] = 1×10^{-2} M; a: *38c*, b: *38b*, c: *29*

These ligand-metal ion ratios

$$k_{obsd} = k_0' + \frac{k_c K[L]_T [M]_T}{(1 + K([L]_T + [M]_T)) - \frac{K}{k_c}(k_{obsd} - k_0')} \quad (7)$$

(n = 2 and 1) are the same in both CTAB and Triton X-100 micelles, and the results of kinetic analysis are shown in Table 5. This and Table 3, 4 indicate that larger k_{obsd} values of bisimidazole complexes than those of monoimidazole complexes are primarily due to favorable K values for complexation.

However, the k_c values of monoimidazole complexes are larger than those of bisimidazole complexes although each of the former complexes have two hydroxyl groups so that statistical correction by factor 2 may be necessary for comparison.

3.3.3.2 Kinetics under the Conditions of [Substrate] ≫ [Ligand]: Burst Kinetics

In the foregoing micellar reactions, it is likely that the reaction proceeds through the acylation of the hydroxyl group of the ligands, and the results indicate that the acylation step is greatly enhanced by complexation with Zn^{2+} ions under micellar

Table 5. Kinetic parameters for the release of p-nitrophenol from p-nitrophenyl picolinate *1*[a]

L – M	surfactant	K	k_c
29-Zn^{2+}	CTAB	1.7×10^5 M^{-2}	3.7×10^3 M^{-1} s^{-1}
(2:1 complex)	Triton X-100	2.68×10^6 M^{-2}	10.5×10^3 M^{-1} s^{-1}
38c-Zn^{2+}	CTAB	2.6×10^4 M^{-1}	0.783×10^3 M^{-1} s^{-1}
(1:1 complex)	Triton X-100	2.6×10^4 M^{-1}	1.37×10^3 M^{-1} s^{-1}
39-Zn^{2+}	CTAB	2.2×10^4 M^{-1}	1.37×10^3 M^{-1} s^{-1}
(1:1 complex)			

[a] 25 °C, pH 7.05; see footnotes in Table 3 and 4 for other conditions.

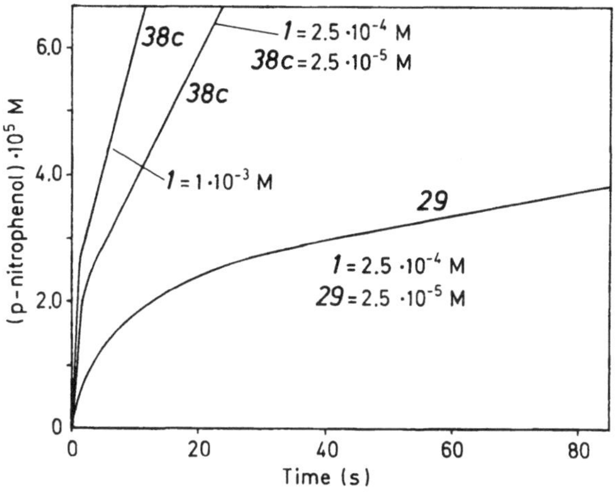

Fig. 10. Burst kinetics under [substrate]$\gg$[ligand-Zn^{2+} ion complex]

conditions. However, a little is known about the subsequent deacylation step since a ligand is always in excess over the substrate. It is clear that a system is not a catalytic one if deacylation is too slow to regenerate the free hydroxyl group of the ligand.

Figure 10 shows typical examples of burst kinetics observed for the reactions of *29*-Zn^{2+} and *38c*-Zn^{2+} ion complexes under the conditions of excess substrate over ligand. Such burst kinetics can be accounted for a two-step reaction involving an acylated intermediate as in Scheme 4, and the rate constants, k_a and k_d, can be obtained based on Eqs. 8–11 [38, 39], where A is the slope of the steady-state line and B is the intercept obtained by extrapolating the steady-state line to time = 0. The k_a should be the same with the k_c in Table 5.

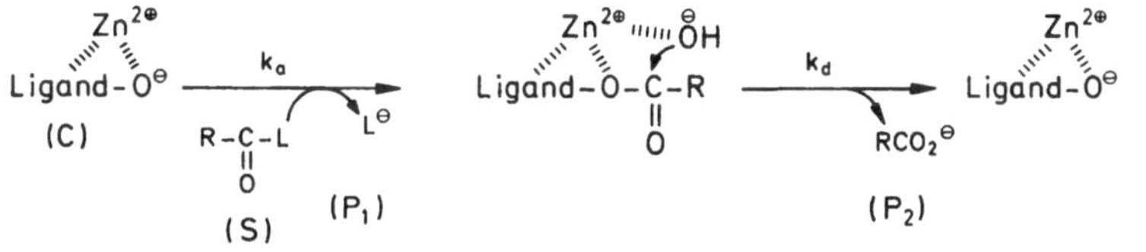

Scheme 4

$$[P_1] = At + B(1 - e^{-bt}) \tag{8}$$

$$\frac{d[P_1]}{dt} = A = \frac{k_a k_d [S]_0 [C]_0}{k_a[S]_0 + k_d} + k_0'[C]_0 \tag{9}$$

$$B = \frac{(k_a[S]_0)^2[C]_0}{(k_a[S]_0 + k_d)^2} \tag{10}$$

$$b = k_a[S]_0 + k_d \tag{11}$$

However, the results shown in Table 6 indicate that the k'_a values calculated by using the ligand concentration ($[\text{ligand}]_T$) instead of the complex concentration ($[C]_0$) are much smaller than the k_c value in the case of *29c*-Zn^{2+}. The $[C]_0 = 1.6 \times 10^{-6}$ M for the *29*-Zn^{2+} complex can be calculated based on $K = 1.7 \times 10^{-5}\ M^{-2}$ (Table 5), but it is difficult to calculate corrected k_a values based on the equations in footnote (c) in Table 6, because the B values are close to $[\text{ligand}]_T = 1 \times 10^{-4}$ M and much larger than $[C]_0$, due to the acylation of almost all ligands.

Table 6. Kinetic parameters in burst kinetics

ligand-Zn^{2+}	pH	$A \times 10^7$ ($M\ s^{-1}$)	$B \times 10^4$ (M)	k'_a [c] ($M^{-1}\ s^{-1}$)	k_a [c] ($M^{-1}\ s^{-1}$)	$k'_d \times 10^4$ [c] (s^{-1})	$k_d \times 10^4$ [c] (s^{-1})
29-Zn^{2+}	6.50	0.136	0.912	5.65	—	1.42	—
	7.05	0.488	0.923	24.4	—	5.08	—
	7.50	1.24	0.924	63.6	—	12.09	—
	8.01	6.16	0.931	352	—	63.8	—
38c-Zn^{2+}	7.03	1.70	0.214	370	577	74	75
	7.96	26.6	0.208	4840	7550	1170	1190

[a] $[\text{CTAB}] = 1 \times 10^{-2}$ M, $[29] = 1 \times 10^{-4}$ M, $[Zn^{2+}] = 1 \times 10^{-3}$ M, $[1] = 5 \times 10^{-4}$ M.
[b] $[\text{CTAB}] = 1 \times 10^{-2}$ M, $[38c] = 2.5 \times 10^{-5}$ M, $[Zn^{2+}] = 1 \times 10^{-3}$ M, $[1] = 2.5 \times 10^{-4}$ M.
[c] k'_a and k'_d are calculated by assuming $[C]_0 = [\text{ligand}]_T$ and k_a and k_d are calculated by using calculated complex concentration $[C]_0$:
$k_a = k'_a([\text{ligand}]_T - B^{1/2}[\text{ligand}]_T^{1/2})/([C]_0 - B^{1/2}[C]_0^{1/2})$
$k_d = k'_d([\text{ligand}]_T^{1/2})/([C]_0^{1/2})$

In the case of the *38c*-Zn^{2+} ion complex, $[C]_0 = 2.4 \times 10^{-5}$ M calculated by using $K = 2.6 \times 10^{-4}\ M^{-1}$ (Table 5) is close to $[\text{ligand}]_T = 2.5 \times 10^{-5}$ M and it is possible to calculate corrected k_a values which are in reasonable agreement with the k_c value in Table 5. Thus ligand *38c* is more appropriate than *29* for the study of mechanism of catalysis under various conditions. The reasons are that *38c* forms a more stable Zn^{2+} ion complex than *29*, and furthermore such complexing ability appears to be retained in the acylated intermediate in the *38c* complex which, however, may be lost largely in the case of *29*. The latter reasoning appears to be supported by much larger deacylation rate constants k_d for *38c* than for *29* complexes. Namely, the deacylation rate is also greatly enhanced by complexation with Zn^{2+} ions.

3.3.3.3 pH Rate Profile of k_c and k_d

Equation 12 applies under the condition of $[L]_T \gg [M]_T \simeq [S]$ for

$$k_{obsd} - k'_0 = k_c[M]_T \tag{12}$$

$$A = k_d[\text{ligand}]_T \tag{13}$$

38c (see Eq. 2 and 7), and Eq. 13 applies for both *39* and *38c* under the conditions of $k_a[S]_0 \gg k_d$ (Eq. 8). The pH-rate profiles of k_c and k_d based on these Eqs., are shown in Fig. 11.

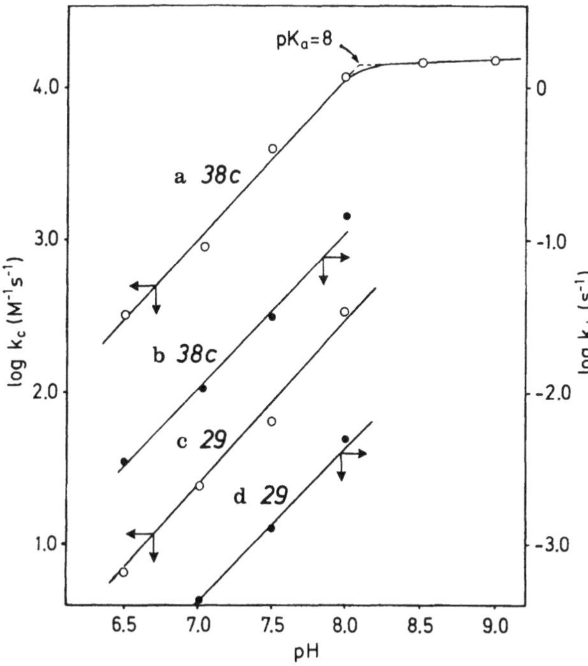

Fig. 11. pH-Rate profile for the acylation and deacylation rate constants (k_c and k_d) for the reactions of *29* and *38c* complexes; (a) and (c) for k_c, [CTAB] $= 1 \times 10^{-2}$ M, [*1*] $= 5 \times 10^{-5}$ M, [Zn^{2+}] $= 1\text{–}4 \times 10^{-5}$ M, [*38c*] or [*29*] $= 5 \times 10^{-4}$ M; (b) and (d) for k_d

The k_c and k_d values given there are satisfactorily in agreement with those listed in Table 5, 6. It is interesting to note that the curve (a) for *38c* shows leveling off near pH 8 which seems to correspond with the pK_a of the hydroxyl group of the *38c*-Zn^{2+} ion complex, indicating that complexation with Zn^{2+} ions in a cationic micelle lower the pK_a of a hydroxyl group considerably. Figure 11 also shows that the k_d values of the *38c* complex are more than 20 times larger than those of the *29* complex.

3.3.3.4 Cu^{2+} Ion Effects

The Table 7 shows Cu^{2+} ion effects. In a CTAB micelle, the Cu^{2+} ion itself without ligand enhances the rate 1900 fold as compared to 14 fold of the Zn^{2+} ion (Table 4).

Table 7. Pseudo-first-order rate constants, k_c and K values in the Cu^{2+} ion catalyzed reaction of *1*

ligand, condition	k_{obsd} [a] (s^{-1})	K (M^{-2})	k_c ($M^{-1}\,s^{-1}$)
CTAB	0.002×10^{-2}	—	—
CTAB+Cu^{2+}	3.78×10^{-2}	—	—
CTAB+Cu^{2+}-*29*	4.91	1.09×10^{8}	3.23×10^{5}
CTAB+Cu^{2+}-*36*	4.10	8.12×10^{7}	3.82×10^{5}
CTAB+Cu^{2+}-*37*	1.01	—	—
CTAB+Cu^{2+}-*38c*	0.153[b]	—	—
CTAB+Cu^{2+}-*44*	0.105	—	—

[a] 25 °C, pH 7.03, [CTAB] $= 1 \times 10^{-2}$ M, [*1*] $= 5 \times 10^{-5}$ M, [ligand] $= 1 \times 10^{-4}$ M, [Cu^{2+}] $= 10^{-4}$ M.
[b] [*1*] $= 2.5 \times 10^{-5}$ M, [ligand] $= 0.5 \times 10^{-4}$ M. [Cu^{2+}] $= 2.5 \times 10^{-4}$ M.

Thus the Cu^{2+} ion is specially active toward a pyridine ester like *1*. This activity further increases about 100 fold in the presence of monoimidazole ligands, e.g. *29* and *36*. Thus, in the presence of 1×10^{-4} M each of Cu^{2+} and *29*, the spontaneous rate is enhanced 2.5×10^5 fold at pH 7.05. This rate enhancement is much larger than the corresponding 1.5×10^3 fold observed in the presence of 2×10^{-4} M of ligand and 1×10^{-3} M of Zn^{2+} (Table 4). Kinetic analyses indicate that these monoimidazole ligands also form 2:1 ligand-Cu^{2+} ion complexes as in the cases of Zn^{2+} ion complexes. The observed K value of the *29*-Cu^{2+} ion complex is 640 times larger than that of the *29*-Zn^{2+} ion complex (Table 5 and 7). The k_c values of the former complex is also 87 times larger than that of the latter complex. In contrast, the activities (k_{obsd}) of the bisimidazole ligands *38c*, and **44** are low. The activity of the *42*-Cu^{2+} ion complex having a primary hydroxyl group is also low. Although the reason is not yet clear, it is likely that the Zn^{2+} and Cu^{2+} ion complexes require different structures, i.e. square planar or tetrahedral, for activity and a more flexible 2:1 complex of monoimidazole ligands is more suitable for the required Cu^{2+} ion complex than the 1:1 complex of bisimidazole ligands.

3.3.3.5 Substrate Specifity

Table 8 indicates substrate specifity as measured by rate enhancement. The most reactive ester is *1* which undergoes rate enhancement of more than 10^4 fold in the presence of 2×10^{-4} M of catalyst at pH 7.05.

Table 8. Substrate specificity as measured by rate enhancement ($k_{catalyzed}/k_{uncatalyzed}$)[a]

catalyst ester $RCO_2C_6H_4NO_2(p)$	*29*-Zn^{2+}		*38c*-Zn^{2+}	
	CTAB	Triton X-100	CTAB	Triton X-100
R = CH_3	29	210	—	11.2
C_2H_5	82	—	—	—
C_5H_{11}	—	1040	—	83.8
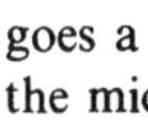 (*1*)	1565	18600	21500	14500

[a] conditions (a) in Table 4.

In contrast to *1*, isomeric p-nitrophenyl nicotinate shows almost no catalysis. Thus, it is clear that substrate coordination to the metal ion complex plays the critical role for an enormous rate enhancement. The lipophilic ester ($R = C_5H_{11}$) also undergoes a large rate enhancement indicating the importance of substrate binding into the micellar phase by hydrophobic interaction. A large rate enhancement can also be seen in lipophilic esters which lack the metal coordination site as given below with the enantioselective micellar reactions (Table 9, 10).

4 Enantioselective Hydrolysis

Enantioselective deacylation of esters in micellar reactions has been extensively studied in order to understand enzyme stereospecificity, and some micellar systems involving chiral histidine-functionalized surfactants have been found to exhibit moderate or high enantioselectivities [40–43]. However, there has been no report, except ours [37], on the micellar catalysis dealing with metal ion catalyzed acyltransfer reactions involving lipophilic ligands with chiral centers.

4.1 Effects of Ligands, Metal Ions, and Surfactants

Lipophilic ligands, e.g. *45–49* having chiral L-2-pyrrolidine-methanol moieties, have been prepared and examined for their catalytic activities in reactions with optically active esters (*50–52*).

Table 9 indicates that the rate enhancement (k^L/k^L_0) is relatively small when Zn^{2+} ions or a ligand is used separately for both *50* and *52* substrates. A large rate enhancement is obtained only when a ligand and the metal ion are used together as in the previous examples (Table 1, 3, 4, 7). Ligands L-*45* and L-*46* are relatively inactive as compared to other ligands having the imidazole moiety. The ligand activation by metal ion is the order of $Zn^{2+} > Co^{2+} > Ni^{2+}$ in all the cases, the same as in non-micellar reactions (Table 1). Rate-enhancing effects (k^L/k^L_0) of L-*47*-Zn^{2+}, L-*48*-Zn^{2+}, and L,L-*49*-Zn^{2+} ion complexes are remarkably large in view of the consideration

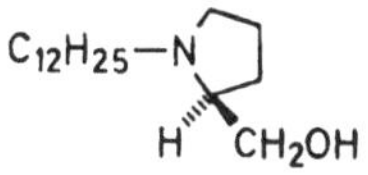

L-*45*

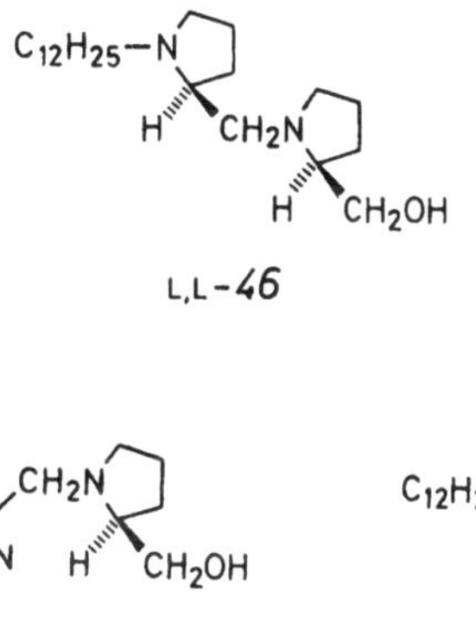

L,L-*46*

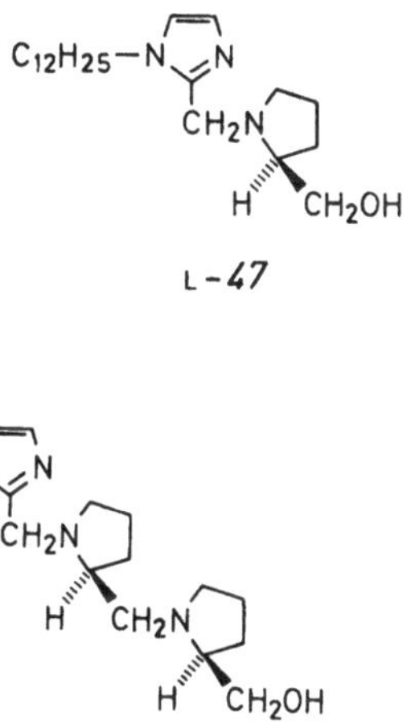

L-*47*

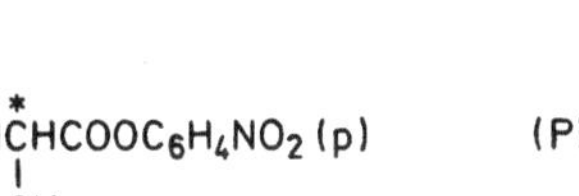

L-*48*

L,L-*49*

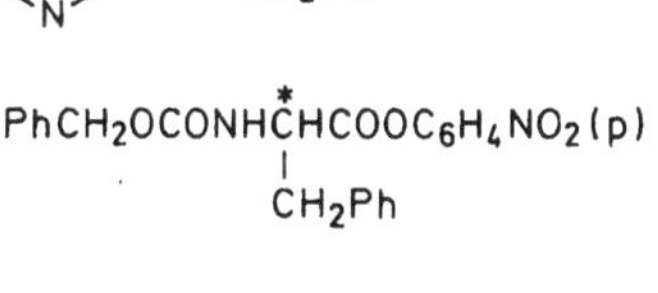

(Pic)–Phe–PNP
50

(Nic)–Phe–PNP
51

$PhCH_2OCONHC^{*}HCOOC_6H_4NO_2$(p)
CH_2Ph

(Z)–Phe–PNP
52

Table 9. Pseudo-first-order rate constants for the release of p-nitrophenol in the reactions of optically active esters in a CTAB micelle

ester	catalyst	$k^L \times 10^3$ (s^{-1})	$k^D \times 10^3$ (s^{-1})	k^L/k_0^L	k^L/k^D
(Pic)-Phe-PNP	none	2.16	—	1.0	—
50[a]	Zn^{2+}	3.42	—	2.1	—
	L-*47*	6.15	3.15	2.8	1.95
	L-*45*-Zn^{2+}	3.79	3.78	1.8	1.0
	L-*46*-Zn^{2+}	4.35	3.50	2.0	1.24
	L-*47*-Zn^{2+}	141	35.7	65	3.96
	L-*47*-Co^{2+}	45.9	26.7	21	1.72
	L-*47*-Ni^{2+}	5.89	5.45	2.7	1.08
	L-*48*-Zn^{2+}	54.5	21.8	25	2.54
	L,L-*49*-Zn^{2+}	14.6	7.26	6.8	1.92
(Z)-Phe-PNP	none	0.0703	—	1.0	—
52[b]	Zn^{2+}	0.0797	—	1.1	—
	L-*47*	0.408	0.359	5.8	1.14
	L-*48*	0.226	0.159	3.2	1.42
	L,L-*49*	0.503	0.209	7.2	2.41
	L-*45*-Zn^{2+}	0.112	0.106	1.6	1.06
	L,L-*46*-Zn^{2+}	0.750	0.230	10.7	3.26
	L-*47*-Zn^{2+}	19.7	10.7	280	1.84
	L-*47*-Co^{2+}	1.2	0.89	17.1	1.35
	L-*47*-Ni^{2+}	0.0997	0.0950	1.42	1.05
	L-*48*-Zn^{2+}	4.23	2.99	60.2	1.41
	L,L-*49*-Zn^{2+}	30.8	7.88	438	3.91
	L,L-*49*-Co^{2+}	1.29	0.355	18.3	3.63
	L,L-*49*-Ni^{2+}	0.146	0.0975	2.08	1.50

[a] [CTAB] = 1×10^{-2} M, [ligand] = 5×10^{-4} M, [M^{2+}] = 1×10^{-3} M, [50] = 1×10^{-4} M, pH 7.29 (2,6-lutidine buffer), 25 °C.

[b] [CTAB] = 1×10^{-2} M, [ligand] = 1×10^{-3} M, [Zn^{2+}] = 2×10^{-3} M, [52] = 1×10^{-4} M, pH 7.51, 25 °C.

that the esters *50* and *52* have no favorable metal ion binding site near the reacting C=O bond unlike in the case of the ester *1*. Among these three complexes, the most active one is the L-*47*-Zn^{2+} ion for the ester *50* ($k^L/k_0^L = 65$), and the L,L-*49*-Zn^{2+} ion for the ester *52* ($k^L/k_0^L = 438$). In parallel with these activities, the L-*47*-Zn^{2+} ion shows the highest enantioselectivity for the ester *50* ($k^L/k^D = 3.96$), and the L,L-*47*-Zn^{2+} ion for the ester *52* ($k^L/k^D = 3.91$).

The ester *51* undergoes almost no catalysis by any complex, although the uncatalyzed rate of *51*, $k_{obsd} = 11.4 \times 10^{-3}\ s^{-1}$ (pH 7.51, 25 °C) is fairly large but smaller than those of the catalyzed rates of *50* and *52* (Table 8). The reason of such non-catalysis is not yet clear.

Table 10 indicates the results obtained in non-ionic micelles and may be compared with Table 9 of cationic micelles.

The uncatalyzed rates in the Triton X-100 micelle are much smaller than those of the CTAB micelle as expected in ionic micellar reactions. However, in the catalyzed reactions much larger rate enhancements occur in the former micelle than in the latter micelle, similarly as in achiral systems (Table 4, 5). In Table 10, ester *50* shows

Table 10. Pseudo-first-order rate constant for the release of p-nitrophenol: in Triton X-100

ester	catalyst	$k^L \times 10^3$ (s^{-1})	$k^D \times 10^3$ (s^{-1})	k^L/k_0^L	k^L/k^D
(Pic)-Phe-PNP	none	0.368	—	1.0	—
50[a]	Zn^{2+}	2.74	—	7.45	—
	L-*47*	0.838	0.728	2.28	1.15
	L-*48*	0.560	0.526	1.52	1.06
	L,L-*49*	0.494	0.452	1.34	1.09
	L-*47*-Zn^{2+}	260	209	707	1.24
	L-*48*-Zn^{2+}	127	105	345	1.21
	L,L-*49*-Zn^{2+}	15.0	14.7	40.8	1.02
(Z)-Phe-PNP	none	0.00328	—	1.0	—
52[a]	Zn^{2+}	0.0117	—	3.56	—
	L-*47*	0.0479	0.0512	14.6	0.94
	L-*48*	0.0385	0.0265	11.7	1.05
	L-*47*-Zn^{2+}	5.00	2.26	1520	2.21
	L-*48*-Zn^{2+}	1.30	0.827	396	1.60
	L,L-*49*-Zn^{2+}	8.12	2.26	2480	3.58
(Z)-Phe-PNP	Cu^{2+}	0.0181	—	5.51	—
52[b]	L-*47*-Cu^{2+}	0.501	0.395	153	1.27
	L-*48*-Cu^{2+}	0.415	0.220	127	1.88
	L,L-*49*-Cu^{2+}	1.16	0.306	354	3.79

[a] [Triton X-100] = 1×10^{-2} M, [ligand] = 1×10^{-3} M, [Zn^{2+}] = 2×10^{-3} M, [ester] = 1×10^{-4} M, pH 7.50 (2,6-lutidine buffer), 25 °C.

[b] [Triton X-100] = 1×10^{-2} M, [ligand] = 1×10^{-4} M, [Cu^{2+}] = 1×10^{-4} M, [ester] = 5×10^{-5} M, pH 7.50, 25 °C.

only low enantioselectivities with all catalysts in contrast to some high values in Table 9. It may be interesting that the L,L-*49*-Zn^{2+} ion catalyst is the most active toward the ester *52* in both CTAB and Triton X-100 micelles.

4.2 K and k_c Values obtained under [ligand] ≫ [substrate]

As described previously, lipophilic monoimidazole ligands form 2:1 complexes with the Zn^{2+} ion (n = 2 in Scheme 2) as active catalysts except for some sterically hindered ligands (Table 3, 5, 7), and bisimidazole ligands form 1:1 complexes (n = 1 in Scheme 2, Table 5). In this chiral system, the latter 1:1 complex accords with kinetic analyses for both L-*47* and L,L-*49* ligands as shown in Fig. 12 and Table 11. These conclusions seem to be reasonable since monoimidazole derivatives have only one imidazole nitrogen, while the other bisimidazole and chiral ligands have more than two nitrogen atoms which can effectively coordinate to the Zn^{2+} ion.

The K values in Table 11 appear to be reasonably large. They correspond to the association between ligand and Zn^{2+} ion involving almost no interaction with the substrate in the pre-equilibrium stage so that they are the same for both L- and D-ester and for different esters (*50* and *52*). The enantioselectivities (k^L/k^D) in Table 9 and 10 involve some contributions from the free ligands in addition to those from the com-

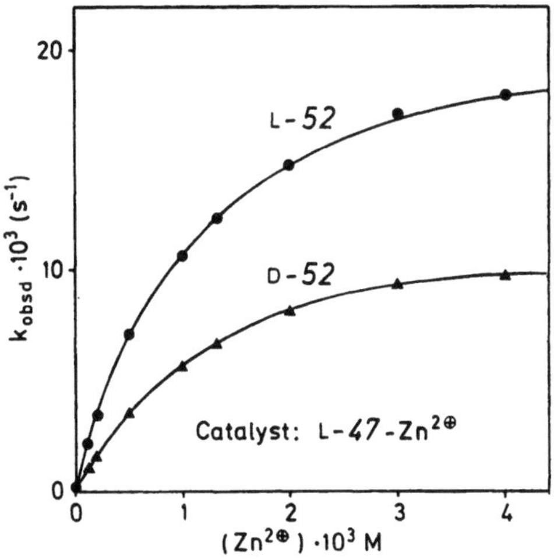

Fig. 12. Plots of pseudo-first-order rate constants for the release of p-nitrophenol from *L-52* and *D-52* as a function of zinc ion concentration. See Table 9 for other conditions

Table 11. Kinetic parameters for the hydrolysis of L- and D-ester in a CTAB micelle[a]

catalyst	L- or D-ester	$K \times 10^{-3}$ [b] (M^{-1})	(Pic)-Phe-PNP(*50*)		(Z)-Phe-PNP(*52*)	
			k_c (M^{-1} s^{-1})	k^L/k^D	k_c (M^{-1} s^{-1})	k^L/k^D
L-*47*-Zn^{2+}	L	4.92	503	5.86	23.1	1.85
	D	5.18	85.8		12.5	
L,L-*49*-Zn^{2+}	L	17.4			31.8	3.84
	D	15.4			8.29	

[a] [CTAB] = 1×10^{-4} M, [ligand] = 0—3×10^{-3} M, [Zn^{2+}] = 0—4×10^{-3} M, [substrate] = 1×10^{-4} M, pH 7.31, 25 °C.

[b] K values are primarily determined by the association between the ligand and the Zn^{2+} ion, and almost the same for both substrates.

plexes. Meanwhile the k^L/k^D values in Table 11 correspond to those of the pure complexes, and hence it may be reasonable that the value of 5.86 is significantly larger than 3.96 in Table 9.

4.3 Burst Kinetics under [substrate] ≫ [ligand]

Figure 13 indicates burst kinetics. As discussed before, such biphasic curves indicate the reaction to occur through two steps involving an acylated intermediate. The initial slopes for the presteady state can be taken as the measure of acylation rates, and the slopes of the later straight line for steady-state can be taken as the measure of deacylation rates.

Table 12. Enantioselectivities in the acylation and deacylation steps in the burst kinetics of the reaction of (Z)-Phe-PNP(*52*)

catalyst	surfactant	acylation k^L/k^D	deacylation k^L/k^D
L-*47*-Zn^{2+}	CTAB	1.8	2.2
	Triton X-100	2.1	0.56
L,L-*49*-Zn^{2+}	CTAB	4.0	1.8
	Triton X-100	3.05	0.39

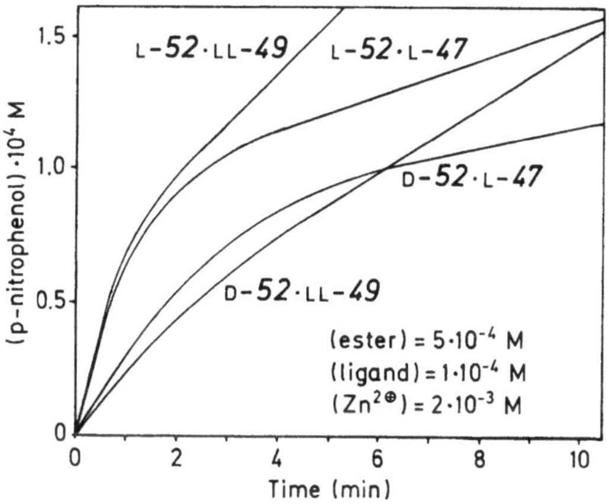

Fig. 13. Burst kinetics: [ester]/[ligand] = 5.0

The ratios of these slopes for L- and D-esters are shown in Table 12. The k^L/k^D values of the acylation step in the CTAB micelle are very close to those in Table 9, as they should be. It is interesting to note that the second deacylation step also occurs enantioselectively. Presumably it is due to the deacylation ocurring by the attack of a zinc ion-coordinated hydroxide ion which, in principle, should be enantioselective as in the hydroxyl group of the ligand. Alternatively, the enantioselectivity is also expected when the free hydroxide ion attack the coordinated carbonyl groups of the acyl-intermediate with the zinc ion. At any rate, the rates of both steps of acylation and deacylation for the L-esters are larger than those for the D-esters in the CTAB micelle. However, in the Triton X-100 micelle, the deacylation step for the D-esters become faster than for the L-esters.

The above enantioselectivities are obviously complex functions of many factors, perhaps even more complex than in natural enzymes. Complexity is partly due to the present co-micellar system in which it is difficult to analyze separately the interaction of the substrate with the achiral micelle, and that of the substrate with the catalyst complex.

5 Concluding Remarks

Several model systems related to metalloenzymes such as carboxypeptidase and carbonic anhydrase have been reviewed. Breslow contributed a great deal to this field. He showed how to design precise geometries of bis- or trisimidazole derivatives as in natural enzymes. He was able to synthesize a modified cyclodextrin having both a catalytic metal ion moiety and a substrate binding cavity (*26*). Murakami prepared a novel macrocyclic bisimidazole compound which has also a substrate binding cavity and imidazole ligands for metal ion complexation. Yet the catalytic activities of these model systems are by no means enzymic.

Our micellar models show unusually high catalytic activities as compared with other related model systems. Foregoing results and discussions may be summarized by referring to a generalized mechanism of catalysis shown in Scheme 5.

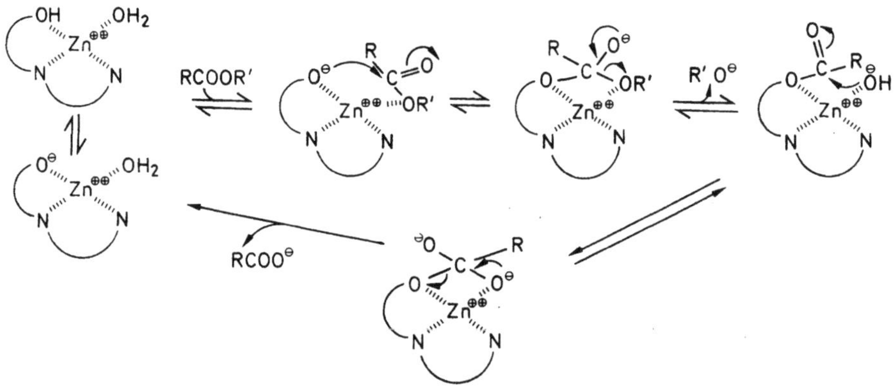

Scheme 5. Summarized mechanism of the catalysis of the micellar model of a metalloenzyme

First, complexation lowers the pK_a of ligand-hydroxyl group which occurs more strongly with stronger complexation. Complexation of lipophilic ligands seems to occur more strongly in non-ionic than in cationic surfactant. In micellar phase, it is likely that an ion-pair between oxyanion and zinc ion is a much more powerful nucleophile than a free hydroxide anion in a bulk water media. The acylation takes place through substrate binding into the micellar phase and subsequent to the formation follows decomposition of the tetrahedral addition intermediate. In all these steps, both substrate lipophilicity and metal ion binding ability appears to be important for the rate enhancement. It is likely that the zinc ion acts as an electrophile to polarize the carbonyl group or to assist the leaving of the p-nitrophenoxide anion, although these possibilities must be further examined by using substrates having weaker leaving groups. Final deacylation steps occur presumably through the attack of a zinc ion-bound hydroxide anion. This step is remarkably fast in the case of the bisimidazole complex *38c*-Zn^{2+} (Table 6). It should be mentioned that the reaction stops at the acylation of the hydroxyl group in the case of Sigman's complex (*3*-Zn^{2+} ion). There are many examples of micellar activation of the hydroxyl group for acylation, but subsequent deacylation is usually very slow. Thus, the zinc ion seems to play as both acid and base catalyst efficiently under neutral conditions.

The study on micellar models is still at the beginning. An amphiphilic ligand which can form micelles by itself has not yet been prepared. It is necessary to obtain complexes of higher stability in order to activate the hydroxyl group strong enough in the reactions of inactive esters or amides. Enantioselectivity must reach higher specifity. Nevertheless it seems to be clear that many features or some important clues have already been disclosed for further refinements of this micellar systems. More details about the present micellar reactions will be reported elsewhere in near future.

6 Acknowledgment

The authors thanks Dr. T. Eiki, Mrs. T. Minami, K. Machiya, I. Tomita, and O. Tanaka for their collaboration in carrying out this research. This research was supported in part by a Grant-in-Aid for Scientific Research from the Ministry of Eduction, Science and Culture, Japan.

7 References

1. Lipscomb, W. N.: Acc. Chem. Res. *3*, 81 (1970)
2. Hartsuch, J. A. and Lipscomb, W. N.: Enzymes *3*, 1 (1971)
3. Kaiser, E. T., Kaiser, B. L.: Acc. Chem. Res. *5*, 291 (1972)
4. Lipscomb, W. N.: Tetrahedron *30*, 1725 (1974)
5. Lipscomb, W. N.: Acc. Chem. Res. *15*, 232 (1982)
6. Lindskog, S. et al.: Enzymes (3rd Ed.) *5*, 587 (1971)
7. Lindskog, S.: Struct. Bonding (Berlin) *8*, 153 (1970)
8. Mildvan, A. S.: Enzymes (3rd Ed.) *2*, 446 (1971)
9. Pocker, Y., Sarkanen, S.: Adv. Enzymol. *47*, 149 (1978)
10. Nostrand, B., Vaara, I., Kannan, K. K.: in Isozyme I, Molecular Structure (Markert, C. Ed.), Academic Press, New York 1975, p. 575
11. Lindskog, S.: Adv. Inorg. Biochem. *4*, 115 (1982), Elsevier
12. Breslow, R., Chipman, D.: J. Am. Chem. Soc. *87*, 4195 (1965)
13. Sigman, D. S., Jorgensen, C. T.: J. Am. Chem. Soc. *94*, 1724 (1972)
14. Eiki, T., Kawada, S., Matsushima, K., Mori, M., Tagaki, W.: Chem. Lett. *1980*, 997
15. Ogino, K., Shindo, K., Minami, T., Tagaki, W.: Bull. Chem. Soc. Jpn. *56*, 1101 (1983)
16. Jencks, W. P.: Catalysis in Chemistry and Enzymology, McGraw-Hill, New York 1969
17. Bender, M. L., Marshall, T. H.: J. Am. Chem. Soc. *90*, 201 (1968)
18. Tang, C. C., Davalian, D., Huang, P., Breslow, R.: ibid. *100*, 3918 (1978)
19. Breslow, R., Hunt, J. T., Smiley, R., Tarnowski, T.: ibid. *105*, 5337 (1983)
20. Huguet, J., Brown, R. S.: ibid. *102*, 7571 (1982)
21. S.-Tilk, H., Cocho, J. L., Frakman, Z., Brown, R. S.: ibid. *106*, 2421 (1984), and the references therein
22. Brown, R. S., Salmon, D., Curtis, N. J., Kusuma, S.: ibid. *104*, 3188 (1982)
23. Tabushi, I., Kuroda, Y., Mochizuki, A.: ibid. *102*, 1152 (1980); Tabushi, I., Kuroda, Y.: ibid. *106*, 4580 (1984)
24. Breslow, R., Overman, L. E.: ibid. *92*, 1075 (1970)
25. Murakami, Y. et al.: J. Chem. Soc., Perkin Trans. 2, *1980*, 1665
26. Murakami, Y.: Topics in Current Chemistry *115*, 107 (1980)
27. Bunton, C. A., Romsted, L. S.: Micellar Effects upon Deacylation, in The Chemistry of Acid Derivatives (Patai, S. Ed.), part 2, Chapt. 17, John Wiley, New York 1979
28. Fendler, J. H.: Membrane Mimetic Chemistry, John Wiley & Sons, New York 1982
29. Menger, F. M.: Acc. Chem. Res. *12*, 111 (1979)
30. Menger, F. M. et al.: J. Am. Chem. Soc. *106*, 1109 (1984)

31. Fendler, J. H., Npme, F., Woert, H. C.: ibid. *96*, 6745 (1974)
32. Seno, S., Shiraishi, S., Araki, K., Kise, H.: Bull. Chem. Soc. Jpn. *48*, 3678 (1975)
33. Sunamoto, J., Kondo, H., Akimaru, K.: Chem. Lett. 821 (1978)
34. Kondo, H., Hamada, T., Yamamoto, S., Sunamoto, T.: ibid. 809 (1980)
35. Melhado, L. L., Gutsche, C. D.: J. Am. Chem. Soc. *100*, 1850 (1978)
36. Eiki, T., Mori, M., Kawada, S., Matsushima, K., Tagaki, W.: Chem. Lett. 1431 (1980)
37. Ogino, K., Tomita, I., Machiya, K., Tagaki, W.: ibid. 1875 (1982)
38. Bender, M. L., Marshall, T. M.: J. Am. Chem. Soc. *90*, 201 (1968)
39. Kunitake, T., Okahata, Y., Sakamoto, T.: ibid. *98*, 7799 (1976)
40. Ihara, Y. et al.: J. Chem. Soc. Perkin Trans. 2, 1741 (1983)
41. Ueoka, R., Murakami, Y.: ibid. 2, 219 (1983)
42. Ohkubo, K., Matsumoto, N., and Ohta, H.: J. Chem. Soc. Chem. Commun, 738 (1982)
43. Murakami, Y. et al.: J. Am. Chem. Soc. *103*, 728 (1981)

Chemical Basis of Ion Transport Specificity in Biological Membranes

Dan W. Urry

Laboratory of Molecular Biophysics, University of Alabama in Birmingham, School of Medicine, Birmingham, Alabama 35294, USA

Table of Contents

This review addresses the issues of the chemical and physical processes whereby inorganic anions and cations are selectively retained by or passed through cell membranes. The channel and carrier mechanisms of membranes permeation are treated by means of model systems. The models are: the planar lipid bilayer for the cell membrane, Gramicidin for the channel mechanism, and Valinomycin for the carrier mechanism.

With respect to the channel mechanism, the phenomenology of channel transport is noted; the molecular structure of the Gramicidin channel is briefly reviewed; the cation binding sites are located within the channel; using Eyring rate theory a free energy profile for ion transit through the channel is developed based on the location of the binding site and the determination of binding and rate constants by physical methods which are independent of the transport mechanism, and it is demonstrated that both binding site location and rate constants must be independently determined in order to achieve the unique description of ionic mechanism. It is shown that inorganic anion vs cation selectivity is the result of the chemical structure of polypeptides combined with conformational energetics of the channel; it is shown that monovalent vs multivalent cation selectivity is the result of the proximity of membrane lipid to the channel proper and properties are proposed for a divalent cation channel; and it is argued that selectivity among monovalent cations is enhanced by the conformation energetics of the channel. Furthermore, a formalism is given which leads to a means of evaluating thermodynamics relative to selectivity among monovalent cations.

With respect to the carrier mechanism, the phenomenology of the carrier transport of ions is discussed in terms of the criteria and kinetic scheme for the carrier mechanism; the molecular structure of the Valinomycin-potassium ion complex is considered in terms of the polar core wherein the ion resides and comparison is made to the Enniatin B complexation of ions; it is seen again that anion vs cation selectivity is the result of chemical structure and conformation; lipid proximity and polar component of the polar core are discussed relative to monovalent vs multivalent cation selectivity and the dramatic monovalent cation selectivity of Valinomycin is demonstrated to be the result of the conformational energetics of forming polar cores of sizes suitable for different sized monovalent cations.

It should be apparent that the principles of selective ion transport are independent of the specific models being treated here and that many of these principles are at variance with what were traditional views on the basis of selective membrane permeation by inorganic ions. Thus, the concept of selectivity among monovalent cations being based on values of hydrated radii is replaced by the demonstration that greater selectivity comes with increased dehydration. The perspective that hydration is the best way to lower ion self energy in order to pass through a protein component in a cell membrane is replaced by demonstration that peptide and ester carbonyls are far better solvators than water and that what is critical is the conformational energetics required to achieve adequate coordination. Furthermore, the earlier prevalent view that the repulsive image force due to the presence of the lipid layer would cause the rate limiting barrier to be in the middle of the membrane is shown to be entirely incorrect for monovalent cations but relevant indeed to multivalent cations. It should also be appreciated that there are other physiocochemical data available from these model systems such as the repulsion between ions at a known distance and separated by a string of water molecules and such as the energetics of lipid membrane deformation. Such information while relevant to the mechanisms of selective permeation of cell membranes has a more general and widespread application.

I Introduction

Traditionally concepts of ion selective permeation of biological membranes have centered on differences in the effective radii of hydrated nuclei. An example of that perspective derives from consideration of the resting membrane potential, E, which in the squid axon is approximated by the Nernst equation

$$E = -\frac{RT}{F} \ln \frac{[K^+]_{in}}{[K^+]_{out}} \tag{1}$$

where F is the Faraday, T is the absolute temperature (°K), R = kN is the Boltzmann's constant, k, times Avogadro's number, N, and $[K^+]_{in}$ and $[K^+]_{out}$ are the molar activities of potassium ion inside and outside, respectively, of the cell. The argument has been that the smaller hydrated radius of K^+, e.g. 2.76 Å for Na^+ and 2.32 Å for K^+ [1], resulted in the membrane being more permeable to this ion. The viewpoint of hydrated nuclei being the permeant species is also consistent with the prevalent notion that hydration provides the most effective means of reducing ion self energy during transit across a barrier of low dielectric constant. Generally, the walls of a transmembrane conduit have been treated as a static continuum, though the conduit itself may be of irregular dimension with localized constrictions, and the ion has been considered with an essentially complete first hydration shell. This has been the case whether the approach was based on a free electro-diffusion model or on the discrete jumps of a rate theory model. Present detailed knowledge of lipid membrane permeation, derived from substantially characterized channel and carrier systems, demonstrates quite the inverse to be the case. The permeant ion is largely, or can even be totally, dehydrated and ion selectivity among monovalent cations derives in large part from the conformational energetics required to achieve carbonyl coordination of the cation. This newer perspective of selective monovalent cation transport across cell membranes derives in greatest detail from studies on the Gramicidin A transmembrane channel and the Valinomycin carrier model systems. These will be discussed in this chapter but first it is useful to review the energetics problem of inorganic ionic species penetrating a lipid bilayer membrane.

The Energetics Problem of Cation Transport Across Lipid Bilayer Membranes: A qualitative perspective of the barrier presented by a lipid bilayer membrane can be obtained from the Born expression [2] for solvation energy, SE,

$$SE = -\frac{N^\circ(ze)^2}{2r}\left(1 - \frac{1}{\varepsilon}\right) \tag{2}$$

where N° is Avogadro's number, z is the unit charge on the ion, e ($= 4.8 \times 10^{-10}$) (esu) is the charge on an electron, r is the ionic radius and ε is the dielectric constant. By this expression the solvation energy for a 1 Å radius monovalent ion, on going from a vacuum to an infinite hydration shell with ε = 81, would be —164 kcal/mole. On going from a vacuum to an infinite lipid with ε = 2.13 [3] would give —88 kcal/mole for the solvation energy. The correction for a finite lipid layer is small as may be demonstrated by considering a two shell model with a lipid shell from 1 Å to 15 Å (i.e. about one-half the thickness of a membrane lipid layer) and an aqueous shell from 15 Å to infinity. The barrier encountered for a monovalent ion to pass from an aqueous medium without hydration into the center of a lipid membrane is enormous, greater than 70 kcal/mole. This barrier is an order of magnitude greater than the apparent barrier for Gramicidin A facilitated Na^+ transport across a lipid bilayer [4] and five times greater than that found for Valinomycin facilitated K^+ transport [5]. Clearly, a polar shell immediately surrounding the ion must be present during transit across the lipid layer. A trishell model can next be considered, i.e. polar-lipid-polar. If the polar central shell were simply water, then there would be an unfavorable energy due to the water-lipid contact surface. This interfacial energy problem is solved by the structure of the carrier or channel as the ion transport structures occur with a polar core and a lipid coat just as the structure of the membrane provides for the lipid to water transition by the polar head groups of the lipid. Thus, it is not unreasonable to consider a trishell model with N-methyl acetamide (NMA), a model for the peptide moiety with $\varepsilon = 179$ as the material of the central polar core. The solvation energy of such a trishell model — with a polar central shell 1 Å to 6 Å with $\varepsilon = 179$, a lipid shell 6 Å to 15 Å with $\varepsilon = 2$ and a polar outer shell 15 Å to infinity with $\varepsilon = 81$ — provides an estimate of barrier height [6] that approaches those determined for carrier and channel transport mechanisms [4,5]. These effects of a lipid bilayer membrane on ion permeation and of the carrier and channel mechanisms which allow cation passage are schematically shown in Fig. 1.

In the Born perspective the hydration shell extends for tens of Angstroms with 10% of the hydration energy coming from the shell extending from 10 Å on. In this view, as the fully hydrated ion approaches the membrane, it begins to loose its hydration shell and the associated solvation energy decreases. This is often referred to as a repulsive image force and the barrier which the lipid membrane presents to an ion is called the electrostatic energy barrier. This barrier is proportional to the square of the charge on the ion as indicated in Fig. 1B. It is now clear that the large hydration shell is not required to achieve the solvation energy and that the calculated solvation energies are too high by at least 50% but the charge squared effect does appear to follow well. These effects are discussed below in terms of the monovalent vs multivalent ion selectivity of channels and in terms of the relative barrier sensed by neutral carrier and neutral carrier-cation complex as they cross the lipid membrane from one side to the other.

The prespective to be gained thus far is that in order to pass through a lipid layer an ion must have an appropriate polar shell provided in large part by the carrier or channel structure which by virtue of its conformation and by also having lipophilic side chains provides for the polar shell to lipid shell transition. While the relative permeability of monovalent vs divalent and trivalent ions can be qualitatively appreciated from the z^2 term in Eqn 2, as indicated in Figure 1B, it is essential to know structural and mechanistic detail in order even qualitatively to understand anion vs cation selectivity and to understand selectivity among monovalent cations.

With the adequacy of lipid bilayer membranes as models for the basic structural motif and hence for the ion transport barrier of biological membranes, studies of channel and carrier ion transport mechanisms across such membranes become of central relevance to transport across cell membranes. The fundamental principles derived from these studies, however, have generality beyond the specific model systems. As noted above and as will be treated below, it is found that selective transport

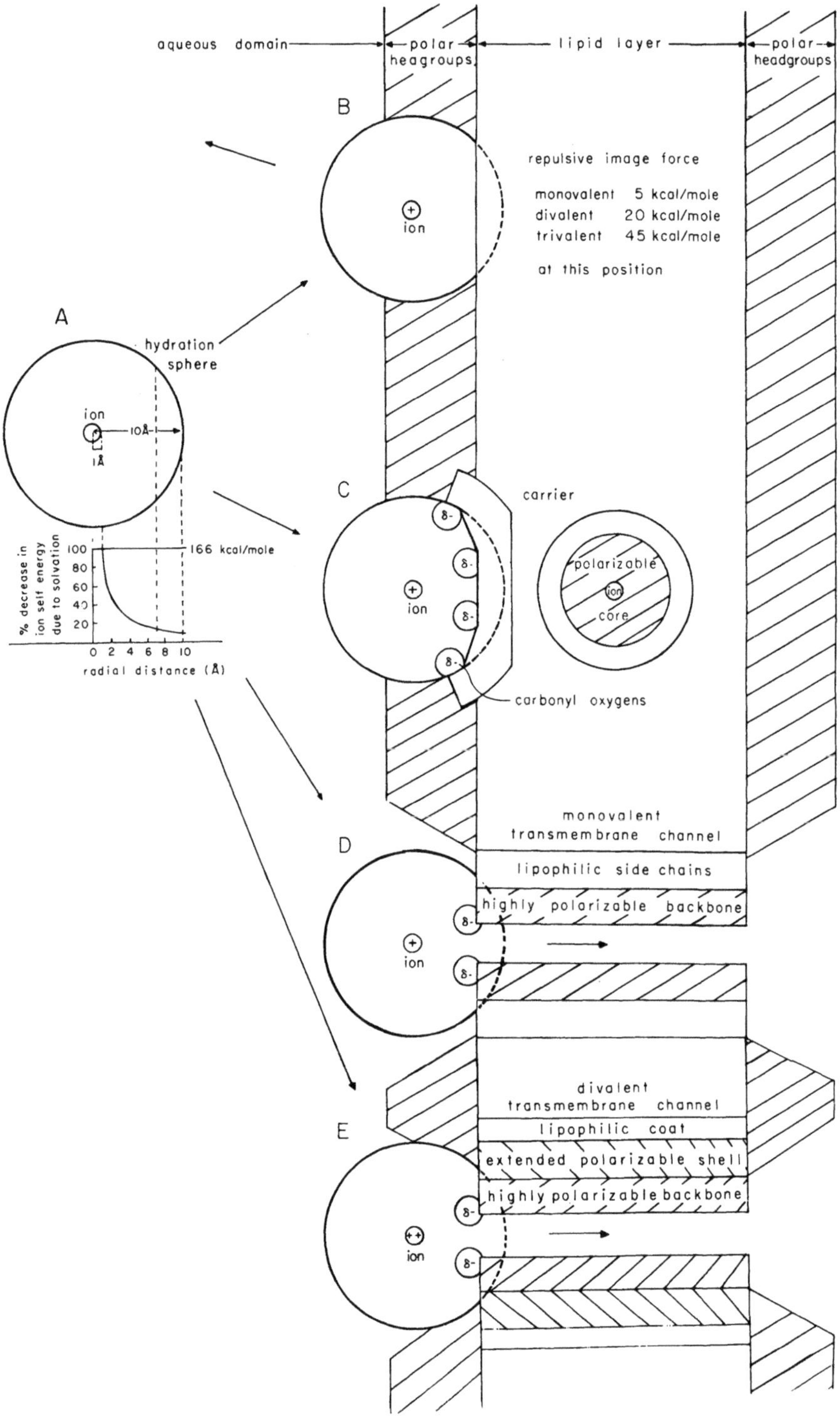
aqueous domain
polar heagroups
lipid layer
polar headgroups
B
ion
repulsive image force
monovalent 5 kcal/mole
divalent 20 kcal/mole
trivalent 45 kcal/mole
at this position
A
hydration sphere
ion
10Å
1Å
100
80
60
40
20
166 kcal/mole
% decrease in ion self energy due to solvation
0 2 4 6 8 10
radial distance (Å)
C
carrier
polarizable
ion
core
carbonyl oxygens
monovalent transmembrane channel
D
lipophilic side chains
highly polarizable backbone
divalent transmembrane channel
E
lipophilic coat
extended polarizable shell
highly polarizable backbone

occurs with largely dehydrated cations and that the channel or carrier impart an ion selectivity due to the energetics of their conformations combined with the energetics of hydration-dehydration. Basic information is obtained on the role of proximal lipid in affecting barrier heights, on the effectiveness of peptide carbonyls as sites for competing with hydration, on the repulsion between ions at known distances with a column of intervening water, on the reason selective transmembrane ionic currents are almost exclusively cationic currents, on the magnitude of ion selective cation currents, and on the usefulness of certain methodologies, such as noise analysis to characterize transport processes. Of most direct interest here is the information that these model systems provide on the basis of ion selectivity. This is to be considered in three stages: 1) the basis of anion vs cation selectivity, 2) the basis of monovalent vs multivalent ion selectivity and 3) the base of selectivity among monovalent ions. For such considerations, prerequisite is discussion of phenomenology, structure and ionic mechanisms.

II The Channel Mechanism of Ion Transport: The Gramicidin Model

The chemical basis for ion transport specificity of the Gramicidin channel is approached by noting the phenomenology of channel transport, by considering molecular structure, by locating the binding sites within the channel, by determining binding constants and rate constants using means independent of the transport phenomeno-

◀

Fig. 1. Born ion self energy and the problem of inorganic ion passage through a lipid barrier.
A. Hydration energy profile, using the Born formalism (Eqn. 1), shows the drop of ion self energy as a function of the radius of a hydration sphere. Note that even with a hydration shell of 10 Å radius not all of the hydration energy is obtained.
B. On approaching a lipid layer the hydration shell becomes increasingly replaced by the lipid; ion self energy increases, and approach is energetically unfavorable. This is referred to as a repulsive (or positive) image force and constitutes an energy barrier to approaching and penetrating a lipid layer. This barrier has a maximum at the center of the membrane. The magnitude of the repulsive force depends on the square of the charge on the ion, i.e. on z^2. While the magntidue of the solvation energy is overestimated by about 50% and the r^{-1} distance dependence is grossly incorrect, the z^2 dependance appears to obtain well. See text for discussion.
C. Hydrated monovalent cation approaching an area of the membrane where an amphiphilic carrier is located with its lipid side in contact with the lipid layer and with polar oxygens directed outward into solution. On close approach to the carrier, water molecules in the first coordination shell become replaced by carrier oxygens. As the ion becomes enclosed, the carrier moves into the lipid layer.
D. Hydrated monovalent cation approaching the carbonyl oxygens of a transmembrane channel. The carbonyl oxygens at the mouth replace water in the first coordination shell. As the ion moves through the channel, it retains one bound water molecule preceding and following it and the walls of the channel provide for lateral coordination. (Parts A through D reproduced with permission from Ref. [6]).
E. Hydrated divalent cation approaching a channel with a slightly larger diameter than in D, but the energy of interaction with the divalent cation is sufficient to deform the channel drawing the walls in to make lateral coordination with the divalent cation. Since the channel is too small for a monovalent cation to pass through with its first hydration shell and since the monovalent cation channel interaction is insufficient to make the channel small enough for lateral coordination of the monovalent cation, the channel is selective for divalent cations. (Part E reproduced with permission from Ref. [68])

logy, and by deriving free energy profiles for ion passage through the channel. With such data and Eyring rate theory, it becomes possible to calculate single channel currents over wide ranges of ion activity and transmembrane potential. Having demonstrated this, effects of dynamic channel structure can be considered. With such background information, issues of anion vs cation selectivity and of monovalent vs divalent ion selectivity can reasonably be discussed, and it is particularly on the issue of selectivity among monovalent cations that dynamic elements of the channel become of special interest in terms of what has been called the peptide libration mechanism.

A Phenomenology of Channel Transport

As first shown by Hladky and Haydon [7,8], it is possible to observe the current due to a single transmembrane channel by using extensions of the planar lipid bilayer approach of Mueller and Rudin [9]. The basic system is shown in Fig. 2 and is commonly referred to as the black lipid membrane (BLM) method. This is because, as the lipid in the hole between the two chambers thins, the areas that have become planar bilayers are seen as black. Additional terms are bilayer lipid membranes or planar lipid bilayer membranes. These lipid bilayer membranes, particularly those which are solvent free, have capacitances which are very close to those of biological membranes.

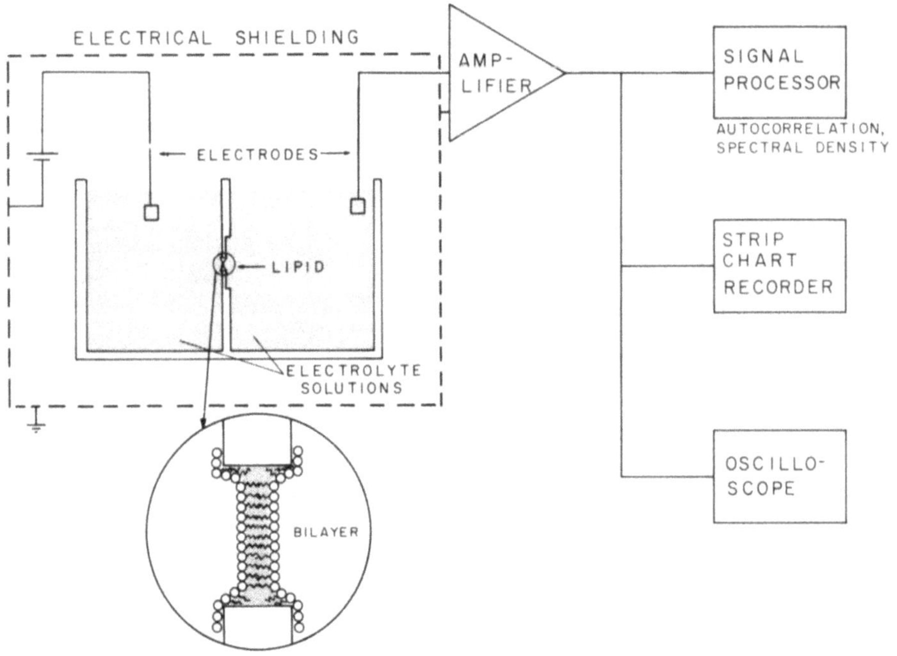

Fig. 2. Experimental set up for characterizing channel or carrier induced currents crossing the small area of a planar lipid bilayer which is schematically shown in enlarged view. Net current derives from either an applied potential or a concentration gradient

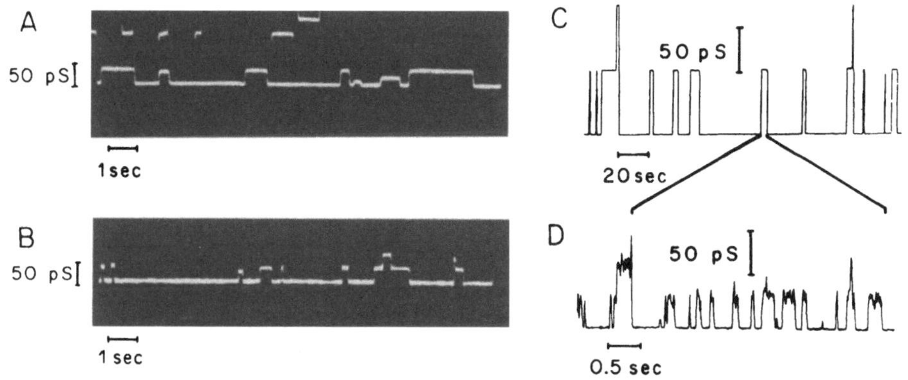

Fig. 3. Step conductance changes across planar bilayer due to Gramicidin A (traces A and C), D · Leu^2 Gramicidin A (trace B) and N-acetyl desformyl Gramicidin A (trace D).
A. and B. Diphytanoyl L-α-lecithin n-decane membranes, 100 mV applied potential, 2 M KCl. Note that, while the height of the conductance steps (single channel conductances) are similar for Gramicidin (above) and D · Leu^2 Gramicidin A (below), the channel lifetimes of the D · Leu^2 Gramicidin A analogue are shorter due to destablization of the head to head junction by the bulky Leu side chain. Reproduced with permission from Ref. [10].
C. and D. Monoolein/hexadecane membranes, 100 mV applied potential 1 M RbCl and 23 °C. Note that, during the time period of an average Gramicidin A channel (above), the N-acetyl desformyl Gramicidin A channel turns on and off many times and that the conductance step is smaller. The N-acetyl methyl replacing the formyl proton also crowds and destabilizes the head to head junction and results in less favorable lateral coordination of the cation at the junction. Reproduced with permission from Ref. [11]

In Fig. 3 [10,11] are traces of the conductance changes across the lipid bilayer due to Gramicidin A (traces A and C), D · Leu^2-Gramicidin A (trace B) and N-Acetyl desformyl Gramicidin A (trace D). A conductance step of 16 picoSiemens (pS) or 16×10^{-12} mhos at 100 mV represents a current of 10^7 ions/sec passing through a single transmembrane channel. It is obvious that these three channel forming molecules exhibit different characteristics. While the height of the conductance step is similar for Gramicidin A and D · Leu^2 Gramicidin A, the duration of the D · Leu^2 Gramicidin A conductance is shorter. In the case of N-acetyl Gramicidin A both the conductance step and the duration are less than for Gramicidin A. The conductance steps can therefore be characterized in terms of their height (single channel conductance) and their duration (channel lifetime). For a given channel structure, membrane composition, transmembrane potential, ionic species and concentration, temperature and to some extent channel concentration, histograms of the distribution of single channel currents (or conductances) and of the distribution of channel lifetimes become fingerprints for a given channel structure. In Fig. 4A is the histogram of single channel conductances of Gramicidin A in diphytanoyl lecithin/n-decane membranes for 1 M, KCl, 130 mV, and 40 °C which resulted from the manual tabulation of over 2,000 events [12]. In Fig. 4B is the histogram of channel lifetimes for the same membrane and salt but at 15 °C and resulting from the manual tabulation of 125 events [12]. Recognizing that we normally get mean values for mM samplings, i.e. $>10^{20}$ species, it is obvious why the histograms are yet somewhat irregular.

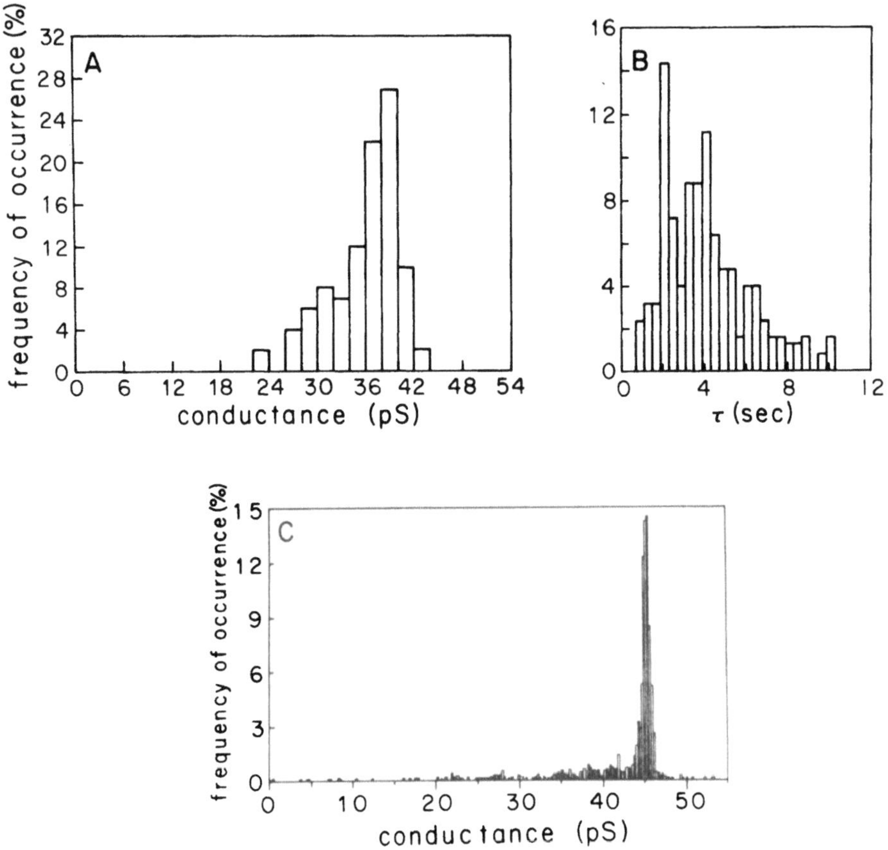

Fig. 4. A. Histogram of single channel conductances, γ, of Gramicidin A in diphytanoyl L-α-lecithin/n-decane membranes for 1 M KCl, 130 mV potential and 40 °C. Note that single channel conductance is simply the single channel current, i_x, divided by the applied potential. Reproduced from reference [12] with permission.
B. Histogram of channel lifetimes for Gramicidin A in diphytanoyl L-α-lecithin/n-decane membranes for 1 M KCl, 130 mV applied potential and 15 °C.
C. Histogram of single channel conductances of Gramicidin A in glyceryl monoolein/hexadecane membranes for 1 M KCl, 103 mV applied potential and 23 °C. Reproduced with permission from Ref. [13]

Computer assisted analysis of single channel data is of great help, as shown in Fig. 4C for Gramicidin A in glyceryl monooleate membranes ar 23 °C, 1 M KCl and 103 mV [13]. Characterization of channel activities, however, can be carried out at higher concentrations using fluctuation analysis methods of autocorrelation analysis [14,15] and of spectral analysis [10,11,16]. Indeed because single channel current data can be readily obtained for these model channels, comparison of this data obtained at low channel concentrations with the results of noise (fluctuation) analysis methods obtained at high channel concentrations provided verification of the latter methods.

B Molecular Structure of the Gramicidin Channel

The primary structure of Gramicidin A (GA) was determined by Sarges and Witkop [17] to be HCO-L·Val^1-Gly^2-L·Ala^3-D·Leu^4-L·Ala^5-D·Val^6-L·Val^7-D·Val^8-L·Trp^9-D·Leu^{10}-L·Trp^{11}-D·Leu^{12}-L·Trp^{13}-D·Leu^{14}-L·Trp^{15}-$HNCH_2CH_2OH$. Two common, naturally occurring analogs are Gramicidin B which is Phe^{11} GA and Gramicidin C which is Tyr^{11} GA. Additionally residue one can also be Ile. These primary structures were also synthetically confirmed by Gross and colleagues [13,19] using the solid phase method of peptide synthesis. These linear pentadecapeptides should not be confused with Gramicidin S (actually of the tyrocidin family of antibiotics) which is a cyclodecapeptide [20].

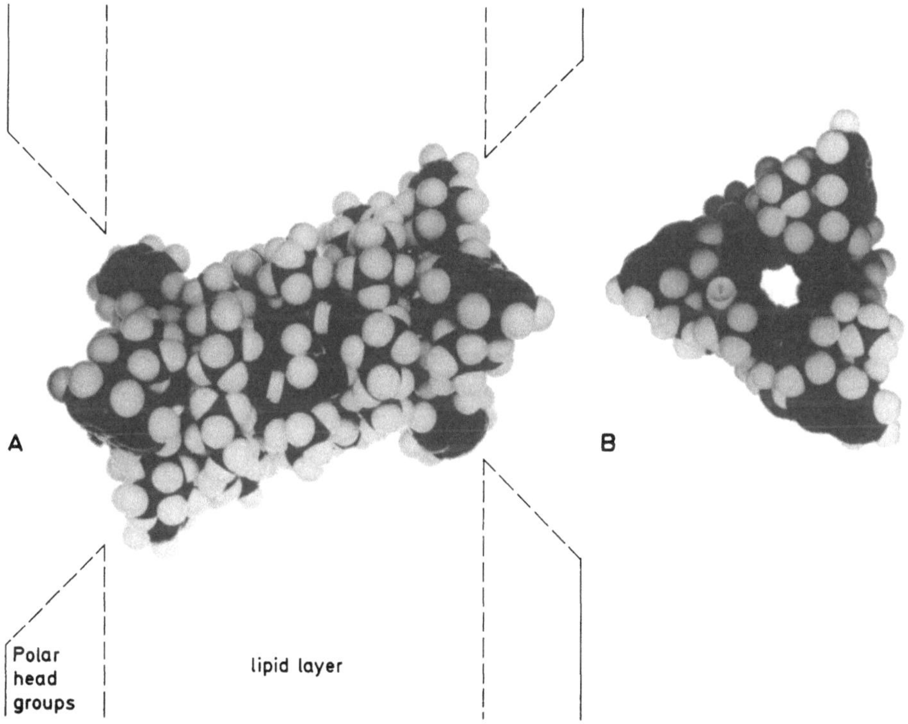

Fig. 5. Space filling model of Gramicidin A transmembrane channel.
A. Side view of channel spanning the lipid layer of a planar lipid bilayer. The structure is comprised of two monomers, each in a left-handed, single stranded $\beta^6_{3,3}$-helical conformation, and joined together at the head or formyl end by means of six intermolecular hydrogen bonds. The two formyl protons are seen at the center of the structure in this view. Replacement of these protons by methyls destabilizes the conducting dimer as shown with N-acetyl desformyl Gramicidin A (Fig. 3D).
B. Channel view showing a channel with a maximal diameter of 4 Å. Also seen in this view are three carbonyl oxygens surrounding the channel and directed outward into solution. These are the peptide carbonyls of the Trp^{15}, Trp^{13} and Trp^{11} residues

The molecular structure of the Gramicidin channel was proposed in 1971 [21,22] to be as shown in Fig. 5. The channel is a dimer with each monomer in a π^6_{LD}-helix (now called a $\beta^{6.3}_{3,3}$-helix) [23] in which there are 6.3 residues per turn and in which the peptide C—O moieties alternately point toward the carboxyl end (L-residues) and toward the amino end (D-residues). The two molecules associate head to head (formyl end to formyl end) by means of six intermolecular hydrogen bonds and the sense of the helix is left-handed. The backbone conformation was independently proposed by Ramachandran and Chandresekharan [24,25] based on a poly L-Ala-D-Ala model which, however, did not make it possible to address the question of helix sense nor of the nature of a dimeric structure. Subsequently, double stranded helices, now referred to as double stranded β-helices [26], were described for the Gramicidins in organic solvents [27] but the original proponents of the double stranded structure [28] and before that Bamberg and coworkers [29] have concluded that the dominant active channel is that given in Fig. 5. It should for purposes of completeness be mentioned that there is yet one group which considers the double stranded β-helices to be the dominant channel structure [30,31]. While it is not the purpose of this review to develop the arguments for molecular structure, it may be noted that the following data on the location of binding sites in the channel are further evidence for the structure in Fig. 5. A separate review focusses on the issue of molecular structure [32].

C Location of Cation Binding Sites in the Channel

The next step in developing a chemical basis for ion selectivity involves an element of the ionic mechanism, the location of the ion binding site. It has been shown that mM concentrations of the channel state can be obtained by heating Gramicidin A in the presence of lysolecithin [33]. This is a sufficient concentration to monitor an isotopically enriched site by carbon-13 nuclear magnetic resonance. By synthesizing the pentadecapeptide with only one of the carbonyl carbons enriched [34], it is possible to determine whether that carbonyl carbon resonance exhibits an ion induced chemical shift in the presence of cations. Carbonyl carbon cation induced chemical shifts have already been observed in organic solvents for Valinomycin [35] and for enniatin B [36], the latter of which is a cyclohexapeptide analog of the Gramicidins.

Using six different enriched sites in six different syntheses — formyl ^{13}C GA, (1-^{13}C)Val^1 GA, (1-^{13}C)Trp^9 GA, (1-^{13}C)Trp^{11} GA, (1-^{13}C)Trp^{13} GA and (1-^{13}C)

▶

Fig. 6A and B. Wire models of the single stranded, head to head dimerized $\beta^6_{3,3}$-helix of the Gramicidin A transmembrane channel and plotted values for the Na^+ and Tl^+ ion induced chemical shifts as a function of location along the channel. Depending on whether the helix sense is left-handed, a., or right-handed, b., the ion binding sites as defined by L-residue carbonyl carbon chemical shifts are separated by different distances. Importantly, however, the D·Leu^{14} carbonyl is at the binding site for the left-handed structure whereas the D·Val^8 carbonyl is at the binding site for the right-handed structure. An observed ion induced chemical shift for D·Leu^{14} C—O and the absence of a shift for the D·Val^8 C—O allows the conclusion of a left-handed helix sense and of the separation between sites of about 23 Å. Reproduced with permission from Ref. [38]

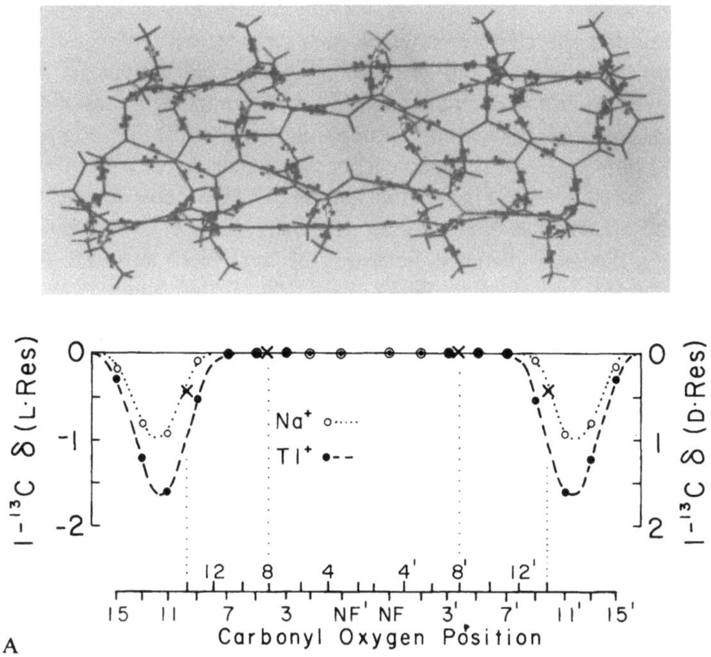

1-^{13}C δ (L·Res)
1-^{13}C δ (D·Res)
Na^{+}
Tl^{+}
12 8 4 4' 8' 12'
15 11 7 3 NF' NF 3' 7' 11' 15'
Carbonyl Oxygen Position

A

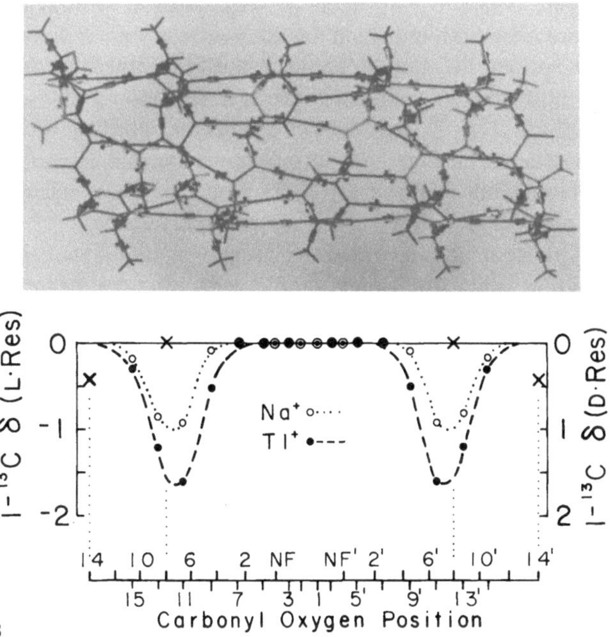

1-^{13}C δ (L·Res)
1-^{13}C δ (D·Res)
Na^{+}
Tl^{+}
14 10 6 2 NF NF' 2' 6' 10' 14'
15 11 7 3 1 5' 9' 13'
Carbonyl Oxygen Position

B

Trp[15] GA — and examining the channel state in each case by carbon-13 nuclear magnetic resonance in the presence and absence of permeable cations, it was found that there are two binding sites related by two-fold symmetry [37)], as shown in Fig. 6 [38)]. The magnitudes of the cation induced chemical shifts for complete site occupancy, about 2 ppm, indicate that the carbonyl oxygens directly coordinate the bare cation [36,37)]. As shown in Fig. 6, the actual distance between the two identified sites depends on whether the helix sense of the channel is left-handed or right-handed. The separation distance between the two sites and the helix sense of the channel depends on whether the D · Leu[14] C—O (left-handed) or the D · Val[8] C—O (right-handed) is located within the binding site. Using synthetic (1-^{13}C)D · Leu[14] GA and (1-^{13}C)D · Val[8] GA, it was possible to show that only the D · Leu[14] carbonyl carbon exhibited a cation induced chemical shift [38)]. It was concluded therefore that the helix sense of the channel is left-handed as initially proposed [21,22)] and that the binding sites are separated by just over 20 Å.

Additional isotopically enriched Gramicidin As have been synthesized in order to map more closely the amino end of the molecule for binding sites; these are (1-^{13}C)L · Ala[3], (1-^{13}C)L · Ala[5] and (1-^{13}C)L · Val[7]. None of these carbonyl carbons exhibited cation induced chemical shifts in the presence of 83 mM TlAc nor in the presence of up to 1 M NaCl [39)]. In addition to being required for detailing the ionic mechanism of transport, identification of the presence and absence of binding sites bears on the question of whether the channel is single stranded or double stranded. The absence of binding along the amino end for the first nine carbonyls of Gramicidin A, the demonstration that the dominant channel state exhibits symmetric current/voltage curves [40)] requiring that the channel have a symmetric free energy profile as occurs with a two fold symmetry axis perpendicular to the channel axis [32)], and the size of the binding site wherein the ion at a fixed location would give the observed chemical shifts are observations which are inconsistent with a double-stranded structure, parallel or antiparallel, but which are beautifully consistent with the structure in Fig. 5. Furthermore, the separation distance between binding sites seen in Fig. 6A, the repeating length of 26 Å [26)], and the helix diameter, which is within a tenth of an Angstrom of the structure in Fig. 5 [23)], have been observed in crystalline K^+ and Cs^+ complexes of Gramicidin A by X-ray diffraction [41)]. The conclusion appears inescapable, whether developed on the basis of ion transport studies [22,29)] or on the basis of independent physical characterizations of phospholipid packaged channels [28,37,38)], or by crystallographic studies on complexes [41)], that the structure in Fig. 6 is the correct structure of the Gramicidin A transmembrane channel.

D Free Energy Profile for Cation Passage through the Gramicidin Channel

1 Use of Eyring Rate Theory

The thermodynamic form of the Eyring rate equation,

$$k' = \frac{kT}{h} e^{-\Delta G^{\neq}/RT} \tag{3}$$

where k, T, h, $\Delta G^{\neq}$ and R are Boltzmann's constant, the absolute temperature (°K), Planck's constant, the Gibbs free energy of activation and the gas constant, respectively, makes it possible to present in a single plot, called the free energy profile, essentially all that is required to describe transport at a given temperature. This simplification, which is a triumph of the Eyring rate theory, makes achieving the free energy profile for passage through the Gramicidin channel an important stage in the development of the understanding of a transport process. The usefulness of the approach is made greater because the required rate constants for each elemental process can be determined by entirely independent physical methods and then applied to the calculation of the transport process as determined by electrical measurements. The objective, therefore, is to calculate the single channel currents such as those reported in Fig. 3 and 4, while using rate constants determined by other unrelated physical methods. In this section, the relevant mathematical formalisms and experimental data will be reviewed. It is a successful, independent calculation of the single channel currents that provides the level of confidence required for a meaningful discussion of the chemical basis of ion selectivity.

In the previous section was given the experimental demonstration of two sites. Here the steady state scheme and equations necessary to calculate the single channel currents are given. The elemental rate constants are thereby defined and related to experimentally determinable rate constants. Eyring rate theory is then used to introduce the voltage dependence to these rate constants. Having identified the experimentally required quantities, these are then derived from nuclear magnetic resonance and dielectric relaxation studies on channel incorporated into lipid bilayers.

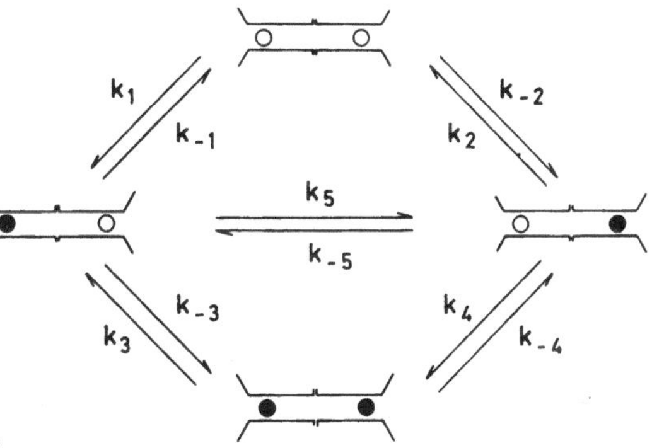

A

$$\dot{\chi}(OO) = -(C'_x k_1 + C''_x k_{-2})\chi(OO) + k_{-1}\chi(XO) + k_2\chi(OX) = 0$$

$$\dot{\chi}(XO) = C'_x k_1 \chi(OO) - (k_{-1} + k_5 + C''_x k_{-3})\chi(XO) + k_{-5}\chi(OX) + k_3\chi(XX) = 0$$

$$\dot{\chi}(OX) = C''_x k_{-2}\chi(OO) + k_5\chi(XO) - (k_2 + k_{-5} + C'_x k_4)\chi(OX) + k_{-4}\chi(XX) = 0$$

$$\dot{\chi}(XX) = C''_x k_{-3}\chi(XO) + C'_x k_4\chi(OX) - (k_3 + k_{-4})\chi(XX) = 0$$

B

Fig. 7. A. Kinetic scheme for two site single filing channel. Ten rate constants are required. In the absence of a transmembrane potential, however, the two-fold symmetry of the channel reduces this to five rate constants. Then Eyring rate theory is used to introduce the voltage dependence as shown in Eq. 6.
B. Steady state eq. for two-site single-filing channel. The values of the rate constants for given transmembrane potential and ion activity allows solving for the probabilities of the occupancy states χ_i. The quantity $\dot{\chi}_i$ is the rate of change with time of the occupancy state which under steady state conditions is zero. The solved for values of χ along with the values of the rate constants and the locations of the binding sites are then used in Eq. 4 or Eq. 5 to calculate the single channel current

2 Expression for the Single Channel Current, i_x

In Fig. 7A is given the steady state scheme for two sites which defines each of the elemental rate constants and in Fig. 7B are the steady state equations for the rate of change with time of the probability, χ, of each of the occupancy states of the channel: oo, xo, ox and xx. C'_x and C''_x are the concentrations of the x ion on the left- and right-hand sides, respectively. The general expression for the current, i_x, due to the ionic species, x, passing through a single channel is

$$i_x = \frac{ze}{2d} \{(d - a_1) C_x(k_1 - k_{-2}) \chi(oo) + [2a_1k_5 - (d - a_1)(k_{-1} + C_xk_3)] \chi(xo) + [(d - a_1)(k_2 + C_xk_4) - 2a_1k_{-5}] \chi(ox) + (d - a_1)(k_3 - k_{-4}) \chi(xx)\} \tag{4}$$

Recognizing that at steady state the net rate over each barrier is the same, the simplest expression for the current is obtained for the central barrier i.e.

$$i_x = ze[k_5\chi(xo) - k_{-5}\chi(ox)] \tag{5}$$

As defined in Fig. 8A, the quantity 2d in Equation 4 is the total length from one side to the other of the channel and a_1 is the distance from the binding site to the central barrier.

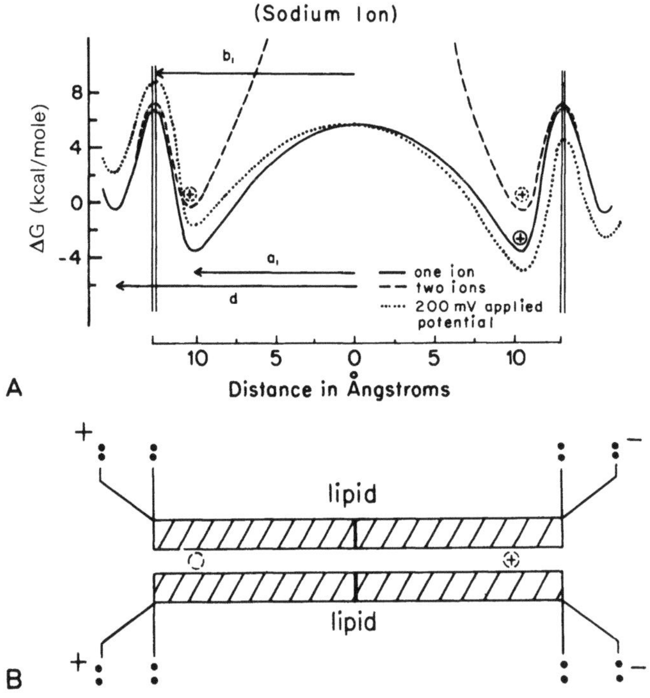

Fig. 8. A. Free energy profile for Na^+ movement through the malonyl Gramicidin channel. Calculated using Eyring rate theory and the locations of binding sites in the channel.
B. Schematic representation of the channel aligned with the free energy profile above and indicating the direction of the field. Reproduced with permission from Ref. [45]

3 Introduction of Voltage Dependence to Rate Constants Determined in the Absence of an Electric Field

Instead of the ten rate constants defined in Fig. 7A, in the absence of an applied field there are only five. These may be defined as the rate constants for the first ion entering the channel, k^{t}_{on}; the off-rate constant for one ion in the channel, k^{t}_{off}; the rate constant for crossing the central barrier, k_{cb}; and for an ion entering and leaving when there is already an ion in the channel, k^{w}_{on} and k^{w}_{off}, respectively. Actually the superscripts, t and w, stand for tight and weak sites, respectively. The difference in binding properties occurs, in spite of the two-fold symmetry and two identical sites, because of the effect of repulsion between two ions. Following the Eyring treatment [42–44], with the free energy change on going from one side to the other of the membrane being zFE (see Eqs. 1 and 2 for definitions), the effect on ions on the positive side of the membrane is to increase their free energy by zFE/2 and on the negative side of the membrane is to decrease their free energy by —zFE/2. At the center of the membrane or midway through the channel, the free energy change due to the applied field is zero. Now defining $X \equiv \exp(zFE/2dRT)$ and taking the potential gradient to be linear, which is a reasonable first approximation so far as the backbone structure of the channel is concerned, we can now write each of the ten rate constants of Fig. 7B in terms of the five experimentally determinable rate constants i.e.

$$
\begin{aligned}
k_1 &= k^{t}_{on}X^{l_1}; & k_{-1} &= k^{t}_{off}X^{-l_2}\\
k_2 &= k^{t}_{off}X^{l_2}; & k_{-2} &= k^{t}_{on}X^{-l_1}\\
k_3 &= k^{w}_{off}X^{l_2}; & k_{-3} &= k^{w}_{on}X^{-l_1}\\
k_4 &= k^{w}_{on}X^{l_1}; & k_{-4} &= k^{w}_{off}X^{-l_2}\\
k_5 &= k_{cb}X^{a_1}; & k_{-5} &= k_{cb}X^{-a_1}
\end{aligned}
\tag{6}
$$

where $l_1 = (d - b_1)$ and $l_2 = (b_1 - a_1)$ with a_1, b_1 and d as defined in Fig. 8A [45]. Therefore, by way of example, the free energy change associated with the rate constant k_1 is

$$\Delta G^{\ddagger}_1 = \Delta G^{t}_{on} + (d - b_1)\, zFE/2dRT \tag{7}$$

For a more complete treatment of the derivations and determination of experimental rate constants (to be discussed briefly below) refer to Ref. 46 for Gramicidin A and Ref. 47 for the malonyl dimer of Gramicidin A. (Malonyl Gramicidin A is formed by deformylation of Gramicidin A and then joining to amino ends together using the malonyl moiety, $—CO—CH_2—CO—$, to form the covalent dimer.)

4 Determination of Binding and Rate Constants by Nontransport Physical Methods

Utilizing ion nuclear magnetic resonance of nuclei with quadrupolar moments and carrying out longitudinal relaxation time studies [48], it is possible to determine binding constants for a tight binding site, $K^{t}_{b} = (k^{t}_{on}/k^{t}_{off})$ and for a weak binding site, $K^{w}_{b} = (k^{w}_{on}/k^{w}_{off})$. Using malonyl GA, this has been done for sodium-23

($K_b^t = 63\,M^{-1}$, $K_b^w = 1.4\,M^{-1}$)[47], lithium-7 ($K_b^t = 14\,M^{-1}$ $K_b^w = 0.5\,M^{-1}$)[49] and for cesium-133 ($K_b^t \simeq 50\,M^{-1}$, $K_b^w = 4\,M^{-1}$)[50]. In the case of sodium-23, transverse relaxation times could also be utilized to determine off-rate constants: $k_{off}^t = 3 \times 10^5/sec$; $k_{off}^w = 2 \times 10^7/sec$[47,51]. Therefore for sodium ion four of the five rate constants have been independently determined. What has not been obtained for sodium ion is the rate constant for the central barrier, k_{cb}. By means of dielectric relaxation studies a rate constant considered to be for passage over the central barrier, i.e. for jumping between sites, has been determined for Tl^+ to be approximately $4 \times 10^6/sec$[52]. If we make the assumption that the binding process functions as a normalization of free energies, recognize that the contribution of the lipid to the central barrier is independent of the ion and note that the channel is quite uniform, then it is reasonable to utilize the value of $4 \times 10^6/sec$ for the sodium ion.

5 Calculation of Single Channel Currents and the Equivalent Free Energy Profile

Utilizing the above five experimentally derived rate constants and Eyring rate theory, the ten rate constants of Eq. 6 are all obtained. With the rate constants known, the probability of each occupancy state, $\chi(ox)$ for example, can be calculated and finally the single channel current can be calculated as a function of molal activity of sodium ion. This is done for a 100 mV transmembrane potential in Fig. 9. It should be emphasized that Fig. 9 represents a calculation of single channel currents

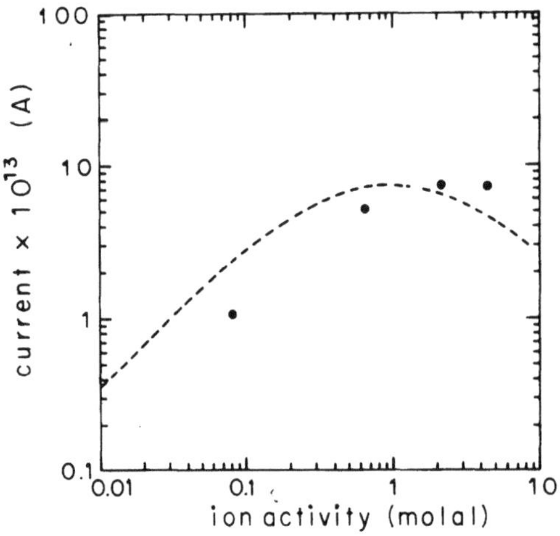

Fig. 9. Calculation of malonyl Gramicidin single channel current due to Na^+ using only experimentally determined rate constants and binding site locations. There are no parameters. The data points are the experimental single channel current values from Ref.[47]. The greater uncertainty in the experimental values is with the value of k_{off}^t. Because it is required to calculate k_{on}^t from the tight site binding constant K_b^t, this means that k_{on}^t is also in question due to $K_b^t = k_{on}^t/k_{off}^t$. Further experimental data in progress suggests that k_{on}^t is a little too large consistent with the higher calculated values at low ion activity. The experimental values used can be considered to be accurate only within a factor of two or three. Accordingly in Fig. 10B, the experimental values have been allowed to vary to achieve a "best fit". The changes on achieving a "best-fit" are within the experimental uncertainties

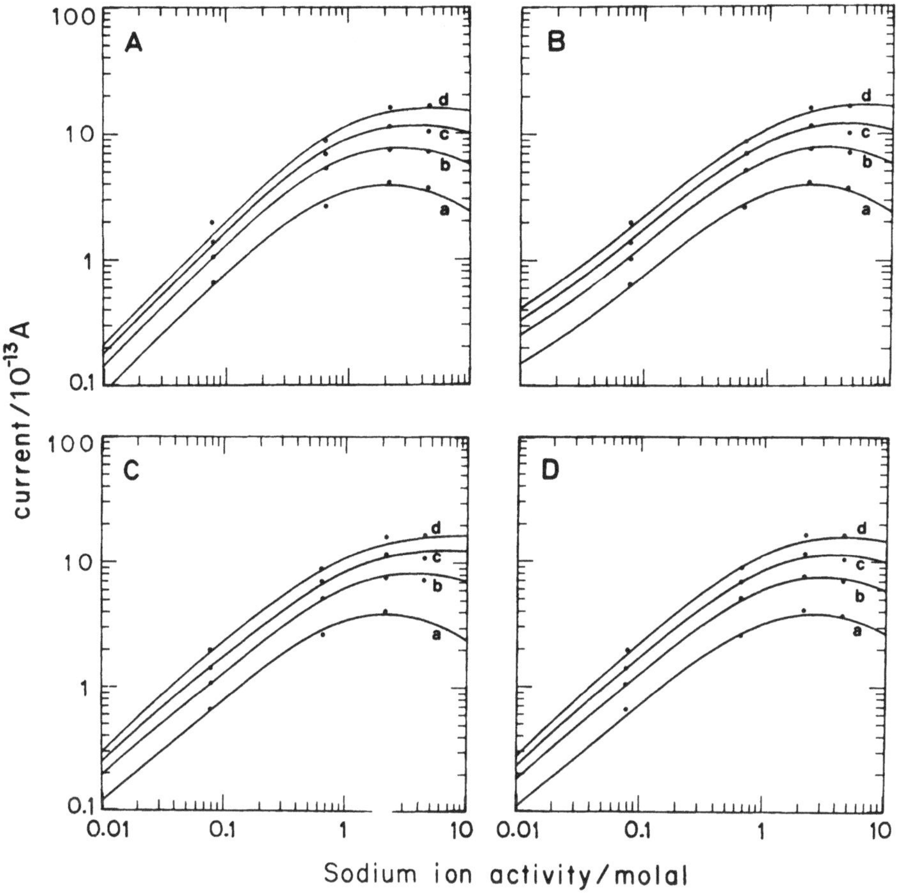

Fig. 10. Calculated sodium ion single channel currents for the malonyl Gramicidin channel and comparison with experimental data points using four different models all of which fit the data well but only one of which, B., is correct. The point to be made is that both the independent determination of rate constants and of the binding site locations are required.

A. Two site model with two equivalent non-interacting binding sites. $K_b = 1.2\ M^{-1}$, $k_{off} = 6.6 \times 10^6$/sec, $k_{cb} = 1.3 \times 10^7$/sec. This model is incorrect because a tight and a weak site have been observed with different off-rate constants as listed in part B.

B. Two site model with equivalent but interacting binding sites. $K_b^t = 200\ M^{-1}$, $k_{off}^t = 1.6 \times 10^5$/sec, $k_{cb} = 1.3 \times 10^7$/sec, $K_b^w = 1.3\ M^{-1}$, $k_{off}^w = 7.9 \times 10^6$/sec. These values are within a factor of two or three of the experimentally determined values and the two sites have been demonstrated to be correct.

C. Three site model with virtual weak sites. This model places the tight site midway through the channel and with the tight site occupied then two weak sites become stabilized just inside the mouth of the channel. With two ions in the channel, one at the central tight site and one at a weak site, a rate constant for the shift of the ions in the direction of the empty weak site is defined as k_{cs}. $K_b^t = 100\ M^{-1}$, $k_{off}^t = 1.1 \times 10^5$/sec, $k_{cs} = 4.8 \times 10^6$/sec, $K_b^w = 1.0\ M^{-1}$, $k_{off}^w = 5.8 \times 10^6$/sec. While the values of the constants are reasonably close to the experimental values, this model is not correct due to the absence of a binding site midway through the channel.

D. General three site model with a parameter, A, for ion repulsion $K_b^t = 75\ M^{-1}$, $k_{off}^t = 2.9 = 10^7$/sec, $K_b^w = 1.0\ M^{-1}$, A = 3.1 kcal-Å/mole. This model is incorrect due to the absense of a binding site in the middle of the channel and due to the value of k_{off}^t.

Part A reproduced with permission from reference 45. Parts B, C and D reproduced with permission from Ref. [47)]

using only experimentally derived rate constants. *There are no parameters.* Realizing, however, that the experimentally determined rate constants can be in error by a factor of two or three and realizing that the lipid in the planar bilayer studies and in the suspension studies has the common phosphatidyl choline head group but has differences in the acyl chains, it becomes of interest to see what happens when a "best fit" is attempted by letting the rate constants vary. Indeed when this is done for the 50 mV (a), 100 mV (b), 150 mV (c) and 200 mV (d) curves of Fig. 10B, the rate constants change by no more than the anticipated error in their determination and the fit falls within the accuracy for determining the most probable single channel current. Accordingly, with these values we can plot the free energy profile as shown in Fig. 8A using Eqs. 3 and 6.

Thus, we have achieved a free energy profile for sodium passage through the malonyl Gramicidin A channel, without parameters, but rather with experimentally determined rate constants which have real meaning and with prior determination of the locations of the binding sites. It is useful to realize that different combinations of occupancy models and of rate constants can also achieve fits to the experimental currents but these (see Figs. 10A, C and D) have no physical reality. With regard to the two site model with non-interacting sites (see Fig. 10A), there is not agreement with the experimental demonstration of a tight site binding constant using sodium-23 NMR nor of the confirmation of this tight binding site by directly observing the channel component with carbon-13 NMR. Additionally, it is reasonable to expect some repulsion between ions on double occupancy. The two three site models (Fig. 10C and D) are calculable using the experimental values for the rate constants but the demonstrated absence of binding midway through the channel excludes these occupancy models. The point to be made is that both independent experimental values for the rate constants and independent demonstration of binding site location appear necessary to arrive at a unique mechanism.

E Comments on the Free Energy Profile

1 Free Energy Change on Binding

Since the sodium ion binding constant, K_b^t, for the first ion to enter the channel is $\sim 63\ M^{-1}$, the free energy change on binding, ΔG_b^t, is

$$\Delta G_b^t = -RT \ln K_b^t \qquad (8)$$

$$= -2.5\ \text{kcal/mole}$$

This demonstrates that the peptide carbonyls are better solvators than water molecules, in spite of the proximity of the lipid. Also the sodium-23 nuclear magnetic resonance chemical shift is upfield by more than 20 ppm [47] on interaction with the channel. The upfield chemical shift indicates a greater electron density surrounding the cation once within the channel, as might be expected for a favorable cation binding site. Analogously, the downfield chemical shift of the L-residue carbonyl carbon resonances, shown in Fig. 6A, indicates electron density loss to the cation. The carbonyls in the region of the binding site, as shown in Fig. 6A, are the carbonyls of residues 9, 11, 13, 14 and 15. While the D · Leu[14] carbonyl is at the binding site, the chemical shift is upfield indicating that only the L-residue carbonyls coordinate the cation and that the hydrogen bonding of the Leu^{14} C—O is weakened and/or the carbonyl oxygen librates outward toward the lower dielectric constant of the lipid [38]. Thus, four carbonyls, two of which (those of Trp^{11} and Trp^{13}) most significantly, induce the cation into the channel with a favorable 2.5 kcal/mole change in free energy. Structurally this allows for two water molecules, one preceding and one following the ion through the channel, to retain direct coordination with the cation.

2 Repulsion Between Ions on Double Occupancy

Binding for the second ion is less favorable with a free energy change of about —0.2 kcal/mole ($= -RT \ln K_b^w$). Since the molecular structure has two-fold symmetry making the binding sites identical, this means that there is a repulsion between ions of about 2.3 kcal/mole at the separation distance of 23 Å, based on a repeat distance of 1.53 Å/dipeptide [53]. Structurally, there is room for no more than five water molecules separating the two ions i.e. about 4 Å per water molecule. With the repulsion between ions being given by $(ze)^2/\varepsilon r$ and with $e^2/a_0 = 627$ kcal/mole given $a_0 = 0.529$ Å, then the effective dielectric constant, ε, with five interceding water molecules would be about 6. This is a reasonable number of considerable interest to electrochemists and physical chemists in general.

3 The Rate Limiting Barrier

Due to the low dielectric constant of the lipid, in the past it has been assumed that the central barrier would be rate limiting for ion penetration across biomembranes. With the Gramicidin channel studies it has been demonstrated that, with the minimal shelding of one thickness of polypeptide backbone, the rate constant for the central barrier was experimentally determined and also fitted to be 4×10^6/sec for malonyl GA [47,52] and was fitted to be 7.2×10^7/sec for GA [46]. With the solution as the reference state of 0 kcal/mole, this gives the free energy values for the central barrier of 5.9 kcal/mole for malonyl GA and 4.5 kcal/mole for GA. On the other hand, the entry barrier, determined from rate constants of 3×10^7/sec for GA and malonyl GA, is just over 7 kcal/mole. Thus, the rate limiting barrier is at the entrance to the channel; it is the partial dehydration step. Two components of that barrier may be considered to be the barrier for replacing a coordinating water molecule with a coordinating carbonyl and the energy to depart from the channel coordinates sufficient to bring the several carbonyls into adequate contact with the cation. These two elements will be discussed in connection with selectivity among monovalent cations.

F Ion Selectivity of the Gramicidin A Transmembrane Channel

With the foregoing developments and with additional results to be discussed in this section, it is possible to demonstrate a qualitative understanding of the chemical basis of ion selectivity. As further studies complete the determination of the elemental rate constants, as further studes adequately determine temperature dependencies, and as further theoretical calculations more accurately describe conformations and ion interactions, a more quantitative picture will emerge. This is particularly important with the more subtle and very interesting problem of selectivity among monovalent cations. As will be presented below, it is the chemical structure and its conformational energetics that gives rise to selectivity for cations over anions; it is the proximity of the lipid which results in exclusion of multivalent cations; and it is dominantly the energetics of conformational dynamics, the energetics of peptide librations, coupled with the energetics of partial dehydration that give rise to selectivity among monovalent cations.

1 Anion vs Cation Selectivity

Whether the species of interest is an inorganic cation or anion, the perspective derived from the channel structure with a channel diameter of no greater than 4 Å (see Fig. 5) is the requirement for partial replacement of the waters of hydration in order to pass through the channel. The candidates for replacing the waters of hydration are obvious: the peptide carbonyl oxygens for cations and the peptide NH hydrogens for anions. The relative effectiveness of such interactions is apparent from Fig. 11 which contains the partial charges on each atom of the peptide moiety [54, 55]. The carbonyl oxygen has a negative charge of —0.384 whereas the positive charge on the hydrogen is less than half that magnitude, 0.176. All other aspects being equal, the energy barrier for permeation of anions through the channel could be expected to be twice that of cations. But all other aspects are not equal. As is shown below, the structure itself favors cation permeation; the arrangement of carbonyl oxygens and the conformational energetics required to achieve ion coordination favor cation entry into the channel.

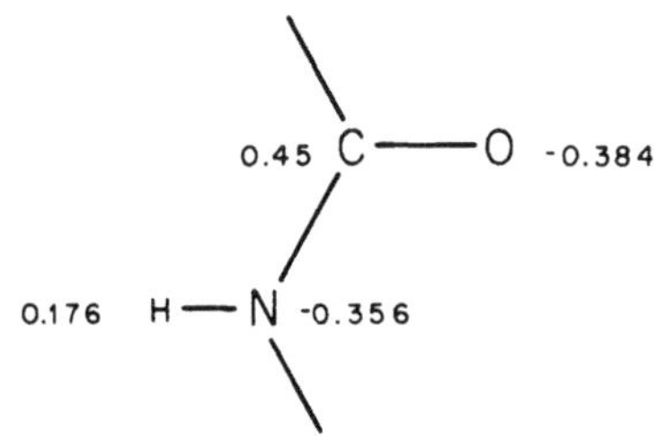

Fig. 11. Peptide moiety indicating the monopoles (net charges) on each of the atoms [54, 55]. This demonstrates that, on the basis of size, charge and accessibility, a polypeptide could more effectively provide selectivity for cations

1.1 Role of Channel Structure

As viewed from solution, the entrance to the channel (see Fig. 5B) is guarded by three carbonyl oxygens; they are the carbonyls of L-Trp[15], L-Trp[13] and L-Trp[11]. Less imposing are the smaller, somewhat recessed peptide NH moieties of the same residues with their lesser positive partial charge. Accordingly, the entrance to the channel presents a negative potential which is attractive to cations and repulsive to anions. The approach of cations is obviously favored. Yet within the channel, and to some extent at the entrance, the potential, whether positive or negative as viewed by an ion on the channel axis, depends on the energetics of peptide libration, the rocking or torsional oscillation of the peptide moiety. The potential for a species on the channel axis depends on whether it is more favorable in terms of conformational energy for the oxygen end of the peptide moiety or the proton end of the peptide moiety to librate into the channel. The peptide libration decreases the channel radius from a maximum of about 2 Å to a radius capable of providing adequate lateral coordination of the bare ion.

1.2 Energetics of Peptide Limited Libration

Our most current energy minimized conformation of the single stranded $\beta^6_{3,3}$-helix is given in Fig. 12A [53]. In this stereoperspective along the channel axis, it is apparent that the carbonyl oxygens of the L-residues are already librated into the channel and the carbonyl oxygens of the D-residues are librated slightly outward. This lines the channel with the negative ends of the carbonyl and creates a

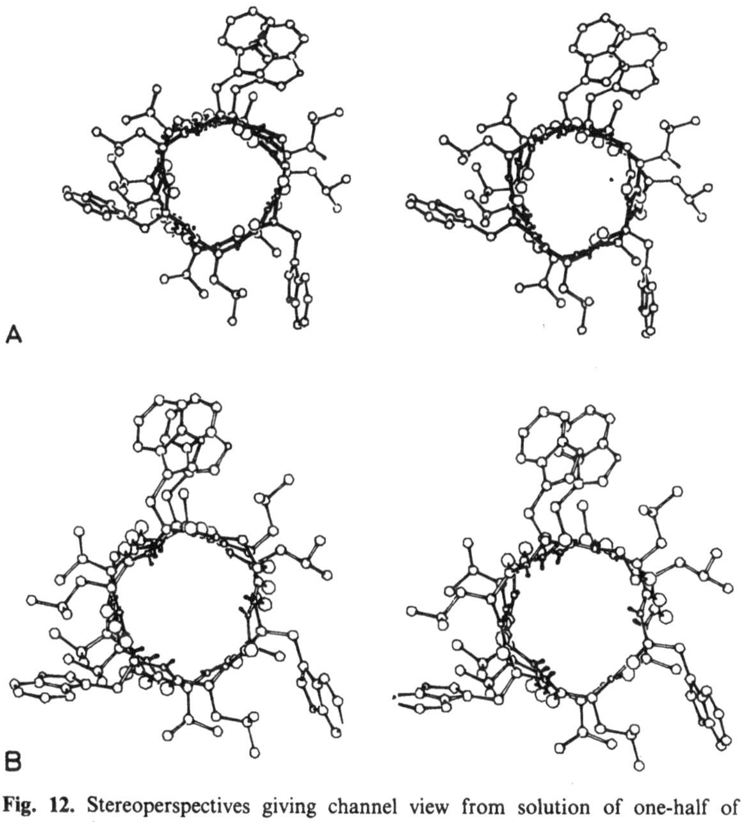

Fig. 12. Stereoperspectives giving channel view from solution of one-half of a Gramicidin A transmembrane channel.
A. In vacuo lowest energy conformation showing the peptide carbonyls to be slightly librated into the channel by about 10°, Reproduced with permission from Ref. [53].
B. In vacuo calculation of a librated state of the Gramicidin channel showing the peptide NH moieties to be librated into the channel. This is a relatively high energy conformation, some 8 kcal/mole higher than in A. C. M. Venkatachalam and D. W. Urry, unpublished results

negative potential for cations. The channel radius for this lowest energy conformation is such that it would be optimal for an ionic radius between that of Rb^+ and Cs^+ whose ionic radii are 1.47 Å and 1.67 Å, respectively.

It is energetically possible for the D-residue carbonyls to librate further outward which brings the L-residue NH into the channel, as shown in Fig. 12B [56]. This structure, though energetically less favorable by about 8 kcal/mole channel, would be suitable for anion permeation. Relative to cation passage through the channel seen in Fig. 12A, anion passage by means of the conformation in Fig. 12B is impaired due to the relatively higher energy of this conformation and due to the lesser effectiveness of the peptide NH hydrogen in providing a partial positive charge for an ionic species in the channel. *Thus, cation selectivity is the result of the energetics of the channel structure and the chemical nature of the peptide moiety.*

1.3 Possibility of Limited Anion Interaction Site

While under most conditions there is no significant anion permeability [57,58], anion exclusion is not absolute. As shown by Eisenman et al. [59,60], there are conditions of high thallium ion occupancy that acetate and chloride may be about 1/10 as permeant as Tl^+. This is consistent with the above structural perspective which may be further considered to call attention to a location within the channel that would provide a localized anion binding site. By virtue of the head to head dimerized single stranded β-helical structure [21,22] and as is apparent by careful examination of the abscissa in Fig. 6A, the entrances to the channel are dominated by L-residue carbonyls which can librate into the channel. Similarly, the formyl ends which occur midway through the channel are dominated by the D-residue carbonyls that can most readily librate outward. Due to the planarity of the peptide moiety, this brings the L-residue NH moieties into the channel. Accordingly, the structure would provide for a local minimum for anions in the region of the head to head junction. This is, of course, superimposed on the positive image force of the lipid which is most sensed midway through the channel. Particularly with Tl^+ in each of the two cation binding sites, an anion placed midway between them may not constitute an energetically unacceptable configuration.

2 Monovalent vs Multivalent Cation Selectivity

2.1 Multivalent Ion Exclusion

The calcium ion conductance of the Gramicidin A channel is very low [8], no value having yet been established. Sodium ion and calcium ion have essentially the same ionic radius, 0.97 Å and 0.99 Å, respectively. Why then does sodium ion pass through at rates of 10^7 ion/sec when there is no measurable conductance for calcium ion? Calcium ion does affect channel function, however. The single channel conductance due to monovalent cations is decreased by the presence of calcium ion [8,61]. Based on analyses of blocking effects, Bamberg and Lauger [61] estimated binding constants for Ca^{2+} and Ba^{2+} binding to a site at or near the mouth of the channel to be of the order of 1 to 10 M^{-1}. Using (1-^{13}C) enriched synthetic Gramicidin As, it has been possible to demonstrate Ca^{2+} [62] and Ba^{2+} [63] binding in the channel (see Fig. 13). On the basis of calcium ion concentration dependence of chemical shift, a binding

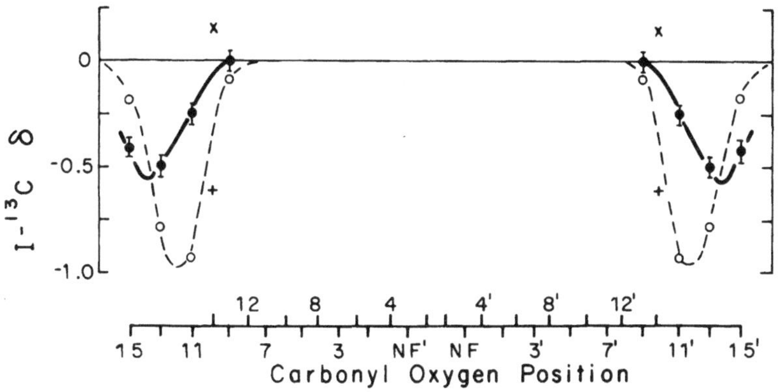

Fig. 13. Location of calcium ion binding site in the Gramicidin A channel —●—. For comparison, the sodium ion binding site is included - - -. Calcium ion binding site is 1 to 2 Å closer to the aqueous solution and pushed away from the lipid layer. Since exchange with the binding site is fast, the lack of Ca^{2+} transport must be due to the repulsive image force of the lipid. Reproduced with permission from Ref. [62]

constant of 1 M^{-1} could be estimated [62]. Additionally as only one resonance was observed for the (1-^{13}C) carbonyl carbon throughout the titration up to 1 M $CaCl_2$ or 1 M $BaCl_2$, it can be deduced that the exchange rate of the divalent ion with the binding site is faster than 10^6/sec. Thus, divalent ions like Ca^{2+} and Ba^{2+} can bind in the channel and can exchange rapidly with the binding site but they cannot pass through the channel at a detectable rate.

It has also been shown in organic solvents that Ca^{2+} has a much more favorable interaction with Gramicidin A than do Na^+ and K^+ [6], yet when the Gramicidin A is in the membrane just the inverse is true. In addition to having a weaker binding constant for interaction with membrane incorporated Gramicidin A, the calcium ion binding site is pushed outward toward the solution by one or two Angstroms (see Fig. 13). A weaker binding constant itself does not lead to less conductance. In Fig. 10A, the two-site model with equivalent noninteracting (weak) sites, the binding constant is 1.2 M^{-1} and yet the conductance is high. Accordingly, if the entry barrier is not rate limiting and the magnitude of the binding constant is not a factor, then the rate limiting barrier for divalent ions must be the central barrier. Specifically, it must be that part of the central barrier that derives from the proximity of the lipid. As shown above, the central barrier for Na^+ conductance through the GA channel is 4.5 kcal/mole above the solution reference state. Additionally, it may be noted that there is a preference for binding to the Trp end of the channel even in organic solvents [64] so that the intrinsic difference in interaction between the ethanolamine (Trp) and formyl ends of the monomer could be about 1.5 kcal/mole or more. This allows an estimate of that part of the central barrier which is due to the repulsive image force of the lipid (see Fig. 1B) to be about 3 kcal/mole for the Na^+ ion. If we now utilize the z^2 term in the Born expression (see Eq. 2) as the essential difference between monovalent and divalent ions for this effect, the contribution to the central barrier for a divalent ion due to the lipid surrounding the Gramicidin channel would be of the order of $z^2 \times 3$ or 12 kcal/mole. It may be noted here that the relative hydration energies follow very well the z^2 dependence of Eq. 2, because the best experimental determinations of the Gibbs free energies of hydration for ions of 1.0 Å radius are approximately 100 kcal/mole, 400 kcal/mole and 900 kcal/mole for monovalent, divalent and trivalent ions, respectively [65–67]. The z^2 effect would reduce the 10^7 ion/sec current for a sodium ion to the order of 10^{10} ions/sec for a calcium ion. *Thus, the proximity of the lipid provides the mechanism for selectivity against multivalent cations.*

2.2 Proposed Properties of a Divalent Ion Channel

With this general reasoning, it has been argued that a channel would require an extended polarizable layer as indicated in Figure 1E in order to allow passage of divalent cations [68]. The problem for a calcium ion channel then becomes one of achieving selectivity over sodium ions, i.e., of impeding passage of sodium ions. As the calcium ion can have very favorable binding constants with peptides and proteins of the order of 10^6 M^{-1} [69, 70], the free energy of binding can be as high as 8.5 kcal/mole. If part of that free energy change were utilized to achieve a conformational change, as great as 5 kcal/mole, there would still be an adequate binding constant for channel transport. The channel binding site could, therefore, require a 5 kcal/mole deformation energy before a 1 Å radius ion would be adequately coordinated. As sodium ion binding constants only reach values of 10^3 M^{-1}, the 4 kcal/mole interaction energy would not be sufficient to bring about the conformational change required for achieving the interaction. Accordingly, a calcium ion channel should have a diameter too small for hydrated sodium to pass through but the conformational energy should be such that the calcium ion can achieve the librational or other conformational change required for coordination but sodium ion cannot. Thus, the studies on the Gramicidin channel provide an insight to a possible means of achieving a divalent cation channel.

3 Selectivity Among Monovalent Cations

The alkali metal ion permeability ratios for Gramicidin channels in erythrocyte lipid/n-decane planar bilayers was determined by Myers and Haydon [57] to be: Li^+ (0.33) < Na^+ (1.0) < K^+ (3.9) < Rb^+ (5.5) < Cs^+ (5.8). This is the same order as for ionic mobility (1) which may also be ratioed to sodium ion to give: Li^+ (0.77) < Na^+ (1.0) < K^+ (1.48) < Rb^+ (1.55) < Cs^+ (1.56). While indeed the order is the same, the significant point is that the range of discrimination has been expanded by almost an order of magnitude. The concern then becomes the mechanism whereby the Gramicidin channel effects this enhancement of selectivity. An important source of selectivity that must be considered is the relative solvation energy of water and peptide. This has been examined by Eisenman and colleagues [71] and found for formamide, N-methyl acetamide and dimethyl formamide to give the selectivity series $Li^+ < Na^+ < Cs^+ < Rb^+ < K^+$. Only the first two are in the correct order. The last three are in an inverse order. The amide selectivity series does not provide an explanation for the selectivity of the Gramicidin A transmembrane channel. Another mechanism must be dominant. The mechanism is called the peptide libration mechanism; it was proposed in the original papers [22, 23] and has already been employed in the above discussion on anion vs cation selectivity. This mechanism takes advantage of the different ionic radii: Li^+ (0.68 Å), Na^+ (0.97 Å), K^+ (1.33 Å), Rb^+ (1.47 Å) and Cs^+ (1.67 Å).

3.1 Peptide Libration Mechanism

The original Fig. depicting the peptide libration mechanism [23] is given in Fig. 14. As drawn, both L- and D-residue carbonyls are shown coordinating a laterally dehydrated potassium ion. About a 20° libration is required to achieve close contact between carbonyl oxygen and cation. As shown in Fig. 12A, our must recent *in vacuo* calculations give a lowest energy conformation in which the L-residue carbonyl oxygens are already partly librated into the channel [55]. Also, as considered in the section treating the location of the ion binding sites, it is seen that only the L-residue carbonyl oxygens coordinate the monovalent cation (see Fig. 6A). A channel axis view from solution of the L-Trp carbonyl oxygens coordinating a sodium ion is shown in Fig. 15 where the extent of libration of the oxygens into the channel is greater than in Fig. 14A for K^+. The lowest energy calculated structure, seen in

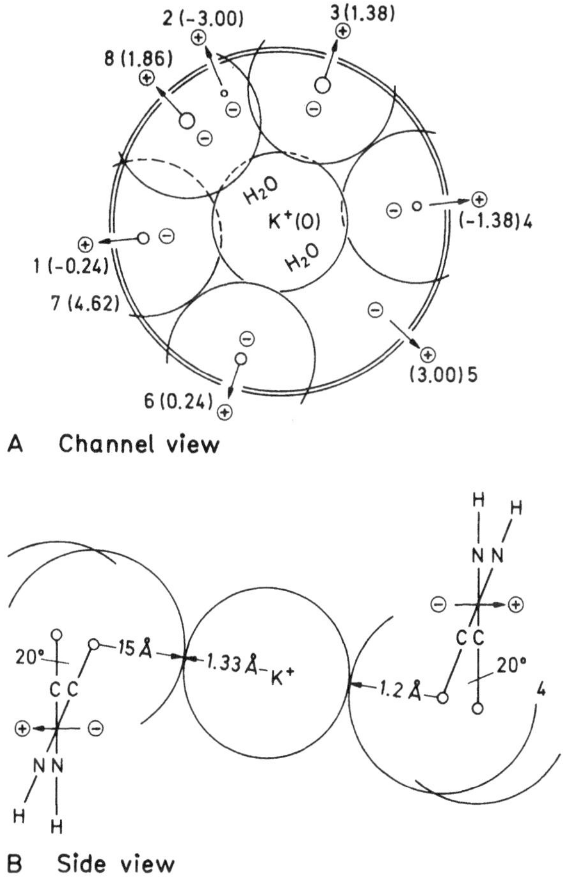

Fig. 14. Peptide libration mechanism as orginally depicted [23]. The rocking or torsional oscillation of the peptide moiety brings the carbonyl oxygen into the channel making the channel suitable for cation interaction. In this depiction, both L- and D-residue carbonyls are shown interacting with K^+. Subsequent studies have shown that only the L-residue carbonyl binds to the cation [38]. A 20° libration is shown as adequate to bring the carbonyl oxygen into close contact with a potassium ion. A lesser libration angle would be required for Rb^+ and Cs^+ interaction whereas a progressively greater libration angle would be required for Na^+ and Li^+. Taking the energy to increase with angle of libration, this becomes a mechanism for discriminating against ions of smaller radius. Reproduced with permission from Ref. [23]

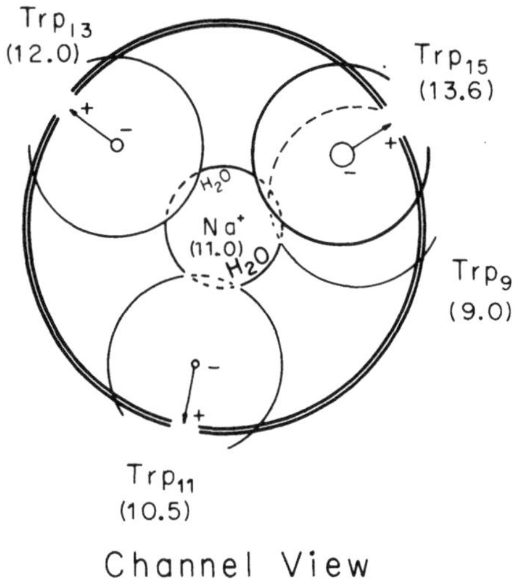

Fig. 15. Librated states of the L-Trp^{9,11,13,15} carbonyls to achieve interaction with Na^+ at the binding site. The numbers in parentheses are the distances of the oxygen atoms and Na^+ from the two fold symmetry axis of the channel. The water molecules are to indicate that water is expected to be bound on each side of the cation as it passes through the channel.

Fig. 12A, has a 10° libration. This gives a channel size which would be optimal for an ionic radius between that of Rb^+ and Cs^+. Therefore enhanced discrimination is not expected between Rb^+ and Cs^+, but the energy required to librate further inward to make contact with smaller ions in the series can be expected to enhance selectivity between these ions. Work is currently in progress to calculate the change in channel energy as a function of libration angle or of the equivalent, the effective channel radius [56]. The implications of a peptide libration mechanism for enhancing ion selectivity can also be pursued experimentally as outlined below.

3.2 Formalism for Evaluating Thermodynamics Relevant to Selectivity Among Monovalent Cations

A general expression for the single channel current equation may be obtained by rewriting Eq. 4 as

$$i_x = ze \sum_r \chi_r \sum_s \bar{\lambda}_{rs} q_{rs} \tag{9}$$

χ_r is the probability of the rth occupancy state, e.g., the xo state. $\bar{\lambda}_{rs}$ is the fractional length across the membrane that an ion moves as the rth occupancy state changes by any of the s possible ways, for example by the jump over the central barrier from the xo state to the ox state which involves the fractional length, a_1/d (see Fig. 8A for definitions of the lengths). The fractional length, $\bar{\lambda}_{rs}$, is a signed quantity being positive when the ion movement is with the potential gradient and negative when it is against the field. The quantity, q_{rs}, is the rate for the process; it may simply be the rate constant, e.g., k_5 as when going from xo to ox or k_{-1} when going from xo to oo; or it could be the product of a rate constant and a concentration as when going from oo to ox where it would be $C_x''k_1$.

Equation 9 and Eq. 4, however, consider only the rates of changes of occupancy states and assume that the channel itself is a constant unchanging entity. But the channel is a dynamic structure. It has different energy states, i.e. different librated states, that are required for conductance of different ions and the probability distributions of the conformations change with temperature. Accordingly, it is necessary to introduce the dynamic nature of the channel into the current equation in order to approach experimentally those channel energy components relevant to ion selectivity.

Whether concern is with the fraction of time that a given molecule is in a particular state or with the fraction of molecules within a large collection of molecules that are in the state of interest, the formalism is essentially the same. When considering single channel currents, a histogram of the frequency of occurrence of conductance states is obtained which shows a dispersity of conductance values and which is necessary to define a mean or most probable single channel current (see Fig. 4A and C). In the case of Gramicidin A channels at low concentration, a pair of monomers may associate to give a channel with a given conductance; the pair may dissociate and reform to give a channel with the same or a different conductance; a new pair of monomers may form a channel with similar or different conductance, or a given channel can change its conductance state without turning off and then on [40]. It is our perspective that the different conductance states represented in the histogram result in the main from the same basic channel conformation, the left-handed, single stranded, head to head dimerized $\beta^6_{3,3}$-helical structure; and that different combinations of side chain orientations and different metastable combinations of φ_L, ψ_L and φ_D, ψ_D torsion angles result in different single channel currents. A proper description of these effects needs to be introduced into the expression for single channel currents. It may be noted that a general formalism for the effects of dynamic channels has been presented by Lauger [72].

Considering a segment of the channel involved in a single rate process, for example the set of residues involved in the entrance barrier, there is the probability, P_l, that the segment is in a librated state suitable for ion entry. Writing the probability of all the other states of the segment as P_0, the ratio of probabilities is equivalent to an equilibrium constant, $K°$, relating all other states to the librated state, i.e.,

$$\frac{P_0}{P_l} = K° \tag{10}$$

Since $P_l + P_0 = 1$ and $P_0 = K°P_l$ then

$$P_l = (1 + K°)^{-1} \tag{11}$$

As each individual rate process is designated by the subscript, rs, the constant should be indentified as K°_{rs} and the rate constant, q_{rs}, should be multipied by the probability that the involved channel segment is in the librated state required for conductance. Accordingly, Eq. 9 becomes

$$i_x = ze \sum_r \chi_r \sum_s \bar{\lambda}_{rs} q_{rs} (1 + K^\circ_{rs})^{-1} \tag{12}$$

If all of the peptide carbonyls were optimally librated for conducting then each P_l would be one and each K°_{rs} would be zero, recovering Eq. 9. Since there is a dispersity of single channel currents, as seen in Fig. 4A and C, this means that each channel state in the histogram differs by at least one value of K°_{rs}. If the index t is used to sum over the different K°_{rs} values for each q_{rs}, the mean single channel current, $\bar{i}_x$, would be obtained by summing over the single channel currents in Fig. 4A or C and dividing by the number of channels N, i.e.,

$$\bar{i}_x = \frac{1}{N}\sum_{t=1}^{N} i_t = \frac{1}{N}\sum_{t=1}^{N}\sum_{r}\chi_r\sum_{s}\bar{\lambda}_{rs}q_{rs}(1 + K^{\circ}_{rst})^{-1} \tag{13}$$

3.3 Temperature Dependence of Single Channel Currents

A simpler phenomenological form of Eq. 13 or 12 is useful. This may be approached by using Eq. 4 or its equivalent, Eq. 9, with the rate constants determined for Na^+ transport. Solving for the $\Delta G_i^{\ddagger}$ using Eqn. (3) and taking $\Delta G_i^{\ddagger}$ to equal $\Delta H_i^{\ddagger}$, that is the $\Delta S_i^{\ddagger} = 0$, the temperature dependence of i_x can be calculated as shown in Fig. 16A. In spite of the complex series of barriers and states of the channel, a plot of log i_x vs the inverse temperature (°K) is linear. Accordingly, the series of barriers can be expressed as a simple rate process with a mean enthalpy of activation $\overline{\Delta H}^{\ddagger}$ even though the transport requires ten rate constants to describe it mechanistically. This

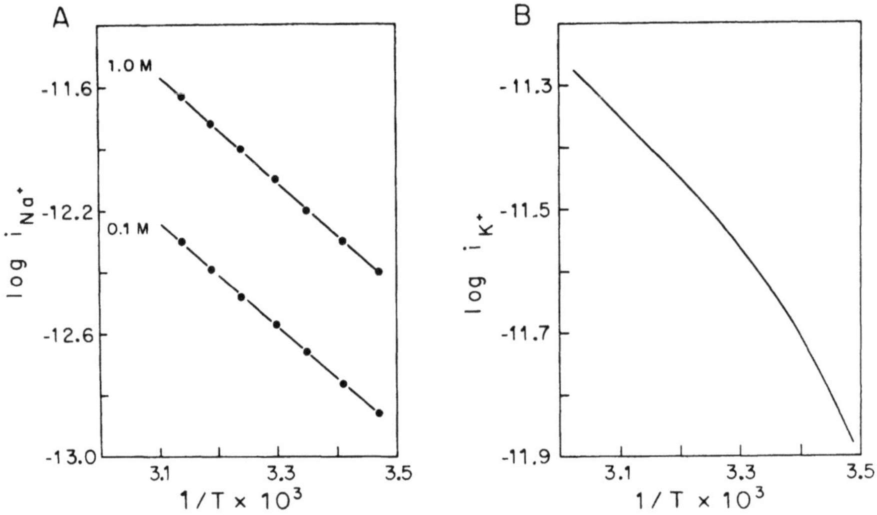

Fig. 16. A. Plot of log i_{Na} as a function of T^{-1} (°K) using the experimental values of the rate constants and the location of the binding sites in Eq. 4. The Gibbs free energy of activation is calculated from Eq. 3; the $\Delta S^{\ddagger}$ are taken to be zero, and the current is calculated by means of Eq. 4. The purpose is to demonstrate that multibarrier channel transport can be seen as single rate process with average values for the enthalpies of activation. Non-linearity of such a plot is then taken to arise form the dynamic nature of the channel.
B. Plot of log i_K vs T^{-1} (°K) introducing the effect of a dynamic librating channel as treated in Eq. 13 and 17. The curvature is due to the $(1 + \bar{K}^{\circ})^{-1}$ term in Eq. 15, and the curve was calculated to fit the temperature dependence of K^+ single channel currents in diphytanoyl lecithin/n-decane membranes, 130 mV, 1 M KCl (73).

means that the thermodynamic form of the Eyring rate eq. can be used, i.e.,

$$i_x = CT\, e^{\overline{\Delta S}^{\neq}/R}\, e^{-\overline{\Delta S}^{\neq}/RT} \quad (14)$$

where C is a constant and therefore that Eq. 12 can be written as

$$\bar{i}_x = CT\, e^{\overline{\Delta S}^{\neq}/R}\, e^{-\overline{\Delta H}^{\neq}/RT}(1 + \bar{K}^{\circ})^{-1} \quad (15)$$

where $\bar{K}^{\circ}$ is an average value for the equilibrium between the librated state for a specific cation conductance and the remainder of the channel states. Since

$$\bar{K}^{\circ} = e^{\overline{\Delta S}^{\circ}/R}\, e^{-\overline{\Delta H}^{\circ}/RT} \quad (16)$$

the natural logarithim of the single channel current becomes

$$\ln i_x = \ln C' + \ln T - \overline{\Delta H}^{\neq}/RT - \ln(1 + e^{\overline{\Delta S}^{\circ}/R}\, e^{-\overline{\Delta H}^{\circ}/RT}) \quad (17)$$

where C′ is a constant that also includes $\overline{\Delta S}^{\neq}/R$. When $\bar{K}^{\circ}$ is non-zero, this term will introduce a non-linearity into the $\ln i_x$ vs T^{-1} (°K) curve. Using values of $\overline{\Delta H}^{\neq} = 3.5$ kcal/mole, $\overline{\Delta S}^{\circ} = -60$ cal/mole degree and $\overline{\Delta H}^{\circ} = -18$ kcal/mole channels in Eq. 17, the curve plotted in Fig. 16B is obtained.

The curvature seen in Fig. 16B approximates that obtained for the temperature dependence of Gramicidin A single channel currents for 1 M KCl at 130 mV using diphytanoyl lecithin/n-decane membranes [73)]. If the energy of peptide libration were to increase monotonically with an increase in the angle of libration, then the curvature in Fig. 16B may be expected to increase as the ionic radius decreases. Studies to check this are in progress. By this approach, the selectivity among monovalent cations can be discussed in mechanistic and thermodynamic terms and compared with theoretical calculations on the conformational energy as a function of angle of libration.

The perspective for selectivity among alkali metal cations, as seen in terms of the Gramicidin channel, is one of a lowest energy state of the channel in which there is little effect on the relative mobilities of Rb^+ and Cs^+ but in which there is progressive retardation of K^+, Na^+ and Li^+ due to the progressively greater conformational energy required to achieve the degree of peptide libration necessary for lateral coordination of the cation.

III The Carrier Mechanism of Ion Transport: The Valinomycin Model

As schematically shown in Fig. 1C, a carrier is an amphiphilic molecule capable of residing at the membrane aqueous interface with its lipophilic side interacting with the lipid of the membrane, with polar moieties directed outward into the aqueous phase, and with the polar moieties of a chemical nature to induce an ion into interaction. In the process of a carrier interacting and complexing with the ion, it envelopes the ion in a polar core while completing the lipophilic coat suitable for becoming entirely surrounded by lipid. The carrier-ion complex can then cross the lipid barrier to the opposite interface and carry out the inverse process of release. Net transport is the result of an ion concentration (activity) gradient or a potential gradient. The carrier mechanism effectively trades the high transport rates of a channel process for the potential of greater ion selectivity. So selective can carriers become that they can be used as the basis for very effective ion selective electrodes [74-76].

Fundamental to the function of a carrier with its polar core of limited size is a re-evaluation of the Born expression and a recognition of its limitations. The plot in Fig. 1A, which uses Eq. 2, indicates that a hydration sphere containing more than 100 water molecules is required to achieve better than 90% of the hydration energy. The landmark experimental studies of Kebarle and colleagues [77,78] on ion-water molecule gas phase equilibria "indicate that the decisive interactions determining heats of hydrations occur during the attachment of the first 8–12 molecules" [78]. These results are also consitent with the theoretical calculations of Clementi and colleagues [79,80] and of the Beveridge group [81,82]. Combining those considerations with the greater heat of solvation achievable from models of the peptide moiety [71,83], it would appear that the essential solvation energy can be achieved by an appropriately designed polar core of limited size and that the electrostatic barrier or repulsive image force due to the lipid need not be a large effect. This will be demonstrated when comparing the rates at which carrier alone and carrier with cation can cross from one side to the other of the lipid membrane.

In what follows, the phenomenology of carrier transport will be briefly reviewed along with the mechanism of the Valinomycin model of carrier transport. The development of the molecular structure of Valinomycin will be considered in some detail, since the key to the dramatic selectivity of Valinomycin is thought to reside in the energetics of the molecular structure. Confidence in an understanding of the molecular structure of the Valinomycin-cation complex becomes tantamount to confidence in the presented basis of ion selectivity.

A Phenomenology of Carrier Transport

The assay systems for carrier transport are more varied than for channel transport. One of the simplest has been called the Pressman cell [84] which is a refinement of the older more cumbersome U-tube [85]. In both cases, a dense organic solvent, such as chloroform or carbon tetrachloride, separates two aqueous phases and transport is measured from one aqueous phase through the organic phase to the second aqueous phase. Somewhat more refined are the ion selective electrodes which use an organic solvent, such as diphenyl ether, to impregnate a disc that separates two aqueous phases. For studies of direct relevance to cell membranes, the planar bilayer system of Fig. 2 is again used. The conductance of planar lipid bilayers, without carriers or channels, is of the order or 10^{-8} mhos/cm^2 [86]. On addition of a carrier, the conductance can increase by many orders of magnitude as ions are transported across the bilayer [9]. Careful characterization of this transport can lead to kinetic mechanism which with structure allow discussion of ion selectivity.

1 Criteria for a Carrier Mechanism

The 1972 review by Lauger [87] of carrier-mediated transport which resulted form work of his Laboratory with his collegues Benz, Ketterer, Neumcke and Stark [88-92] stands as a cogent description of Valinomycin-mediated transport. The following brief statement of phenomenology and kinetic mechanism draws largely from that source. As shown in Fig. 17A for 1 M KCl, the membrane conductance is linear with Valinomycin concentration giving a slope of one. In Fig. 17B for 10^{-7} M Valinomycin in the aqueous phase, conductance is also linear with KCl concentration again with a slope of 1 [88]. The saturation effect above 10^{-1} M KCl is dependent on the Valinomycin concentration. The most obvious interpretation of these results is a carrier mechanism with a 1:1 ratio of carrier to cation. A dramatic verification of the carrier mechanism is due to Krasne, Eisenman and Szabo [93] in studies of membrane conductance as a function of freezing and melting of the lipid bilayers. Using glyceryl dipalmitate:glyceryl distearate (1:1)/n-decane membranes that undergo a melting-freezing transition between 40° and 42 °C, they demonstrated with 10^{-7} M Valinomycin that on melting membrane conductance increased four orders of magnitude from 10^{-7} mhos/cm² to 10^{-3} mhos/cm². A drop was also observed on going from liquid to frozen membrane. Valinomycin mediated conductance required a fluid membrane whereas, in the same report it was shown that Gramicidin conductance continued with a reasonable temperature dependence of conductance. Carrier-mediated conductance requires a fluid lipid layer whereas channel-mediated conductance does not.

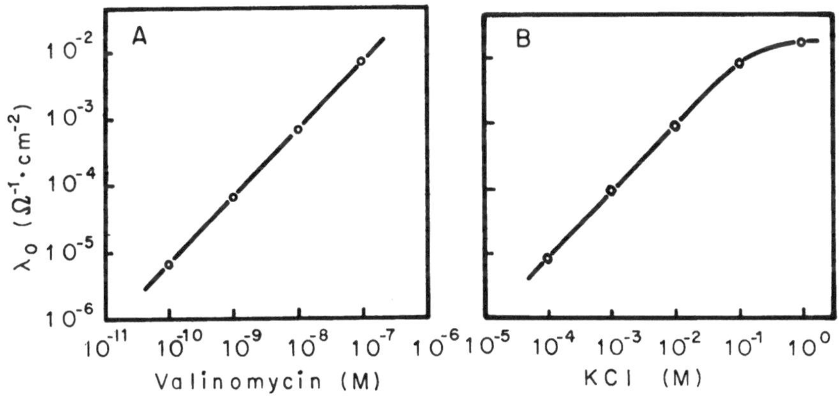

Fig. 17. Membrane current, λ_0, of a dioleoyllecithin membrane, A. as a function of Valinomycin concentration at 1 M KCl and B. as a function of KCl concentration at 10^{-7} M Valinomycin. The linearity and slope of one indicates a carrier mechanism with a 1:1 carrier to cation stoichiometry. Reproduced with permission from Ref. [87]

2 Kinetic Scheme for Carrier Mechanisms

The kinetic scheme applicable to the Valinomycin carrier system is given in Fig. 18 where S is the carrier and MS^+ is the carrier-cation complex. There are five unknown parameters, the four rate constants and N_s, the interfacial concentration of

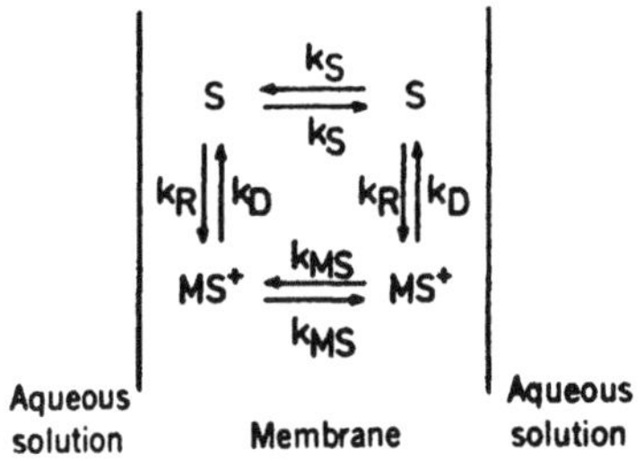

Fig. 18. Kinetic scheme for the Valinomycin carrier mechanism of transport. Reproduced with permission from Ref. [87]

free carrier molecules. Steady state conductance measurements with variables of cation and carrier concentrations and of voltage provide values for three of the five unknowns and amplitude and time course of conductance change resulting from electric field relaxation studies provide for the other two [88,89,91,92]. For the Valinomycin-K^+ system in a phosphatidyl ethanolamine membrane, the rate constants are estimated to be [87]

$$
\begin{aligned}
k_R &\simeq 5\times 10^4/\text{M sec}\\
k_D &\simeq 5\times 10^4/\text{sec}\\
k_S &\simeq 2\times 10^4/\text{sec}\\
k_{MS} &\simeq 2\times 10^4/\text{sec}
\end{aligned}
\tag{18}
$$

And when the aqueous concentration of Valinomycin is 10^{-7} M, N_S is 1.2×10^{12} Valinomycin molecules/cm^2 membrane.

Two interesting points that can be made from these values are the turnover number which is the maximum rate of transport per carrier, and the similarity of magnitude of k_S and k_{MS}. The turnover number, f, is given by the expression

$$f = \left(\frac{1}{k_S} + \frac{1}{k_{MS}} + \frac{2}{k_D}\right)^{-1}$$

and for the Valinomycin-K^+ system is of the order of 10^4 potassium ions per second. This may be compared with 10^7 to 10^8 ions/sec by the channel mechanism. It was noted above that the total hydration energy is achieved by less than twelve water molecules and that the solvation power of carbonyls is greater than that of water. These factors combine to argue that the electrostatic image force for a monovalent cation held within a suitable polar core may not be a significant factor. Thus, it may not be surpising that the lipid barrier translocation rate constants for neutral carrier and ion charged carrier are similar and that k_{MS} can even be as much as 10 times greater than k_S [5]. As will be shown below the structure of the Valinomycin-K^+ complex is such that the repulsive image force would be expected to be considerably less.

B Molecular Structure of the Valinomycin-K^+ Complex

The primary structure of Valinomycin was determined by Shemyakin etal.[94] to be the cyclododecadepsipeptide [(L-Lactyl-L · Valyl-D · Hydroxyisovaleryl-D · Valyl)]$_3$. The secondary structure of the Valinomycin-K^+ complex was derived independently by Ivanov et al.[95] based primarily on infrared studies and by this Laboratory[96] using proton nuclear magnetic resonance. A coarse X-ray structure verified the solution secondary structure and gave the correct direction that the chain wrapped around the ion[85]. The first complete and correct solution conformatin of the Valinomycin-K^+ complex, based on the proton magnetic resonance data and general arguments of the β-turn energetics, appeared shortly thereafter[97,98]. It included an important element in terms of the energetics and stability of the conformation which was the orientation of the isopropyl side chains. The space-filling models as originally published[97,98] are given in Fig. 19. The common reference in the literature is to the "bracelet" conformation[96]. It seems appropriate in this review to stay with the term polar "core" conformation[97,98] as the ion resides in a polar core and the energetics of core size is relevant to ion selectivity.

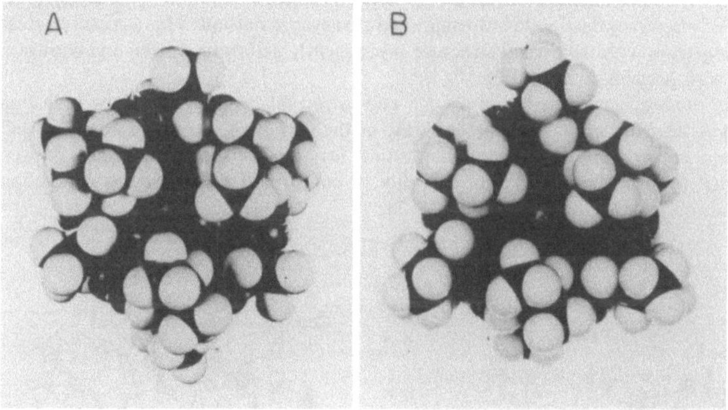

Fig. 19. Space filling models of the molecular structure of the Valinomycin-potassium ion complex as originally determined.
A. Top (D·Val-L·Lac) side. Reproduced with permission from Ref.[97].
B. Bottom (L·Val-D-HyV) side. Reproduced with permission from Ref.[98].
Note that the αCH-βCH torsion angle is trans in all cases (D. W. Urry and N. G. Kumar, Biochemistry, *13*, 1829–1831, (1974). A gauche orientation gives so much steric crowding that the complex will not form[99]

Using a combination of conformation energy calculations and proton nuclear magnetic resonance data, a complete argument was presented for the solution structure of the Valinomycin-K^+ complex which provided a listing of atomic coordinates[99]. That result, plotted in Fig. 20A, is of direct interest to the focus of this review for it was used to calculate conformational energy as a function of the size of the polar core. As will be presented below, the size of the polar core presents a structural argument for the striking selectivity of Valinomycin transport.

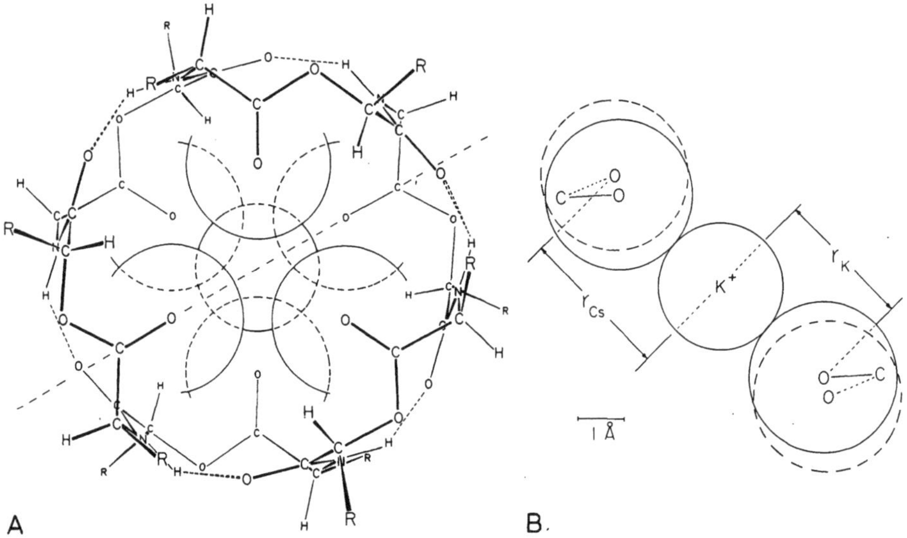

Fig. 20. A. Conformation of the Valinomycin-cation complex derived for solution using a combination of proton magnetic resonance data and conformational energy calculations. This structure agrees within tenths of an Angstrom with the crystal structure subsequently determined (100) and shown in Fig. 21. Reproduced with permission from Ref. [99].
B. Plot of the relative orientations of carbonyl oxygen and cation taken along the plane indicated by the dashed line in A. Depending on the librated state of the carbonyls, the polar core size will vary. The solid oxygen spheres are at an optimal position for the K^+ ion. The dashed oxygen spheres are thc optimal position for a Cs^+ ion. The plot of conformational energy as a function of the oxygen atom-cation distance, r, is given in Fig. 24.

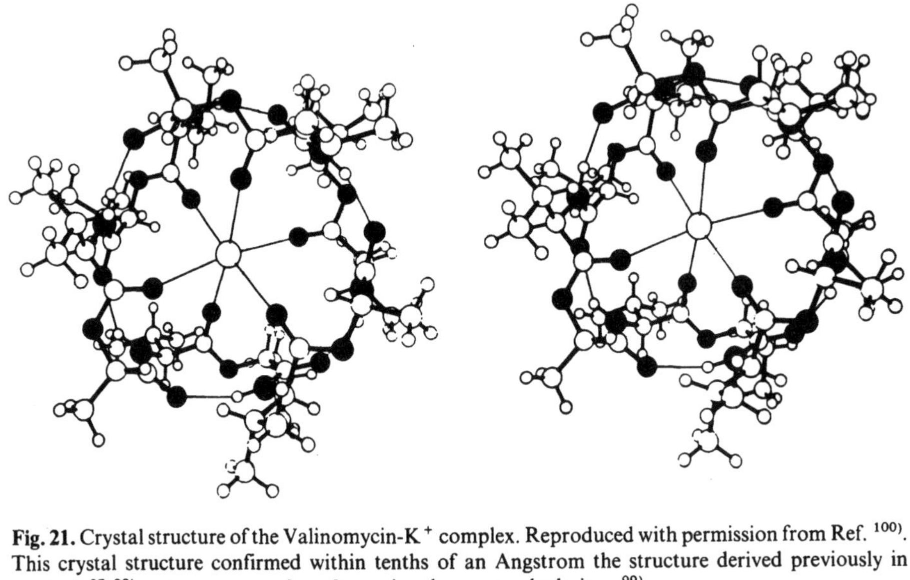

Fig. 21. Crystal structure of the Valinomycin-K^+ complex. Reproduced with permission from Ref. [100]. This crystal structure confirmed within tenths of an Angstrom the structure derived previously in solution [97,98] and by means of conformational energy calculations [99].

It is helful, therefore, to appreciate that the subsequent detailed X-ray structure, shown in Fig. 21 [100], constituted an important verification of the earlier derived solution conformation with no atom out of place by as much as 0.5 Å. Given the twelve residue depsipeptide with 24 backbone torsion angles, six side chain torsion angles and the very large number of possible conformations that were a priori possible, this is a satisfying result, particularly since it was achieved over a decade ago using potential functions and computational methods of an even earlier vintage. The success in deriving the solution conformation of the Valinomycin-K^+ complex provides confidence in the arguments of the associated conformational energetics which are presented here as relevant to ion selectivity.

Another useful element of structure determination is obtained from the carbon-13 nuclear magnetic resonance of K^+ complexation with Enniatin B and with Valinomycin [101, 36]. In Fig. 22A is a plot of the downfield chemical shift of the carbonyl carbon resonances during titration with K^+. As only one resonance is observed, exchange of ion with ionophore is fast, $>10^6$/sec. Of particular interest is the magnitude of the chemical shifts. As the structure of Enniatin B has an alternating L, D sequence, $[\overline{\text{(N-Me-L·Valyl-D·Hydroxyisovaleryl)}}]_3$ [102], it may be considered to be a cyclic analogue of Gramicidin A. As such it provides an estimate of the magnitude of carbonyl carbon resonance chemical shift that could be expected for Gramicidin A (see Fig. 6 and 13) and allows for the argument for lateral coordination of a bare ion in the channel. This is because both crystal [103] and solution [104] structures show direct coordination of carbonyls by K^+, as shown in Fig. 23. In Fig. 22B are the carbonyl carbon spectra for complexation of Valinomycin by K^+. Here the exchange is slow such that throughout the titration the two states of the carbonyls, free and complexed, are resolved. The large magnitude of downfield shift is evidence of an even more favorable ion binding site in the Valinomycin polar core than in Enniatin B or in Gramicidin A. Furthermore, with the resonances assigned the shifts themselves demonstrate which carbonyls are coordinating the ion [36, 101]. This is yet another piece of evidence confirming the Val-K^+ conformation just as it was added evidence for the Gramicidin A channel as discussed above.

C Ion Selectivity and the Energetics of Forming Polar Core Structures

1 Anion vs Cation Selectivity

The transmembrane potential derived from a concentration gradient is calculable by means of the Nernst equation. If K^+ were the only permeable ion then the membrane potential would be given by Eq. 1. With an ion activity (concentration) gradient for K^+ of 10:1 from one side to the other of the membrane at 20 °C, the membrane potential that develops on addition of Valinomycin approaches a limiting value of 58 mV [87]. This is what is calculated from Eq. 1 and indicates that cation over anion selectivity is essentially total. As the conformation of Valinomycin in nonpolar solvents in the absence of cation is similar to that of the cation complex [105], it is quite understandable that anions have no location for interaction. One could with the Valinomycin structure construct a conformation in which a polar core were formed with six peptide N—H moieties directed inward in place of the C—O moieties but this would leave no hydrogen bonding capabilities. The ring of hydrogen bonded peptide moieties would be replaced by a ring of esters with repulsive C—O···O (alkyl) interactions. Accordingly, essentially complete selectivity of inorganic cations over inorganic anions is explicable in terms of the structure of Valinomycin.

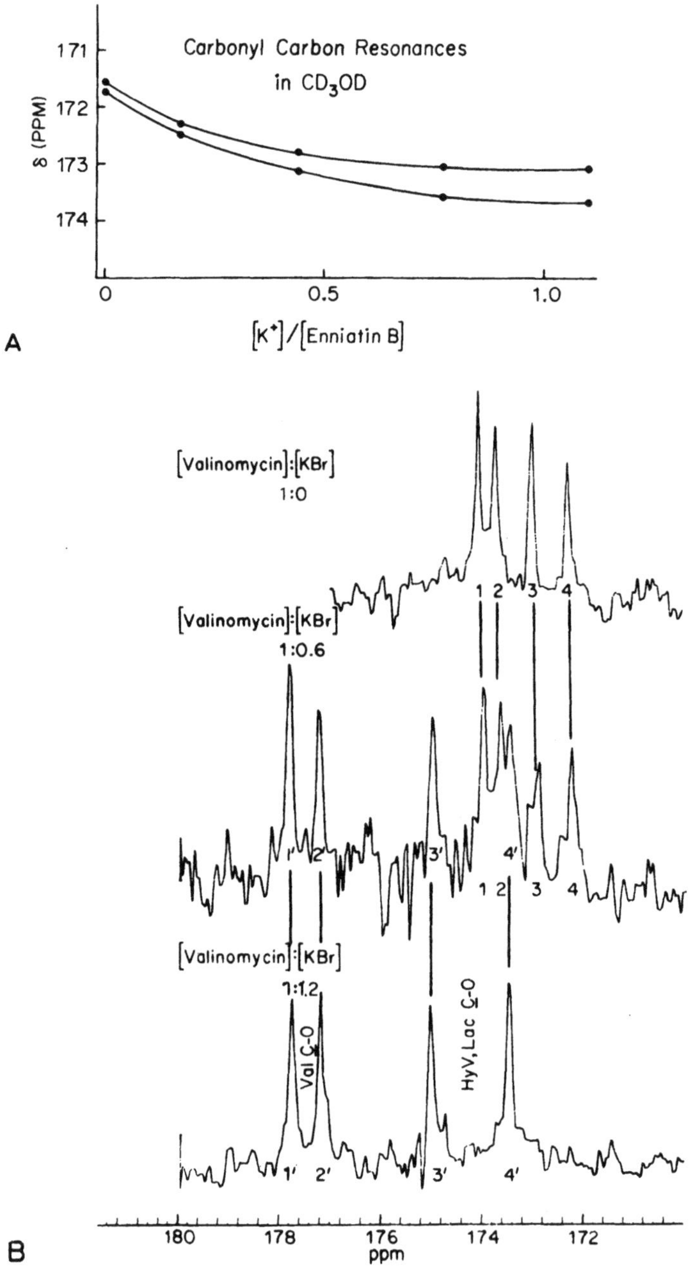

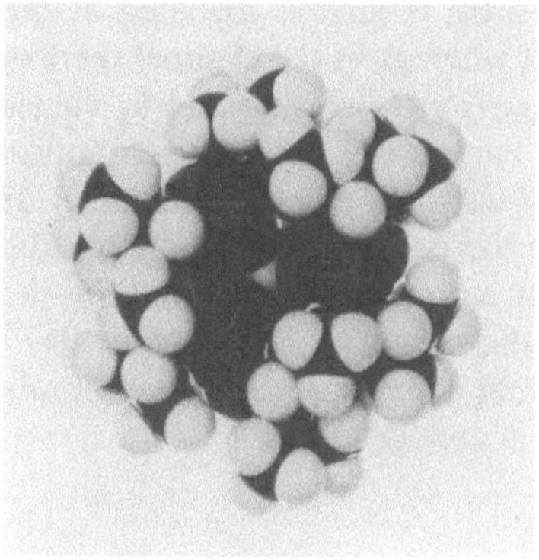

Fig. 23. Space filling model of the Enniatin B—K^+ complex after the crystal structure [103]. Since the carbonyl moieties coordinating the cation are similar for Enniatin B and Valinomycin, the difference in selectivities must arise due to the energetics of the conformations required to achieve coordination of the cation

2 Monovalent vs Multivalent Cation Selectivity

This reviewer has not found well-described experimental data which places limits on the relative Valinomycin transport across planar lipid bilayers of inorganic monovalent and multivalent cations. It is, of course, anticipated that the z^2 effect would dramatically reduce multivalent conductance and perhaps make it nondetectable in a manner analogous to that discussed above for the Gramicidin channel. Comparing the Gramicidin channel and the Valinomycin polar core, however, it would appear that the Valinomycin core would more effectively shield charge from the lipid. The Gramicidin channel has a relatively more limited coordinating capability and utilizes the one layer of peptide backbone separating it from the lipid. Valinomycin on the other hand has a polar core formed by six well-coordinated carbonyls (as is also apparent from the magnitude of the ion induced carbonyl carbon chemical shifts shown in Fig. 22B) and in addition there is a lateral ring of hydrogen bonded peptide moieties forming a layer of polypeptide backbone. By this reasoning, a relative Valinomycin vs Gramicidin channel divalent cation

◄

Fig. 22. Carbon-13 nuclear magnetic resonance of carbonyl carbon chemical shifts on complexation of K^+.
A. KBr titration of Enniatin B in methanol. Since exchange of cation between solution and carrier is relatively rapid only one signal is seen per chemically distinct carbonyl. The titration shows the magnitude of the chemical shift observed. Since Enniatin B may be considered a cyclic analogue of Gramicidin A, these chemical shifts indicate the magnitudes of chemical shifts that can be expected in the Gramicidin A channel (see Fig. 6 and 13) for direct interaction of carbonyl with cation.
B. KBr interaction with Valinomycin carbonyls in trifluoroethanol. Here the exchange is slow on the nuclear magnetic resonance time scale such that at 60% of the way through the titration, for example, complexed and free carbonyls are both seen. It is the Val carbonyls that are interacting with the cation. The shifts of the HyV and Lac carbonyls are due to the intramolecular hydrogen bonding shown in Fig. 20 and 21. The large chemical shifts of the Val carbonyl carbon resonances reflect the favorable cation interaction with a bond angle of about 120° (see Fig. 20B). An angle closer to 90° obtains for the Enniatin B—K^+ complex (see Fig. 23). Reproduced with permission from Ref. [101]

conductance would favor Valinomycin. However, there is limited shielding at the top and bottom of the Valinomycin-cation complex whereas in the channel there is a string of water molecules. Similarly from a comparison of the k_{MS} and k_S values for Valinomycin monovalent cation transport [87,106] with the estimated contribution of the repulsive image force to the central barrier of the Gramicidin channel as discussed above, it would appear that the effect of the proximity of lipid could be less of a limiting factor for Valinomycin than for Gramicidin transport. In the Valinomycin based ion selective electrode systems, the selectivity for K^+ over Ba^{+2} with their almost identical ionic radii (1.33 Å and 1.34 Å, respectively) is essentially the same as the selectivity of K^+ over Na^+ which is greater than 10^4 [107].

3 Selectivity Among Monovalent Cations

The conductances of alkali metal ions due to Valinomycin (10^{-7} M) using 10^{-2} M salts and dioleoyl phosphatidyl choline membranes have been reported by Benz [106]. From that data the following conductance ratios are obtained: Li^+ (0.39) < Na^+ (1.0) < K^+ (1200) < Rb^+ (3900) > Cs^+ (760). These may be compared with those of Enniatin B (5×10^{-6}) from the same source and under the same conditions (106): Li^+ (0.67) < Na^+ (1.0) < K^+ (15) > Rb^+ (12) > Cs^+ (5.9). This comparison is particularly intersting as both carriers coordinate the cations by the same number of similar carbonyl moieties: six ester carbonyls for Valinomycin and 3 ester and 3 N-methyl peptide carbonyls for Enniatin B. Accordingly, differences arising from the carbonyl selectivity series cannot be considered as a basis for the large differences in selectivities nor can the number of aliphatic carbons per residue. The greater number of aliphatic carbons per residue of Enniatin B neither increases conductance nor selectivity [106]. Thus, it appears necessary to consider the conformational energetics required to achieve an optimal polar core for carbonyl coordination of the cation.

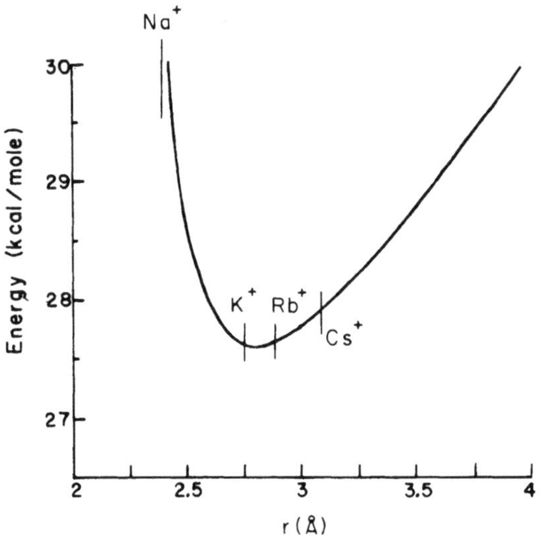

Fig. 24. Calculation of the conformational energy of Valinomycin as a function of the size of the polar core which contains the ion. This uses the structure of Fig. 20. The verticle lines are the optimal core sizes for the indicated ions. Based on the conformational energy component, selectivity for K^+ and Rb^+ would be similar and Cs^+ less favored. Na^+ is off the curve suggesting that this conformation cannot form a polar core small enough to complex Na^+ by means of this conformation. Adapted with permission from Ref. [99].
It seems quite apparent that the dramatic selectivity of K^+ over Na^+ is the result of the conformational energy required to achieve adequate coordination of the ion

Starting from the conformation shown in Fig. 20, the carbonyl moieties can be librated inward and outward changing the size of the polar core as shown in Fig. 20B. As was done over a decade ago [99], the conformational energy can be calculated as a function of libration which changes the distance from the center of the structure where the ion is located to the carbonyl oxygen center. A plot of the conformational energy versus the carbonyl oxygen-metal atom distance is shown in Fig. 24 [99]. On interaction of the metal atom with the carbonyl oxygen there is a bonding that leads to a distance between ion center and oxygen center that is slightly less than the sum of the ionic radius and the oxygen van der Waals radius. In the Valinomycin-K^+ crystal the average shortening is 0.075 Å [100]. Using this figure, O—Me distances are indicated in Fig. 24 as lines crossing the energy curve. Interestingly K^+ and Rb^+ straddle the conformational energy minimum and Cs^+ is slightly higher. Most significantly, however, is that Na^+ is off the curve where the energy rise on this scale appears almost verticle. This indicates that the polar core cannot become small enough to coordinate Na^+ without a substantial increase in conformational energy. Indeed in organic solvents Na^+ forms 1:2 cation to Valinomycin sandwich complexes [108] rather than the 1:1 complexes involved in K^+ transport across lipid bilayer membranes. It would appear that the dramatic selectivity between Na^+ and K^+, exhibited by Valinomycin, is due in large part to the conformational energy required to form the conducting polar core conformation.

The experimental values that would be of particular interest for comparison with the conformational energy calculations would be $k_R(Na^+)$ and $k_R(K^+)$ or the ratio k_R/k_D for each ion. A complete kinetic analysis of Na^+ transport by Valinomycin, however, is not available, presumably due to sensitivity problems arising from the low inherent conductance for sodium. A complete kinetic analysis of alkali metal ion complexation by Valinomycin in methanol is available [109,110] and the relative binding constants closely follow the relative conductance with relative values of K^+ (1700) $<$ Rb^+ (3600) $>$ Cs^+ (560). The discrimination against sodium is greater with Na^+ being 0.1 on this scale. In this study by Grell, Funck and Eggers [109,110], an intermediate complex was observed. A comparison of the relative affinities for the intermediate complex contains the relative selectivity with respect to methanol. While the order is the same as for transport across lipid bilayer membranes, the selectivity is low for the intermediate complex favoring K^+ over Na^+ by only a factor of 2.3. On the other hand, opening from the completed solution conformation indicates that complexation favors K^+ over Na^+ by a factor of 1800. These results constitute experimental verification that the conformational energetics involved in forming a polar core of a size suitable for the ion is the dominant factor responsible for the monovalent cation selectivity displayed by Valinomycin transport.

IV Acknowledgements

This work was supported in part by NIH Grant GM-26898. The author particularly wishes to thank present and past members of the Laboratory of Molecular Biophysics whose cumulative efforts of more than a decade have made this review possible.

V References

1. Cotton, F. A., Wilkinson, G.: Advanced Inorganic Chemistry, A Comprehensive Text, p. 321, Interscience Publishers, 1962
2. Born, M.: Z. Physik. *1*, 45 (1920)
3. Fettiplace, R., Andrews, D. M., Haydon, D. A.: J. Membrane Biol. *5*, 277 (1971)
4. Bamberg, E., Lauger, P.: Biochim. Biophys. Acta *367*, 127 (1974)
5. Benz, R., Lauger, P.: J. Membr. Biol. *27*, 171 (1976)
6. Urry, D. W.: Ann. NY Acad. Sci. *307*, 3 (1978)
7. Hladky, S. B., Haydon, D. A.: Nature *225*, 451 (1970)
8. Hladky, S. B., Haydon, D. A.: Biochim. Biophys. Acta *274*, 294 (1972)
9. Mueller, P., Rudin, D. O.: Biochem. Biophys. Res. Commun. *26*, 398 (1967)
10. Bradley, R. J., Prasad, K. U., Urry, D. W.: Biochim. Biophys. Acta *649*, 281 (1981)
11. Szabo, G., Urry, D. W.: Science *203*, 55 (1979)
12. Urry, D. W., Alonso-Romanowski, S., Bradley, R. J.: in preparation
13. Prasad, K. U., Trapane, T. L., Busath, D., Szabo, G., Urry, D. W.: Int. J. Pept. Protein Res. *19(2)*, 162 (1982)
14. Kolb, H.-A., Lauger, P., Bamberg, E.: J. Membrane Biol. *20*, 133 (1975)
15. Zingsheim, H. P., Neher, E.: Biophys. Chem. *2*, 197 (1974)
16. DeFelice, L. J.: Int. Rev. Neurobiol. *20*, 169 (1977)
17. Sarges, R., Witkop, B.: Biochemistry *4*, 2491 (1965)
18. Fontana, A., Gross, E.: Peptides, Proc. of the 12th Eur. Peptide Symp., p. 229, 1972
19. Noda, K., Gross, E.: Chemistry and Biology of Peptides, (ed. Meienhofer, J.), p. 241, Ann Arbor Science Publishers, Inc., Ann Arbor, Michigan 1972
20. Hunter, F. E., Jr., Schwartz, L. S.: Antibiotics, Vol. I: Mechanism of Action (ed. Gottlieb, D., Shaw, P. D.), p. 636, Springer-Verlag New York Inc., 1967
21. Urry, D. W.: Proc. Natl. Acad. Sci. USA *68*, 672 (1971)
22. Urry, D. W. et al.: ibid. *68*, 1907 (1971)
23. Urry, D. W.: Coformation of Biological Molecules and Polymers — The Jerusalem Symposia on Quantum Chemistry and Biochemistry, V, (eds. Bergman, E. D., Pullman, B.), p. 723, Jerusalem, Israel Academy of Sciences 1973
24. Ramachandran, G. N., Chandrasekharan, R.: Progress in Peptide Research — Volume II (ed. Lande, S.), p. 195, Gordon and Breach, Science Publishers, Inc., New York 1972
25. Ramachandran, G. N., Chandrasekaran, R.: Indian J. Biochem. Biophys. *9*, 1 (1972)
26. Urry, D. W. et al.: Ann. NY Acad. Sci. *264*, 203 (1975)
27. Veatch, W. R., Blout, E. R.: Biochemistry *13*, 5257 (1974)
28. Weinstein, S. et al.: Proc. Natl. Acad. Sci. USA *76*, 2402 (1979)
29. Bamberg, E., Apell, H.-J., Alpes, H.: A. J. *74*, 2402 (1977)
30. Sychev, S. V., Ivanov, V. T.: Membranes and Transport Vol. 2, (ed. Martonosi, A. N.), p. 301, Plenum Press, New York 1982
31. Ovchinnikov, Y. A., Ivanov, V. T.: Conformation in Biology (eds. Srinivasan, R., Sarma, R. H.), p. 155, Adenine Press, Guilderland, New York 1982
32. Urry, D. W.: The Enzymes of Biological Membranes (ed. Martonosi, A. N.), Plenum Press, New York, in press
33. Urry, D. W., Spisni, A., Khaled, M. A.: Biochem. Biophys. Res. Commun. *88(3)*, 940 (1979)
34. Urry, D. W. et al.: Int. J. Pept. Protein Res., *21*, 16 (1983)
35. Ohnishi, M. et al.: Biochem. Biophys. Res. Commun. *46*, 313 (1972)
36. Urry, D. W.: Nuclear magnetic resonance and the conformation of membrane-active peptides. In: Enzymes of Biological Membranes, Vol. 1, (ed. Martonosi, A.), p. 31, Plenum Publishing Corp., New York 1976
37. Urry, D. W., Prased, K. U., Trapane, T. L.: Proc. Natl. Acad. Sci. USA *79*, 390 (1982)
38. Urry, D. W., Walker, J. T., Trapane, T. L.: J. Membr. Biol. *69*, 225 (1982)
39. Urry, D. W., Trapane, T. L., Prasad, K. U.: Science, in press
40. Busath, D. Szabo, G.: Nature *294*, 371 (1981)
41. Koeppe, R. E., II et al.: Nature (London) *279*, 723 (1979)
42. Zwolinski, B. I., Eyring, H., Reese, C. E.: J. Phys. Chem. *53*, 1426 (1949)

43. Parlin, B., Eyring, H.: Ion Transport Across Membranes (ed. Clarke, H. T.), p. 103, Academic, New York 1954
44. Eyring, H., Urry, D. W.: Ber. Bunsenges. Phys. Chem. *67*, 731 (1963)
45. Urry, D. W.: On the Molecular Structure and Ion Transport Mechanism of the Gramicidin Transmembrane Channel. In: Membranes and Transport, Vol. 2, (ed. Martonosi, A.), p. 285, Plenum Publishing Corporation, New York 1982
46. Urry, D. W., et al.: Proc. Natl. Acad. Sci. USA *77*, 2028 (1980)
47. Urry, D. W., et al.: J. Membr. Biol. *55*, 29 (1980)
48. James, T. L., Noggle, J. H.: Proc. Natl. Acad. Sci. USA *62*, 644 (1969)
49. Urry, D. W., et al.: J. Phys. Chem. in press
50. Urry, D. W., et al.: in preparation
51. Venkatachalam, C. M., Urry, D. W.: J. Magn. Resonance *41*, 313 (1980)
52. Henze, R., et al.: J. Membr. Biol. *64* (3), 233 (1982)
53. Venkatachalam, C. M., Urry, D. W.: J. Comput. Chem., in press
54. Momany, F. A., et al.: J. Phys. Chem. *78*, 1595 (1974)
55. Momany, F. A., et al.: ibid. *79*, 2361 (1975)
56. Venkatachalam, C. M., Urry, D. W.: J. Comput. Chem., in press
57. Myers, V. B., Haydon, D. A.: Biochim. Biophys. Acta *274*, 313 (1972)
58. Bamberg, E., Kolb, H.-A., Lauger, P.: in The Structural Basis of Membrane Function (ed. Hatefi, Y.), p. 143, Academic Press, New York 133–167
59. Eisenman, G., Sandblom, J., Neher, E.: Biophys. J. *22* (2), 307 (1978)
60. Eisenman, G., Sandblom, J., Neher, E.: Metal-Ligand Interactions in Organic Chemistry and Biochemistry, Part 2, (eds. Pullman, B., Goldblum, N.), p. 1, D. Reidel, Dordrecht-Holland
61. Bamberg, E., Lauger, P.: J. Membrane Biol. *35*, 351 (1977)
62. Urry, D. W., et al.: J. Biol. Chem. *257*, 6659 (1982)
63. Urry, D. W., Trapane, T. L., Prasad, K. U.: Int. J. Quantum Chem.: Quantum Biology Symp. No. *9*, 31 (1982)
64. Hinton, J. F., Turner, G. L., Millett, F. S.: J. Magnetic Res. *45*, 42 (1981)
65. Rosseinsky, D. R.: Chem. Rev. *65*, 467 (1965)
66. Stokes, R. H.: J. Am. Chem. Soc. *86*, 979 (1964)
67. Noyes, R. M.: ibid. *84*, 513 (1962)
68. Urry, D. W.: Frontiers of Biological Energetics, Vol. 2 (eds. Dutton, P. L., Leigh, J., Scarpa, A.), p. 1227, Academic Press, New York 1978
69. Potter, J. D., Gergely, J.: J. Biol. Chem. *250*, 4628 (1975)
70. Cox, J. A., Wnuk, W., Stein, E. A.: Calcium Binding Proteins and Calcium Function (eds. Wasserman, R. H., Corradion, R. A., Carafoli, E., Kretsinger, R. H., MacLennan, D. H., Siegel, F. L.), p. 266, Elsevier/North Holland, New York 1977
71. Krasne, S., Eisenman, G.: Membranes, Vol. 2, Lipid Bilayer And Antibiotics (ed. Eisenman, G.), p. 273, Marcel Dekker, Inc. New York 1973
72. Lauger, P.: J. Membrane Biol. *57*, 163 (1980)
73. Urry, D. W., Alonso-Romanowski, S., Bradley, R. J.: in preparation
74. Stefanac, Z., Simon, W.: Chimia (Switzerland) *20*, 436 (1966)
75. Pioda, L. A. R., Stankova, V., Simon, W.: Anal. Lett. *7*, 665 (1969)
76. Simon, W., Morf, W.: Membranes — A Series of Advances, Vol. 2, p. 29, (ed. Eisenman, G., Dekker, New York 1973
77. Dzidic, I., Kebarle, P.: J. Phys. Chem. *74*, 1466 (1970)
78. Arshadi, M., Yamdagni, R., Kebarle, P.: ibid. *74*, 1475 (1970)
79. Kistenmacher, H., Popkie, H., Clementi, E.: J. Chem. Phys. *61*, 799 (1974)
80. Kistenmacher, H., Popkie, H., Clementi, E.: ibid. *58*, 1689 (1973)
81. Mezei, M., Beveridge, D. L.: ibid. *74* (1), 622 (1981)
82. Mezei, M., Beveridge, D. L.: ibid. *74* (12), 6902 (1981)
83. Balasubramanian, D., Misra, B. C.: Biopolymers *14*, 1019 (1975)
84. Pressman, B. C.: Fed. Proc. *32*, 1698 (1973)
85. Pinkerton, M., Steinruf, L. K., Dawkins, K.: Biochem. Biophys. Res. Commun. *35*, 512 (1969)
86. Mueller, P., et al.: J. Phys. Chem. *67*, 534 (1963)

87. Lauger, P.: Science *178*, 24 (1972)
88. Stark, G., Benz, R.: J. Membr. Biol. *5*, 133 (1971)
89. Ketterer, B., Neumcke, B., Lauger, P.: ibid. *5*, 225 (1971)
90. Neumcke, B., Lauger, P.: Biophys. J. *9*, 1160 (1969)
91. Lauger, P., Stark, G.: Biochim. Biophys. Acta *211*, 458 (1970)
92. Stark, G., et al.: Biophys. J. *11*, 981 (1971)
93. Krasne, S., Eisenman, G., Szabo, G.: Science *174*, 412 (1971)
94. Shemyakin, M. M., et al.: Experientia *21*, 548 (1965)
95. Urry, D. W.: J. Phys. Chem. *72*, 3035 (1968)
96. Ohnishi, M., Urry, D. W.: Biochem. Biophys. Res. Commun. *36*, 194 (1969)
97. Urry, D. W., Ohnishi, M.: Spectroscopic Approaches to Biomolecular Conformation, p. 263, (ed. Urry, D. W.), American Medical Association Press, Chicago, Illinois 1970
98. Ohnishi, M., Urry, D. W.: Science *168*, 1091 (1970)
99. Mayers, D. F., Urry, D. W.: J. Am. Chem. Soc. *94*, 77 (1972)
100. Neupert-Laves, K., Dobler, M.: Helv. Chim. Acta *58*, 432 (1975)
101. Urry, D. W.: Enzymes of Biological Membranes, Vol. 1, (ed. Martonosi, A.), p. 31, Plenum Publishing Corp., New York, New York 1976
102. Plattner, PL. A., et al.: Helv. Chim. Acta *46*, 927 (1963)
103. Dobler, M., Dunitz, J. D., Krajewski, J.: J. Mol. Biol. *42*, 603 (1969)
104. Shemyakin, M. M., et al.: J. Membr. Biol. *1*, 402 (1969)
105. Bystrov, V. F., ett al.: Eur. J. Biochem. *78*, 63 (1977)
106. Benz, R.: J. Membr. Biol. *43*, 367 (1978)
107. Fiedler, U., Ruzicka, J.: Anal. Chim. Acta *67*, 179 (1973)
108. Ivanov, V. T.: Ann. N.Y. Acad. Sci. *264*, 221 (1975)
109. Grell, E., Funck, Th.: J. Supramol. Structure *1*, 307 (1973)
110. Funck, Von Th., Eggers, F., Grell, E.: Chimia *26*, 637 (1972)

Ion-Selective Bulk Membranes as Models for Biomembranes

Active Ion Transport as a Consequence of Stationary State Situations at Asymmetric Biomembranes

Karl Cammann

Department of Analytical Chemistry, University of Ulm, 7900 Ulm, FRG

Table of Contents

Ion-selective bulk membranes are the electro-active component of ion-selective electrodes, which sense the activity of certain ions by developing an ion-selective potential difference according to the Nernst equation at their phase boundary with the solution to be measured. The main differences to biological membranes are their thickness and their symmetrical structure. Nevertheless they are used as models for biomembranes.

The general theoretical treatment of ion-selective membranes assumes a homogeneous membrane phase and thermodynamic equilibrium at the phase boundaries. Obvious deviations from a Nernstian behavior are explained by an additional diffusion potential inside the membrane. However, allowing stationary state conditions in which the thermodynamic equilibrium is not established some hitherto difficult to explain facts (e.g., super-Nernstian slope, dependence of the selectivity of ion-transport upon the availability of co-ions, etc.) can be understood more easily.

A consequence of this theoretical approach which includes kinetic parameters is the establishment and coupling of certain ion fluxes across the phase boundary (equality of the sum of cathodic and anodic partial currents leading to a mixed potential). If a similar approach can be applied to asymmetric biological membranes with different thermodynamic equilibrium situations at both surfaces, the active ion transport could also be understood.

A Introduction

I General

Ion-selective bulk membranes are the electro-active component of ion-selective electrodes. They differ from biological membranes in many aspects, the most marked being their thickness which is normally more then 10^5 times greater, therefore electroneutrality exists in the interior. A further difference is given by the fact that ion-selective membranes are homogeneous and symmetric with respect to their functioning. However, because of certain similarities with biomembranes (e.g., ion-selectivity order, etc.) the more easily to handle ion-selective membranes were studied extensively also by many physiologists and biochemists as model membranes. For this reason research in the field of bio-membranes, and developments in the field of ion-selective electrodes have been of mutual benefit.

Ion-selective electrodes are electrochemical half-cells in which a potential difference arises across the phase boundary: electro-active membrane/measuring solution. This potential difference ideally depends only on the activity of one kind of ions and follows the Nernst equation. Most ion-selective electrodes are in principle built up like the well-known pH glass-membrane electrode, i.e., with a suitable internal solution and an electrode of second kind (e.g., Ag/AgCl) to conduct the variable potential difference at the outer phase boundary to the high impedance voltmeter. Besides glass membranes, other membranes with an ion-selective behavior have been developed. Liquid electro-active phases are normally supported by more or less electro-inactive membrane filters, porous ceramic materials or embedded together with an appropriate plasticizer in a PVC matrix forming rigid membranes. But also pressed pellets and/or single crystals are used. Table A1 in the appendix gives a survey of commercially available ion-selective electrodes.

If the measuring circuit is completed with a proper reference electrode, an ion-selective EMF response which can be described within certain limits by an extended Nernst equation is obtained [1,2]:

$$E = E_0 + \frac{R \cdot T}{z_M \cdot F} \ln \left[a_M + \sum_{a_I} K^{pot}_{M,I} (a_I)^{z_M/z_I} \right] \tag{1}$$

with: E_0 = EMF of the cell with a standard solution $a_M = 1$ M, $a_I = 0$
- R = gas constant
- T = temperature in Kelvin
- z_M = charge of the ion to be measured
- z_I = charge of the interfering ion
- F = Faraday constant
- a_M = activity of the ions measured
- a_I = activity of the interfering ions
- $K^{pot}_{M,I}$ = selectivity coefficient between the measured ion M and an interfering ion I.

The so-called potentiometric selectivity coefficient $K^{pot}_{M,I}$ reflects the non-ideal behavior of ion-selective membranes and determines the specificity of this electro-analytical method. For the development of further ion-selective membranes it is therefore of vital importance to know which factors influence this coefficient. The answer should be obtainable through the right theoretical approach.

Ion-selective electrodes are increasingly used in analytical chemistry. The possibility of differentiating ions of the same element by their valence status (e.g., NH_4^+ *vs.* NO_3^-, S^{2-}, *vs.* SO_4^{2-}, Cl^- *vs.* ClO_4^-, etc.) enables new analytical approaches. In many instances time-consuming sample preparations such as filtrations, distillations, etc. can be avoided. Figure 1 — without being exhaustive — shows a survey of some species which can be determined with the help of ion-selective electrodes with the accuracies given in Table 1. The instrumentation is equivalent to the electrometric pH instrumentation and therefore well established and simple. Often, a few microliters of samples solution (e.g., a wetted spot of filter paper) is sufficient for an analysis. Flow-through cells with a minimal dead volume are equally possible. Ion-selective micro-electrodes for Na^+, K^+, Ca^{2+}, Li^+, Cl^-, etc. are unchallenged for physiological measurements in biological cells. Their broad analytical range — from saturated solutions down to concentrations of less than 10^{-5} M (in ion-buffered solutions occasionally to $<10^{-17}$ M) — together with their rapid response of a few seconds made them a very valuable tool in the field of analytical chemistry in the laboratory as well as in industrial process control. Industrial monitors for more than a dozen species (e.g., Cl_2, SO_2, NH_3, NO_x, H_2S, CN^-, Cl^-, Na^+, K^+, Ca^{2+}, water hardness,

Table 1. Total error

analytical method	error with monovalent ions	error with divalent ions	remarks
direct potentiometry (analogous to pH-measurements)	1–5% mV-error × 4 = % error	2–10% mV-error × 8 = % error	error is smaller if concentration of standard is close to the sample concentration; activity measurements in general limited by the accuracy of the activity-standard
standard addition or subtraction with known electrode slope	1–4%	2–8%	determination of total concentration, if ionic strength, degree of complexation, and slope remain constant
standard addition or subtraction with unknown electrode slope (double addition or subtraction or addition and subsequent dilution)	ca. 8%	ca. 16%	in general only recommended when working near the detection limit
Gran-plot (multiple standard addition with $10^{\Delta E/S}$ vs. concentration)	0.1–3%	0.1–3%	less accurate if interfering ions are present; best results if many points between 30 and 80% of a titration curve are evaluated and discordant points are eliminated
titration with inflection point evaluation	ca. 1%	ca. 1%	accurate endpoint determination only with symmetrical titration curves
tangent crossing method	<1%	<1%	also higher valent ions can be indirectly determined with high accuracies

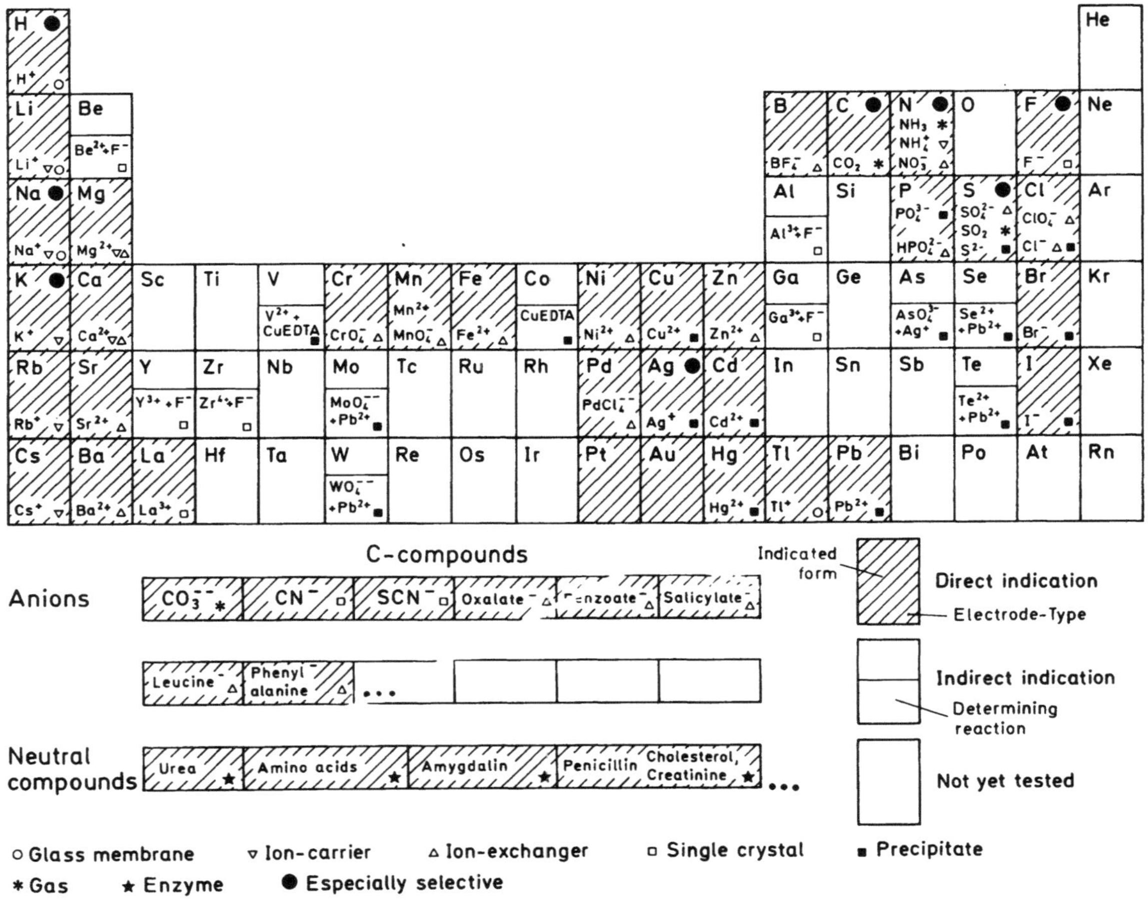

Fig. 1. Survey of the ions and selected neutral compounds which can presently be determined with the help of ion-selective electrodes

etc.) are already commercially available. During recent years more than five new clinical electrolyte analyzers for Na^+, K^+, Ca^{2+}, and Cl^- based on ion-selective electrodes were introduced. Even companies famous for their traditional flame photometric analyzers entered the ion-selective electrode market.

The number of applications of ion-selective electrodes is steadily increasing. This is partially due to the fact that special electrodes selective for certain ions have been constructed by the user, and partly because substances are determined indirectly

Table 2. Short survey of the behavior of ion-selective electrodes and selected fields of application

Advantage of ion-selective electrodes
- Simple instrumentation (~$ 500)
- Information over the state of oxidation and degree of binding
- Determination of the activity of the free ion (important in physiology)
- Measurement without sample consumption
- Measuring range between $\sim 10^{-17}$ M and saturated solutions with constant accuracy
- Delivery of a continuous electric signal with response times in the seconds range
- Easy miniaturization

Drawbacks of ion-selective electrodes
- Selectivity not always sufficient
- Direct read-out is less accurate with higher valent ions
- EMF-Drift may require frequent standardization

Fields of Application for ion-selective electrodes
- Routine-analytical work in the laboratory (direct or as end-point indicators; application frequency in industry ~30%)
- Clinical analyzers for Na^+, K^+, Ca^{2+}, pH, pCO_2, etc.
- Process analyzers
- Enzyme electrodes for the determination of biochemical important substrates
- Microelectrodes used in physiology for Na^+, K^+, Ca^{2+}, Li^+, Cl^-, etc.
- Physico-chemical research (single ion activity coefficient)
- Potentiometric ionophore-modulation immunoassay

via a stoichiometrically known chemical reaction where an ion, for which an ion-selective electrode exists, is released or removed. Typical examples are the enzyme sensors (Table A2 in the appendix) in which a substrate is transformed into measurable species with the aid of one or more enzymes. For details the reader is referred to special monographs and reviews [3-5]. A novel aplication is the potentiometric immunoassay for the determination of antibodies [5a]. This new technique shows the specificity and sensitivity of the radioimmunoassay but does not require a radioactive labeling.

This short review cannot be comprehensive as of the exponential increase of the literature dealing with new procedures and applications. Table 2 summarizes selected data and lists fields of application. For a more inclusive view on the subject the interested reader is referred to existing monographs [6-16] or to the critical biennial review in Analytical Chemistry. As the readers of this general volume on membranes most likely are acquainted with electro-analytical sensors, this article will be limited to the introduction of a theoretical approach which might be helpful also to researchers in the bio-membrane field.

II Historical Development

The history of ion-selective membranes as the active part of ion-selective electrodes began with the discovery by Cremer in 1906 — reported in a physiological journal [17] — that membranes made from certain sorts of soft glass showed a transmembrane potential if solutions of different pH were separated by them. Haber and Klemensiewicz [18] studied this effect in more detail and published their results in 1909 demonstrating a near Nernstian H^+-response of certain sorts of Tnuringian glasses. But already in 1908 Haber showed that ionically conducting solid electrolytes develop a transmembrane potential in relation to the concentration of the ions also constituting the solid membrane phase. Fused discs, pressed pellets, and also salts embedded in paraffin were used as membranes. With respect to the instrumentation, some of these high-impedance measurements were extremely difficult to perform without distortion of the electrochemical equilibrium at the corresponding phase boundary. In 1908 Haber and Beutner [19] also published results with liquid water-immiscible phases, e.g., nitrobenzene, whose behavior resembles that of salt phases. Acidic liquids responded to cations. and basic liquids to anions.

Such phase boundary potentials were already described in 1902 by Nernst and Riesenfeld [20] who investigated "oil cells" saturated with a common compound. A classical review of this research dealing with membrane potentials was published in 1922 by Michaelis [21].

In 1923 it was shown [22,23] that pH glass membranes also responded to other monovalent cations like silver, sodium, and potassium, especially at high pH values (alkali error). MacInnes and Dole [24] in 1930 optimized the glass composition which afforded a considerably error-free pH range and minimum resistance (high impedance electrometers were extremely delicate at that time). The so-called MacInnes glass made by Corning Glass Works with the designation 015 consists of 72% SiO_2, 22% Na_2O, and 6% CaO. In 1934, Lengyel and Blum [25] published their results on the interrelation of glass composition and the so-called alkali error and showed that this error drastically increased when certain amounts of trivalent metal oxides (e.g., Al_2O_3 or B_2O_3) were added to the glasses. At high pH values those glasses responded in a Nernstian manner to monovalent cations like silver, sodium, and potassium. This was the beginning of the development of membranes sensitive also to other ions than H^+. But before this phenomenon was investigated further, Sokolof and Passynsk [26] made some lithia-based pH-glasses which showed Nernstian responses up to pH 12.5 and a minimum alkali error in solutions with a high Li^+-concentration only.

Parallel to this development of glass membranes Beutner [27] and Osterhout [28], from 1920 to 1935, investigated liquid phases with a protoplasma-like behavior. However, none of the systems was selective and no phase showed the slightest theoretical Nernstian response. In 1929 Michaelis [29] published his classical studies on collodion membranes which showed a nearly Nernstian behavior toward some alkali ions. This led Teorell [30] and Meyer and Sievers [31] to the development of a theory viewing membranes as ion-exchange bodies (fixed-charge theory).

As early as 1937, Kolthoff and Sanders [32] used silver halide discs as ion-sensitive membranes for the corresponding anion. They found that interferences through redox-systems were negligible with those membranes compared to electrodes of second kind. In the following years many ion-conducting electrolytes (mostly solids) were tested in membrane form either as pieces, pressed pellets, or embedded in an inert support material (e.g., paraffin) in an effort to dicover ion-specific membranes which could be of some analytical use. Here, the classical work of Marshall and collaborators [33-35] should be mentioned along with that of Sollner and coworkers [36-38]. In the years 1939–1954 Marshall studied natural minerals and Sollner used differently treated collodion to obtain cation-sensitive (oxidized collodion) or anion-sensitive (protamine containing) membranes. But unfortunately all these attempts to build highly ion specific membranes failed. The selectivity was rather poor and a general analytical applicability was therefore not found. An exception is the sodium selective NAS 11–18 glass membrane described by Eisenman, Rudin, and Casby [39].

The first stage in the development of the so-called ion-selective electrodes was the finding of Pungor and collaborators [40-41] in 1964 that silicon rubber was an ideal support material for various ionic conducting salts, e.g., silver halides which, embedded in the silicon gave fairly selective membranes for Ag^+ and the corresponding halide. Electrodes of this type were still manufactured until 1982 by the Hungarian company Radelkis. But the real break-through was only in 1966 when Frant and Ross filed their patents for the fluoride- and sulfide-selective electrodes based on a LaF_3 single crystal and a Ag_2S pressed pellet, respectively [43,44]. Subsequently, several ion-selective membranes were developed in just a few years which led to the currently available electrodes and to the expression "renaissance in potentiometry".

Only a few milestones should be mentioned: The calcium-selective membrane of Ross [44] consisting of calcium di-(n-decyl)phosphate in di-(n-octylphenyl)phosphonate and that of Shatkay and co-workers [45] consisting of thenoyltrifluoroacetone in a PVC-tributylphosphate matrix, as well as the use of Aliquat S 336 for anion-sensitive membranes studied by Coetzee and Freiser [46]. Štefanac and Simon [47] constructed the first highly selective potassium electrode using valinomycin as the electro-active compound. Moody, Oke, and Thomas [48] optimized a calcium-selective membrane with the use of PVC as an inert matrix; with this technique other membranes also can be improved over the common liquid membrane constructions. Excellent ion-selective membranes for Ca^{2+}, Ba^{2+}, Na^+, Li^+, etc. made from neutral non-cyclic carriers in a PVC matrix were developed by Simon's group [49-52], which remains very active and leading in this field. It is self-evident that the progress in the field of ion-selective membranes was also strongly influenced by the progress in other areas, e.g., instrumentation (electrometer amplifier), bio-membrane research, semiconductor research, etc.

B Theoretical Approaches for Ion-Selective Membranes

I Thermodynamic Equilibrium at the Phase Boundary

The most popular theoretical description of the potentiometric behavior of ion-selective membranes makes use of the three-segmented membrane model introduced by Sollner [53], Teorell [30,54], and Meyer and Sievers [31,55]. In this model the two phase boundaries and the interior of the membrane are treated separately. Here, the total trans-membrane potential consists of two phase boundary potentials (treated as Donnan-potentials [56,57]) and, in case of different mobilities of the various ions in the interior, of an additional diffusion potential inside the membrane. The individual potential contributions can be evaluated if a series of fundamental model assumptions is accepted [58]:

1. At each phase boundary there exists a thermodynamic equilibrium between the membrane surface and the respective adjacent solution. The resulting thermodynamic equilibrium potential can then be treated like a Donnan-potential if interfering ions are excluded from the membrane phase [59-61]. This means that the ion distributions and the potential difference across each interface can be expressed in thermodynamic terms.
2. The mobilities of all particles within the membrane are invariant.
3. The chemical standard potentials of all particles within the membrane are invariant with space and time.
4. Pressure and temperature gradients do not exist across the membrane.
5. The system is in a zero-current steady-state.
6. The membrane is homogeneous with respect to a direction parallel to the phase boundaries.
7. The same solvent is used for the electrolyte solutions on either side of the membrane.
8. The effect of solvent flow across the membrane is negligible.
9. The local activity coefficients are the same for all ions in the membrane, or the individual activity coefficients in the membrane are the same for all cations and anions.

If a diffusion potential occurs inside the membrane, the relation between mass transport and electrochemical potential gradient — as the driving force for the diffusion of ions — has to be examined in more detail. This can be done by three different approaches:

a) Proper use of the Nernst-Planck flux equation [62,63].
b) Application of Eyring's absolute rate theory [64].
c) Application of the laws of irreversible thermodynamics [65].

Since such theoretical considerations also describe the important ion transport through biological membranes, all three approaches are extensively used and were originally developed for the description of thin membranes. For the description of thick ion-selective membranes the first approach is generally used although the others often lead to very similar or identical results. The interested reader is referred to the excellent monograph of W. Morf [58]. Nonetheless, the results should be mentioned here. The main purpose of the theory should be an accurate description of the behavior of real ion-selective membranes. With respect to their analytical usefulness the selectivity is of great importance. Therefore, a proper theory should enable the calculation of the selectivity coefficient, preferable with independently measurable physico-chemical parameters.

According to the nine assumptions and approach a) for the diffusion potential inside the membrane the selectivity coefficient K_{M-I}^{Pot} can be expressed by other parameters. Table 3 shows the results for the different kinds of membranes [66]. In some cases the expressions for K_{M-I}^{Pot} contain ion-mobilities inside the membrane

Table 3. The selectivity coefficient K^{pot}_{M-I} expressed by other parameters [66]

membrane type	K^{pot}_{M-I}
solid membranes	
slightly soluble inorganic salts, e.g., $Ag_2S + AgCl$ etc.	$S.P._{MX}/S.P._{IX}$
ion-exchange membranes, i.e., glass membranes	$(u_I/u_M)\, K^{eq.}_{MI}$
liquid membranes	
charged ion exchangers	
a) dissociated complexes	$(u_I k_I)/(u_M k_M)$
b) strongly associated complexes	
α) mobility of free sites is small	$(u_I + u_S)\, k_I/(u_M + u_S)\, k_M$
β) mobility of free sites is large	$[u_{IS}(K^{stab}_{IS})_m k_I]/[u_{MS}(K^{stab}_{MS})_m k_M]$
electroneutral carrier	k_I/k_M
when $u_{IS} \cong u_{MS}$	$(K^{stab}_{IS})_{aq}/(K^{stab}_{MS})_{aq}$

$S.P._{MX}$, $S.P._{IX}$ = solubility products of MX and IX,
M = measured ion, I = interfering ion, S = site in the membrane;
u_i = mobility of particle i in the membrane;
K^{eq}_{MI} = ion-exchange equilibrium constant for the reaction
$I^+_{solution} + M^+_{membrane} \rightleftarrows I^+_{membrane} + M^+_{solution}$,
k = single ion distribution coefficient;
$[K^{stab}]_{m,aq}$ = stability constant of the complexes in the membrane phase and in the aqueous phase

phase; this indicates that kinetic parameters are at least occasionally necessary for an accurate description of the potentiometric behavior of ion-selective membranes. By scientific theory we may ask how to measure the potential-dependent individual ion mobilities inside the membrane as well as the other parameters independently and not via potentiometric measurements, followed by a method of curve fitting. Furthermore, the results in Table 3 do not indicate the given dependence of the selectivity coefficient on the concentration of the ion to be measured and the interfering ion, as well as on the method used for the empirical determination. An IUPAC commission [67] therefore suggests the fixed interference method.

The three-segment potential theory can hardly describe membranes with blocked interfaces [68], coated wire electrodes [69], Růžička selectrodes® [70], or ion-selective field effect transistors [71,72] without internal solution. The latter operate with extremely thin layers of the electro-active compound (e.g., SiO_2 or Si_3N_4 show a pH-response with layers around 25 Å [73]). Furthermore, a variable selectivity depending on conditioning [69] or a change from a sodium selectivity to a potassium selectivity when the glass membrane is electrolyzed cathodically in a potassium salt solution [74], cannot be deducted from the expressions in Table 3. This holds true also for a super Nernstian slope sometimes obtained even after appropriate corrections for a varying liquid junction potential. In recent years several artificial systems which exhibit active and passive ion transport with mobile carriers in bulky liquid membranes have been reported. For a review see [75]. Here, the ion transport selectivity depends upon the fact wether the ion of interest is studied alone or together with one or more co-ions.

Finally some assumptions could not be verified as, e.g., the complete co-ion exclusion necessary for the treatment of the phase boundary potential as a Donnan potential, or the constant ion mobility through glass membranes with hydrated layers [76].

II No Thermodynamic Equilibrium at the Phase Boundary

If the separate solution method is used for the determination of the potentiometric selectivity coefficient, it is tacitly assumed that for the ion to be measured and for the interfering ion, EMF vs. log. activity curves display identical slopes. This can indeed be found in many cases. Because both slopes then approach the theoretical value, the conclusion can be drawn that the interfering ions also show a reversible behavior. The latter indicates that they rapidly pass the phase boundary [58] to reach their thermodynamic equilibrium in a relatively short time. However, if the ion to be measured is present in the sample solution together with such interfering ions, parallel charge transfer reactions cannot be excluded. In such a case the establishment of the thermodynamic equilibrium of the ion to be measured depends not only on the fact that its charge transfer kinetics is high, but to an even higher degree on the ratio of the charge transfer kinetics between the ion to be measured and the interfering ions.*

General electrochemical experience, however, has shown that parallel electrode reactions necessarily lead to a mixed potential lying in between the quilibrium potentials of each parallel reaction. Furthermore, it has been found — and it is a well established fact in corrosion research — that the mixed potential can be predicted if the current-voltage curve for each of the parallel electrode reaction is known. Current voltage curves can be described by the Butler-Volmer equation [77,78]. One important parameter of this equation is the exchange current density i_0 [79] which reflects the phase boundary kinetics. A high exchange current density is necessary for a good potentiometric sensor [80-84] and is equivalent to a non-polarizable electrode, i.e., a small external current flowing across the phase boundary should not change the equilibrium potential. In a current-voltage curve such a behavior is demonstrated by a very steep slope, nearly parallel to the current axis (as is the case, e.g., for good reference half-cells).

In Fig. 2 an example is given to illustrate how the mixed potential of a silver wire in an oxygen-containing electrolyte can be found. Of main importance is the fact that in the final analysis a reaction takes place, although there is no external current: Dissolved oxygen oxidizes the silver at a constant rate given by the current density at the mixed potential point. The mixed potential is the result of a stationary state situation; it changes if one of the parallel reactions, i.e., the oxygen-reduction-rate is altered. If the latter is diffusion controlled, a change of the stirring speed will alter the mixed potential. Stirring sensitivity with ion-selective electrodes at a high concentration of the ion to be measured therefore indicates a mixed potential at the membrane.

---

* The rate of this charge transfer is not necessarely identical with the rate constant of ion complexation in homogeneous solution which may be diffusion limited! [75]

a

b

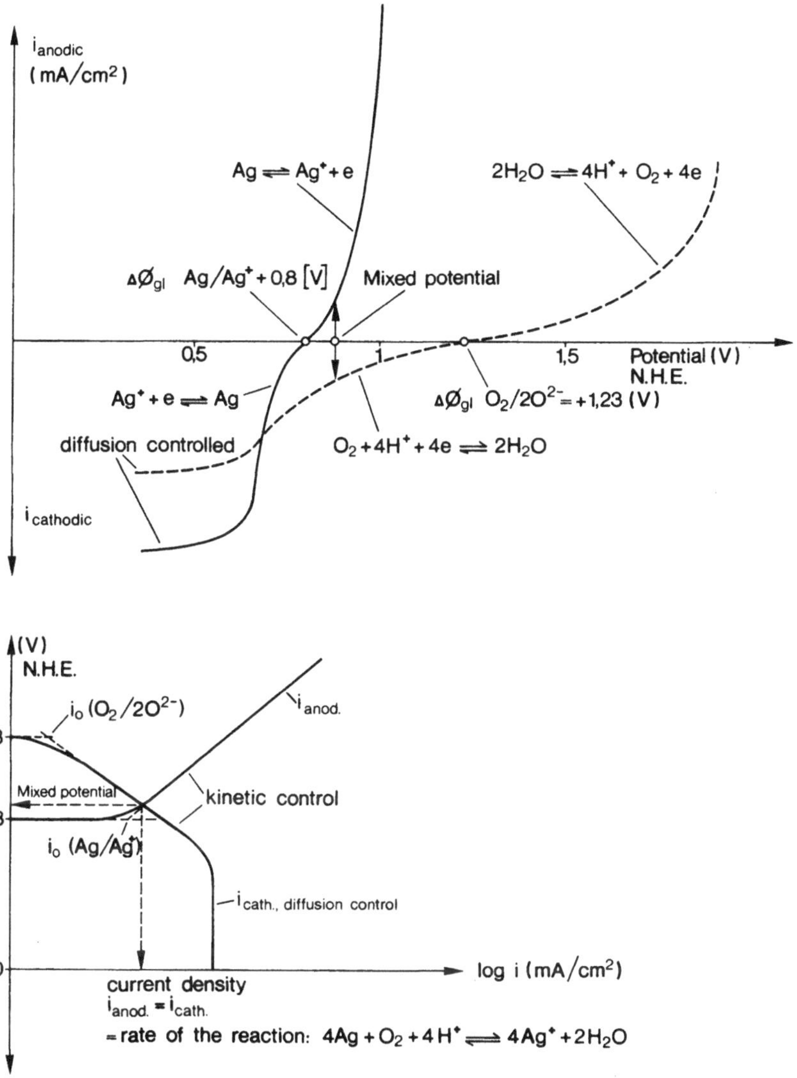

Fig. 2. Current-voltage curves illustrating the mixed potential

a In a silver ion and oxygen-containing electrolyte the thermodynamic equilibrium-potential $\Delta\Phi_{gl}$ for the Ag^+/Ag-system cannot be reached; instead a mixed potential is established where the cathodic partial current density of the oxygen reduction equals the anodic partial current density of the silver dissolution (= corrosion of silver at a constant rate controlled mainly through the kinetics of the O_2-reduction, which is exaggerated here for clarity).

b In an Evans-diagram (electrode potential *vs.* a reference electrode as ordinate versus logarithm of the current density as abscissa) the two partial current densities leading to the mixed potential in a) are drawn in the same quadrant. Their intersect allows an easy location of the mixed potential. Simultaneously the slopes of the log i-voltage curves (= Tafel plots) allow some conclusions concerning the electrode reaction mechanism. Here it is assumed that the cathodic partial current density is still under kinetic control; N.H.E. = normal hydrogen electrode; the electron e bears a negative charge not extra indicated throughout this article

Another example of the importance of the exchange current density and the underlying current-voltage characteristic for the behavior under no external current flow (potentiometric condition) are the amalgam electrodes. In principle, all ions forming an amalgam in the reduced state can be determined in single electrolyte solutions with the corresponding amalgam electrodes; but their practicability (e.g., selectivity, redox-sensitivity, poisoning) was very limited so that they were scarcely used for analytical purposes [3b]. However, the use of liquid amalgams for quantitative cementation of a variety of ions as a special case of liquid-liquid extraction offers many advantages especially in trace analysis and is often performed [85-88]. During the cementation process an amalgam of a less noble element, e.g., Zn(Hg) is used to reduce and extract more noble species, e.g., Cu^{2+} according to the interfacial reaction

$$Cu^{2+} + Zn(Hg) \rightleftarrows Cu(Hg) + Zn^{2+}$$

The cementation process starts with a large difference of the equilibrium potentials [e.g., $Cu^{2+}/Cu(Hg)$ and $Zn^{2+}/Zn(Hg)$] and will arrest when the potentials of both systems are equal, i.e., when (at 298 K) [85-88]

$$E^o_{Cu} + 0.029 \lg a_{Cu^{2+}} = E^o_{Zn} + 0.029 \lg a_{Zn^{2+}}$$

Given

$$E^o_{Cu} = +0.34 \text{ V}$$
$$E^o_{Zn} = -0.76 \text{ V}$$

one can calculate the ratio of activities of both metals in the solution and thereby the extraction coefficient under this condition. Under the assumption that the E^0-values remain constant it will be

$$\frac{a_{Zn^{2+}}}{a_{Cu^{2+}}} = 10^{38}$$

This value does not express the actual result since side and/or parallel reactions (e.g., H^+ or O_2 reduction) are not considered, but it does demonstrate the completeness of the cementation process and the effectiveness of this liquid-liquid extraction. During this extraction no external current flows through the phase boundary Hg (amalgam)/solution thereby establishing a potentiometric condition. The question of the potential difference at the phase boundary can be answered by constructing the experimentally accessible current-voltage curves for the reactions:

$$Cu^{2+} + 2e \rightleftarrows Cu(Hg)$$

and

$$Zn(Hg) - 2e \rightleftarrows Zn^{2+}$$

In an Evans diagram [89] the mixed potential can easily be found and also be verified by measuring the open circuit potential of a zinc-amalgam electrode in a Cu^{2+}-ion solution. Even the complication by the simultaneous presence of another reducible species, e.g., Pb^{2+} can be graphically demonstrated for different limiting conditions

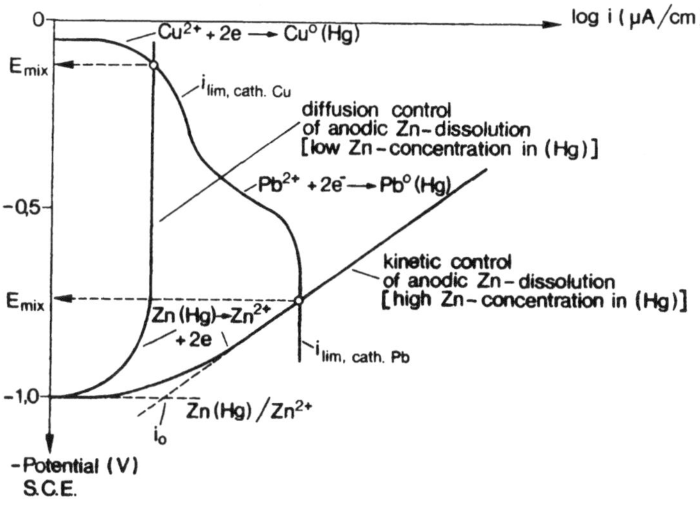

Fig. 3. Evans-diagram for the cementation of Cu^{2+} and Pb^{2+} with zinc amalgam of different zinc content. If the zinc concentration in the mercury employed for this special extraction technique is low, the anodic zinc-dissolution current density may be diffusion controlled and below the limiting cathodic current density for the copper reduction. The resulting mixed potential will lie near the half-wave potential for the reaction $Cu^{2+} + 2\,e \rightleftarrows Cu^0(Hg)$ and only Cu^{2+} ions are cemented into the mercury.

If — during this process — the Cu^{2+}-concentration decreases, the mixed potential will shift along the cathodic partial current density curve (like a polarographic curve in this example) toward the equilibrium potential of the zinc amalgam, in case the amalgam reservior is large enough.

If the zinc amalgam concentration is high, both Cu^{2+}- and Pb^{2+}-ions are simultaneously cemented from the very beginning and the mixed potential lies near the equilibrium potential of the zinc amalgam.

This figure demonstrates that also under potentiometric conditions (= no external current flow) electrochemical net reactions occur. The EMF of the zinc-amalgam in a given Zn^{2+}-ion solution depends on the current-voltage characteristic of other ions (in this example, Cu^{2+} and Pb^{2+} are interfering ions with respect to the Zn^{2+} equilibrium potential) at the amalgam electrode. EMF drifts are thus explainable.

(Fig. 3). In this case the cathodic current-voltage curve is given by the corresponding polarogram for Cu^{2+} and Pb^{2+}. The non-equilibrium stationary state potential during the cementation of both ions now depends on the concentration ratio Cu^{2+}/Pb^{2+} and the value of the kinetically or diffusion controlled (limited) anodic current for the zinc-dissolution. If the latter is greater than the sum of the diffusion limited cathodic currents of copper and lead, the mixed potential at the beginning of the cementation process lies closer to the half-wave potential of zinc, and Cu^{2+} and Pb^{2+} are simultaneously reduced at a rate according to their wave heights. If the limited anodic current is smaller than the diffusion current for the copper reduction the mixed potential at the beginning will lie closer to the half-wave potential of the latter and in the beginning only copper will be cemented. Because of the diminishing Cu^{2+}-concentration the anodic current will soon be larger (large amalgam reservior) than i_{lim}, cathodic Cu and the cementation of Pb^{2+} will start with a corresponding shift in the mixed potential. The limited anodic current for the zinc dis-

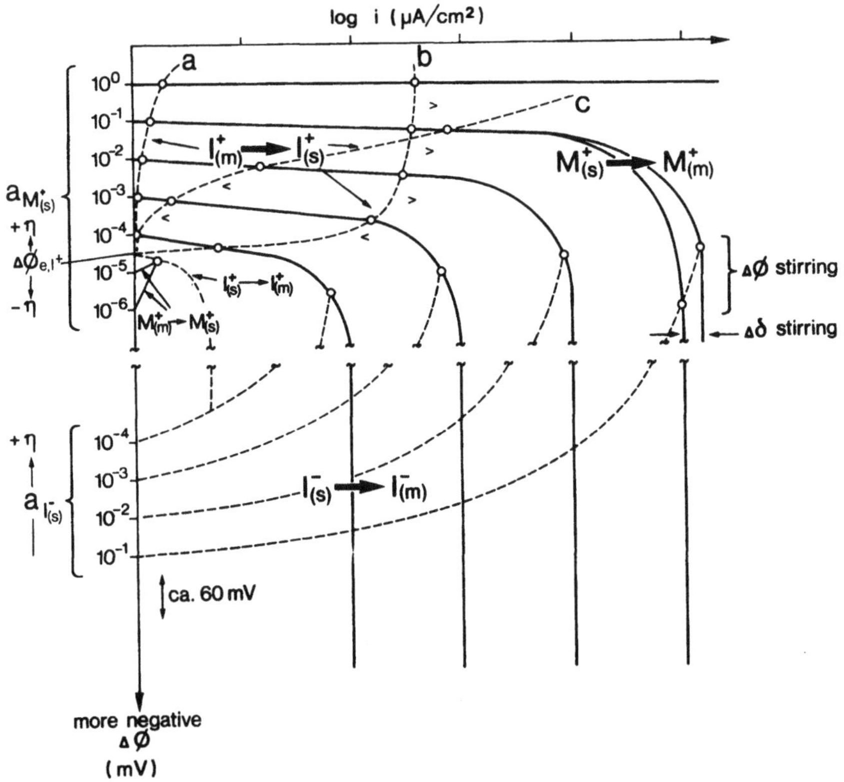

Fig. 4. Evans-diagram of a cation selective membrane; circles indicate mixed potentials. Unbroken lines describe the current-voltage curves of the measured ion M^+ at different activites. Broken lines are the corresponding current-voltage curves for interfering ions if they are simultaneously present. In the upper part, an interfering cation I^+ is assumed to have an equilibrium galvani potential (without M^+ ions being present) lying between the potentials of 10^{-4} and 10^{-5} M solutions of the measured ion. This is equivalent to assuming a selectivity coefficient of about $K_{M^+-I^+} \cong 3 \cdot 10^{-5}$. If present at a concentration exceeding 10^{-4} M, M^+ will then cause a positive overpotential η for the equilibrium reaction of I^+, driving the reaction $I^+_{(m)} \to I^+_{(s)}$, the release of interfering ions from the membrane phase (m) into the solution phase (s). The corresponding current in the opposite direction is the transfer of measured ions from the solution into the membrane: $M^+_{(s)} \to M^+_{(m)}$, since the equilibrium galvani potential of the interfering ions $\Delta\Phi_{e,I^+}$ acts like a negative overpotential η for the equilibrium of the measured ions. The current caused by the interfacial reaction $I^+_{(m)} \to I^+_{(s)}$ cannot be distinguished from the current caused by the flow of some less lipophilic anions in the opposite direction $I^-_{(s)} \to I^-_{(m)}$ not shown here, since it would require concentration-dependent partial current-voltage curves (as demonstrated with the strongly lipophilic anion) rendering the Figure fully into a maze. Depending on the exact course of the current-voltage curve of the interfering ions, different cases a)–c) may be encountered: Course a), the interfering ion I^+ is present in the membrane phase only in minute amounts so that the current of the reaction $I^+_{(m)} \to I^+_{(s)}$ is very small compared to the exchange current density of the measured ion M^+; the resulting response is near Nernstian. Course b), the interfering ion I^+ is present in the membrane phase in higher amounts than in the previous case and the current-voltage curve for positive overpotentials shows a considerable current with a diffusion-limited plateau. As the signs (⟨or⟩) indicate, the calibration curve shows a sub-Nernstian response for low and a super-Nernstian slope for high M^+ activities. Course c), the current-voltage behavior of the interfering ions in case of a positive overvoltage acts like a kinetically controlled curve resulting always in a sub-Nernstian slope of the calibration curve of M^+. With M^+-activities of 10^{-5} and 10^{-6} M, the opposite reactions, as indicated in the figure, occur at the membrane surface leading to one and the same mixed potential which defines the detection limit.

solution depends on the zinc concentration in the mercury and on the diffusional conditions inside the mercury and on the solution side.

In the mixed potential theory of ion-selective electrodes [80-84,106] the ion-selective membrane/sample solution interface is treated as being similiar to the better known metal/electrolyte-solution interface, especially with respect to the kinetics of the interfacial reactions involved and their effect upon an ideal Nernstian behavior. It has been shown by Gerischer [90] that the interfacial model developed so far for metallic electrodes can also be used with some modifications to describe the analogous situations at semiconductor/electrolyte-solution interfaces. Because ion-selective membranes are sometimes semiconductors (Ag_2S pressed pellet membranes) or insulators with a larger band gap (glass membranes), it was pointed out as early as 1973 [11] that parallel charge transfer processes lead to deviations from a thermodynamic equilibrium and, consequently, to mixed potentials, which depend very much on the kinetics of the interacting interfacial reactions. Thus, in order to study the ion-exchange kinetics at ion-selective membranes, some measurement techniques developed for metallic electrodes to determine the important exchange current density, i_0, were also used in the former case. The author succeeded in 1975 in obtaining consistent results by applying different techniques for Ag_2S-, glass-, and valinomycin-based liquid membranes [80,84] and also for LaF_3-membranes [83], not being aware of the research of Guastalla [91] and Gavach [92,93] in the field of electrochemistry at the interface of two immiscible electrolyte solutions (ITIES), which also demonstrated the similarity with metallic interfaces. However, knowing the uncertainties involved (amount of Frumkin- and double layer charging corrections, measurement problems in obtaining a reliable determination of a small charge transfer resistance in series with a much larger — and sometimes slowly changing — ohmic bulk resistance), the results were carefully considered as being only "apparent exchange current density estimates". Surprisingly, these early results agree very well with recent determinations [94-98] using more elaborate techniques.

Of fundamental importance in understanding the electrochemistry of ion-selective membranes and also of biomembranes is the research in the field of voltammetry at ITIES mainly pioneered by Koryta and coworkers [99-101]. Koryta also demonstrated convincingly that a treatment like corroding metal electrodes is possible [102]. For the latter, the description in the form of an Evans-diagram is most appropriate; Fig. 4 shows schematically some mixed potentials, which are likely to arise at cation-selective membranes if interfering ions disturb an ideal Nernstian behavior [82]. Here, the vertical axis describes the galvani potential differences (absolute po-

In the lower part of the figure, an interfering lipophilic anion I^- is considered. Because of its negative charge, the enormous positive overpotential of the measured ions drives the reaction $I^-_{(s)} \rightarrow I^-_{(m)}$, the extraction of the anion together with M^+ into the membrane phase. It is assumed that the reaction $M^+_{(s)} \rightarrow M^+_{(m)}$ is diffusion controlled, which means that any kind of stirring may reduce the diffusion layer thickness by $\Delta\delta$, thereby resulting in a higher partial current density. A consequence of these effects is that the mixed potential can vary within relatively wide limits. All mixed potentials in the lower part of the figure follow the typical calibration curve of a valinomycin-based potassium electrode with KSCN-solutions, e.g. beginning in a sub-Nernstian positive slope and ending in a super-Nernstian fashion with a negative slope of about 70 mV/decade.

tential difference) across the phase boundary: ion-selective membrane/sample solution, with the latter taken as reference. This means that whenever a cation gains energy leaving the aqueous solution and entering the membrane phase, a positive potential difference (according to $\Delta G = -n \cdot F \cdot \Delta\Phi$) is created. The opposite holds for anions. Because the absolute potential difference across an interphase cannot be measured, an absolute scale has been omitted in Fig. 4. But changes in that potential difference (e.g., according to the Nernst-equation) are indicated. It is further assumed that the standard thermodynamic equilibrium potential difference $\Delta\Phi_{e,I^+}$ for the interfering cation lies between 10^{-4} and 10^{-5} M of that of the ion to be measured (thus indicating a selectivity coefficient around 10^{-4}). For concentrations $>10^{-4}$ M of M^+ the ion to be measured $\Delta\Phi_{e,I^+}$ acts like a negative overvoltage forcing the reaction $M^+_{(s)} \rightarrow M^+_{(m)}$ [(s) stands for solution, (m) for membrane], for smaller concentrations of M^+ it acts like a positive overvoltage forcing this reaction in the opposite direction. At the same time, the equilibrium potential difference of the ions to be measured $\Delta\Phi_{e,M^+}$ acts accordingly upon $\Delta\Phi_{e,I^+}$ like a positive or negative overvoltage. The interfering ions do not determine the galvani potential difference at the membrane/solution interface until their exchange current density (current at $\Delta\Phi_e$ parallel to the log i axis) is comparable to that of the ion to be measured. This leads to a reduced Nernstian behavior for the latter at low concentrations. If $\Delta\Phi_{e,I^+}$ would always be more positive than $\Delta\Phi_{e,M^+}$ the membrane phase thermodynamically would favor the interfering ions (e.g., Zn^{2+} at some Ca^{2+}-sensitive, Rb^+ at K^+-sensitive membranes). The reaction $I^+_{(s)} \rightarrow I^+_{(m)}$ would always be driven by the negative overvoltage η established through $\Delta\Phi_{e,M^+}$. During longer periods the concentration of interfering ions would increase in the membrane phase at the expense of the concentration of the ions to be measured. Since the exchange current density is concentration dependent, it will change accordingly. A drifting potential difference will result as well as a failure to show a response in a Nernstian manner to the ion to be measured. Since such interfering ions can be present at least in trace amounts in every real sample solution the limited lifetime of certain ion-selective membranes and the influence of interfering ions thereon can be understood. Of course, the loss of electro-active substance from the membrane phase may also limit the life time [103].

The circles in Fig. 4 represent stationary state conditions (mixed potentials). The reaction $I^+_{(m)} \rightarrow I^+_{(s)}$ is viewed here as an anodic current flowing across the phase boundary because $\Delta\Phi_{e,M^+}$ creates a positive overvoltage η for this reaction. This partial current-voltage curve can also been sustained by the interfacial reaction of interfering anions I^- in the opposite direction $I^-_{(s)} \rightarrow I^-_{(m)}$. In order to distinguish between both possibilities additional analyses have to be performed. The cathodic current is here the transfer of measuring ions from the solution phase into the membrane phase $M^+_{(s)} \rightarrow M^+_{(m)}$ driven by the negative overvoltage created through $\Delta\Phi_{e,I^+}$. The stationary state exists when the net current flow is zero, that is when the anodic current density is equal to the cathodic current density. In the case of the circles of row a) the exchange current density of the ion to be measured is several orders of magnitude higher than the current density at the mixed potential so that the equilibrium potential difference of M^+ is not disturbed by this small current flow relative to i_0.

At anodically polarized metal electrodes (corroding metal) the metal dissolution follows a η vs. log i law over several orders of magnitude since the concentration on the metal side can be regarded as exceptionally high. However, with cathodic polarization this kinetically controlled range depends on the concentration of the ion in the solution and is much smaller leading soon to a diffusion controlled situation. With ion selective membranes the latter situation can arise also with anodic polarization. The course of the anodic polarization curve depends on the availability and mobility of the ion under study in the membrane phase. This is demonstrated in Fig. 4 with the circles of row b) and c) for the anodic current through $I^+_{(m)} \rightarrow I^+_{(s)}$. In case c) the resulting potential differences between the mixed potential points in course of several concentration decades is always sub-Nernstian $\left(\text{slope} < \frac{2{,}3 \cdot R \cdot T}{z \cdot F}\right)$.

In case b) a potentiometrically measured calibration curve for ion M^+ would show a sigmoidal form, starting with a sub-Nernstian slope leading to a super-Nernstian $\left(\text{slope} > \frac{2{,}3 \cdot R \cdot T}{z \cdot F}\right)$ behavior. This might be the case when the interfering cation concentration and mobility in the membrane phase is relatively high or some slightly lipophilic anions is present in the sample solution. In the former case a conditioning of the membrane with the ion to be measured would transfer row b) into row a).

Figure 4 also demonstrates qualitatively the EMF vs. activity curve in case of a lipophilic anion (e.g., a KSCN solution measured with a valinomycin-based potassium electrode) characteristized here with the anodic branches of the current-voltage curves with the interfacial reaction $I^-_{(s)} \rightarrow I^-_{(m)}$. If the membrane is not conditioned with SCN^- anions the calibration curve will start at low KSCN concentrations with a reduced cation-sensitive slope of about 25 mV/decade changing at about 10^{-2} M to a SCN-sensitivity of about 70 mV/decade. This is because during the experiment the membrane becomes increasingly conditioned with the lipophilic anion leading to an increasing exchange current density for that ion until the latter is similar to that of the cation. At high concentrations of KSCN experiments have shown that the membrane potential becomes very sensitive to stirring. This is expressed in the model by assuming a diffusion limited current for the $M^+_{(s)} \rightarrow M^+_{(m)}$ reaction with a stirring-dependent change in the diffusion layer thickness δ. If both reactions $M^+_{(s)} \rightarrow M^+_{(m)}$ and $I^-_{(s)} \rightarrow I^-_{(m)}$ (equivalent to a MI-salt extraction) were to be diffusion controlled at similar current densities, a totally unstable mixed potential would occur leading to unreproducible measurements [84]. It should be mentioned that in case of a picrate conditioned valinomycin membrane the exchange current density for the picrate anion is normally higher than that for potassium leading to an anion response over the total concentration range. The addition of extremely lipophilic anions, such as tetraphenylborate to the membrane phase improves the behavior of carrier-based cationic sensitive membranes [58] at least toward less lipophilic anions like SCN^-. It is assumed that this is the case because negative charges in the membrane reduce the transfer rate of anions. On the other hand, together with the lipophilic anion a corresponding amount of the ion to be measured has been incorporated into the membrane phase leading to a higher exchange current density coupled to an improved potentiometric behavior for the latter.

Figure 4 also demonstrates that through the influence of the different equilibrium potentials upon each other, which is equivalent to externally applied overvoltages, the corresponding interfacial reaction rates at the mixed potential point depend very much on the shape of the partial current-voltage curves involved [104]. If the charge transfer rate is expressed by an electronic analog, the charge-transfer resistances of lipophilic cations and anions are not necessarily additive, as was found to be true for thin lipid bilayer membranes [105].

Summarizing Fig. 4, it should be noted that at the mixed potential points two different electrochemical (i.e., charge-transfer) reactions are coupled. The driving force is the overpotential given by the difference between the individual thermodynamic equilibrium potential differences and the mixed potential. As long as the overpotential remains constant, both reactions run at the same rate. Because of the limited mass capacity of both phases, a net change in the composition of both or only one phase occurs. In the long run this leads to a slowly drifting potential difference, common to most ion-selective membranes.

In the mixed potential model, pure solutions containing only the ion to be measured give the fastest responses since the tendency to form a mixed potential is minimal (e.g., trace impurities in the sample solution or the membrane). However, if a relatively high concentration of an interfering ion is present the establishment of a mixed potential takes more time despite the fact that it lies near the thermodynamic equilibrium potential difference of the ion to be measured. The new stationary state may lead to a new diffusion condition for the interfering ion. Thus, the response time may follow a characteristic $\sqrt{t}$ law. It should be noted that the diffusion current is not necessarily parallel to the voltage axis since migration (movement in an electric field) is involved. This occurs because a large excess of a supporting electrolyte is normally missing. If the time response follows an exponential law, a kinetically controlled, mixed potential may be established. It should also be mentioned that $i \times R$ drops within the membrane surface can alter the shape of the individual current-voltage curves. The former may be effected by so-called mediators or modifiers added empirically to PVC membranes to improve the potentiometric response.

The explicit mathematical treatment for such stationary-state situations at certain ion-selective membranes was performed by Iljuschenko and Mirkin [106]. As the publication is in Russian and in a not widely distributed journal, their work will be cited in the appendix. The authors obtain an equation (s. (34) on page 28) similar to the one developed by Eisenman et al. [6] for glass membranes using the three-segment potential approach. However, the mobilities used in the stationary-state treatment are those which describe the ion migration in an electric field through a diffusion layer at the phase boundary. A diffusion process through the entire membrane with constant ion mobilities does not have to be assumed. The non-Nernstian behavior of extremely thin layers (i.e., ISFET) can therefore also be described, as well as the role of an electron transfer at solid-state membranes.

C Stationary-State Potential Differences at Asymmetric Biological Membranes

Although the author of this review is far from being an expert in the field of biomembrane research, some literature studied so far seems to indicate that the molecular mechanisms of the active ion transport are still not known [107].

"Despite the immense literature about the biochemical properties of the transporting enzyme — the Na^+/K^+ activated membrane ATPase — the intrinsic mechanisms or the mechanistic details by which Na^+ and K^+ ions are transferred through the cell membrane against their concentration gradient still remain unknown" [108,109]. The phenomenology of electrical events occurring on excitable cell membranes can be described in detail by the Hodgkin-Huxley equation [111] "but the attempts to construct physically reliable microscopic models have been rather speculative" [112].

The model proposed by Jardetzky [110] remains consistent with most of the experimental evidence [107], but, unfortunately, not with all.

Otherwise it has been shown that the accumulation of electrolytes by many cells runs at the expense of cellular energy and "is in no sense an equilibrium condition" [113] and that the "use of equilibrium thermodynamic equations (e.g., the Nernst-equation) is not allowed in systems with appreciable leaks which indicate a kinetic steady-state" [114]. In addition, a superposition of partial current-voltage curves was used to explain the excitability of biological membranes [112]. In interdisciplinary research the adaptation of a successful theory developed in a neighboring discipline may be beneficial, thus an attempt will be made here, to use the mixed potential model for ion-selective membranes also in the context of biomembrane surfaces.

Studies on the Na^+/K^+-activated transport enzyme for sodium and potassium indicated that for intact membranes an asymmetric situation exists [115]. It is postulated that the enzyme has specific inner and outer sodium and potassium binding sites and that the Na:K affinity ratio for the inner sites differs from the K:Na affinity ratio for the outer sites. The latter ratio has been estimated to be $\sim 100:1$. If such a situation is transferred to a stationary-state situation at an asymmetric membrane, the result could be as illustrated in Fig. 5. According to the different affinities (ΔG for the ion transfer from the solution into the membrane or the macromolecular enzyme) on the outer membrane side, potassium ions establish a more positive equilibrium galvani potential difference than sodium ions. The reverse holds true for the inner side. The establishment of a mixed potential in the region between the two equilibrium potentials consequently results in the ion fluxes indicated in Fig. 5. It is assumed that certain anions are passively distributed across the membrane. In the proposed model, it is absolutely necessary that sodium and potassium (or equivalent ions, e.g., Rb^+ can replace K^+) are both present on either side. The potential dependence of the pump current [108] becomes understandable through the differences of the potassium and sodium equilibrium potentials, which lead to different current densities at the mixed potential point. Hyperpolarization after injecting sodium ions into the cell interiors [108] can be explained by an increase of the internal equilibrium potential difference for sodium ions and a corresponding shift of the mixed potential on the interior side of the membrane. In the mixed potential model the current densities for $\overleftarrow{K}$ and $\overrightarrow{Na}$ are equal (1:1) in the stationary state. If a different ratio

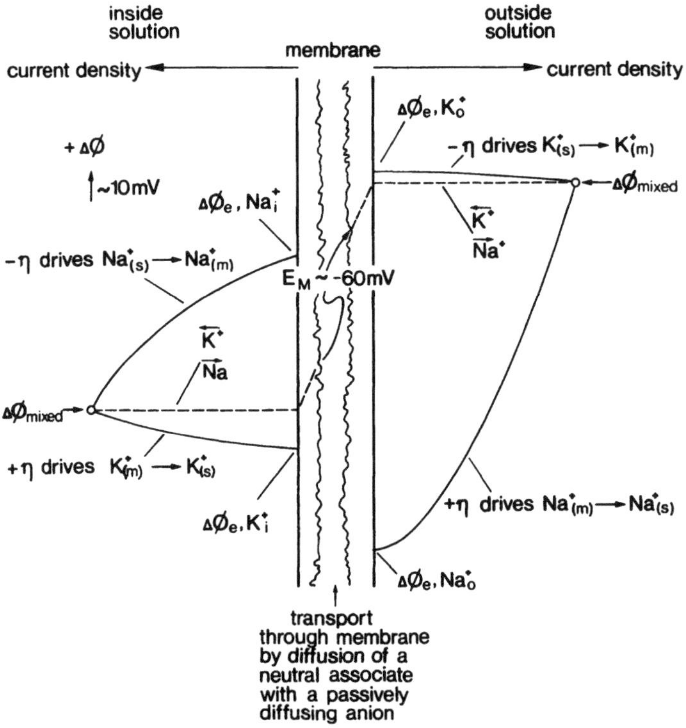

Fig. 5. Tentative mixed potential model for the sodium-potassium pump in biological membranes; the vertical lines symbolyze the surface of the ATP-ase and at the same time the ordinate of the virtual current-voltage curves on either side resulting in different Evans-diagrams. The scale of the absolute potential difference between the ATP-ase and the solution phase is indicated in the upper left corner of the figure. On each side of the enzyme a mixed potential (= circle) between Na^+, K^+ and also other ions (i.e. Ca^{2+}!) is established, resulting in a transmembrane potential of around − 60 mV. This number is not essential; it is also possible that this value is established by a passive diffusion of mainly K^+-ions out of the cell at a different location. This would mean that the electric field across the cell-membranes is not uniformly distributed.

According to the Evans-diagram treatment, the asymmetric surfaces of the Na-K-ATPase show different equilibrium galvani potentials $\Delta\Phi_e$ for Na^+ and K^+-ions inside (i) and outside (o). These galvani potentials act as a corresponding overpotential η driving the indicated surface reactions, where the symbol (s) stands for solution phase and (m) for membrane (more acurately: enzyme) phase. The ions accumulated at the enzyme surfaces then move through the membrane (maybe in-between the phospholipid layer and the ATPase surface and/or along β-sheet domains known to have a relatively large dielectric constant) by passive diffusion along their concentration gradient.

It is very likely that ATP is necessary for creating and sustaining the asymmetric behavior besides the energetic aspect. Quabein and other substances can block certain enzyme positions and thus alter the exchange current density of one or more of the involved ions, resulting in other mixed potentials with different net ion fluxes. (According to the Post scheme $\Delta\Phi_c$, Na_i^+ is established in the enzyme conformation state E1, $\Delta\Phi_c$, K_i^+ then corresponds to E2-ATP, $\Delta\Phi_c$, K_o^+ corresponds to E2 and $\Delta\Phi_c$, Na_o^+ to E1-P). It should be noted that the course of the individual current-voltage curves depends upon several factors (e.g., charge transfer kinetics, ion migration, i × R drop, diffusion layer thickness etc.). Given the concentration differences between Na^+ and K^+, it is very likely that one curve is always diffusion limited.

According to the exact position of the equilibrium galvani potential for Ca^{2+}-ions on the voltage axis, stimulation or inhibition of the Na^+—K^+ exchange can be understood. The effect of anions (e.g., $MgATP^{2-}$) may be similiar as shown in Fig. 4.

(e.g., 2:3) is found, additional reactions must be involved, e.g., in the hydrolysis of ATP there are hydrogen ions involved which also could participate in the stationary-state situation at both membrane surfaces and/or a corresponding anion-flux across the interface occurs simultaneously (e.g. Na^+ + $MgATP^{2-}$ "extraction" — s. Fig. 4 — onto the inner membrane/ATPase surface).

It is possible that the stationary-state situations leading to an active ion transport occur only in localized regions of the membrane, i.e., at ATPase molecule units with diameters of about 50 Å and a length of 80 Å. The vectorial ion currents at locations with a mixed potential and special equipotential lines would appear phenomenologically like ionic channels. If the membrane area where the passive diffusion occurs is large, it may determine the rest potential of the whole cell.

D Conclusion

In this review article an attempt has been made to convey the idea that parallel electrochemical processes like those occurring at metal electrodes (i.e., corrosion, analytically used amalgams) are also possible at ion-selective membranes. A superposition of partial current-voltage curves for the ion to be measured and for the interfering ions allows the exact localization of the resulting mixed potential and explains in an easily surveyable manner the different behaviors, as there are: sub- and super-Nernstian response, EMF-drift, effect of conditioning and/or co-ions, failure due to more lipophilic interfering ions, stirring sensitivity of the EMF, non-ideal functioning of ISFETs, response-times of very thick membranes with or without the presence of interfering ions, EMF-drift, life-times, etc. Of course, the partial current-voltage curves have not been measured empirically due to the great difficulties arising from the high membrane impedances; however, the total current-voltage curves to which the individual add up at ITIES [99–102], extremely thin glass membranes, bilayer membranes, and at cellular bio-membranes, have been measured [116,117,108,118]. On the other hand, the semiconducting properties of lipid films have been demonstrated [119–121] and the kinetics at semidonductor surfaces have been described theoretically [90]. Since ion-selective membranes are similar to biological membranes and at least some of them also show semiconductor properties [11,122] it is justified to apply the laws of electrochemical kinetics in both cases. Thus, "conductances across a membrane are equivalent to kinetic rate constants, permeability through carriers is equivalent to catalytic processes, and different surface reactions preceding or following the transport process are equivalent to similar reactions in the electrode process" [117]. An electronic equivalent circuit of a three layer lipid membrane has already been proposed [123].

Concerning the comparison between the three-segment potential theory, which assumes a thermodynamic equilibrium at the phase boundaries, and the proposed mixed potential theory it should be mentioned that both are identical in describing the EMF versus concentration behavior. The former requires such parameters as the individual ion mobilities within the membrane and single ion extraction coefficients both often obtained a posteriori (through curve fitting of the corresponding EMF versus concentration curve). The latter requires the individual current-voltage curves of all involved parallel reactions. The graphical presentation may be more distinct in showing that for a stationary state condition, ions are indeed moved at a constant rate across the phase boundary.

The proposed model for the so-called sodium-potassium pump should be regarded as a first tentative attempt to stimulate the well-informed specialists in that field to investigate the details, i.e., the exact form of the sodium and potassium current-voltage curves at the inner and outer membrane surfaces to demonstrate the excitability (e.g. N, S or Z shaped) connected with changes in the conductance and ion fluxes with this model. To date, the latter is explained by the theory of Hodgkin and Huxley [111] which does not take into account the possibility of solid-state conduction and the fact that a fraction of Na^+ in nerves is complexed as indicated by NMR-studies [124]. As shown by Iljuschenko and Mirkin [106], the stationary-state approach also considers electron transfer reactions at semiconductors like those of ionselective membranes. It is hoped that this article may facilitate the translation of concepts from the domain of electrodes in corrosion research to membrane research.

E Appendix

Stationary state situations at ion-selective membranes after Iljuschenko and Mirkin [106]

The presupposition is that parallel electrochemical reactions (i.e., ion or electron transfer) occur across the phase boundary, if the measured ions and interfering ions are both present in the solution. A redox process in which electrons pass the phase boundary is also considered an interfering electrochemical reaction.

For the stationary state the mixed potential $\Delta\Phi_{mix}$ can be described as (see also Figs. 1–4):

$$\Delta\Phi_{mixed} = \Delta\Phi_{e,A} + \eta_A = \Delta\Phi_{e,B} + \eta_B = \ldots \quad (2)$$

with: $\Delta\Phi_{e,A}$; $\Delta\Phi_{e,B}$; ... = equilibrium galvani potential for species or electrochemical reactions A, B, ...

η_A; η_B; ... = overpotential for the electrochemical reaction of A, B . . . at the phase boundary including the right sign

The sum of all cathodic partial currents across the phase boundary equals the sum of all anodic partial currents at the mixed potential; therefore, a further condition is:

$$\Sigma i_i = 0 \quad (3)$$

with: i_i = current density of the involved species. The establishment of such a stationary state can create ion-fluxes across the interface and can result in changes in the composition of the involved phases at least near the phase boundary.

If these changes are small (i.e., a current of 1 μA corresponds to an amount of about 10^{-10} mol/s only) and the capacity factor (e.g., charge density) in the involved phases is large, the mixed potential can be stable over a long period of time and can also be very reproducible. On the other hand, the limited lifetime of certain ion-selective membranes (e.g., liquid membranes, PVC-based membranes) can be explained by a change in their composition leading to greater deviations from the Nernst equation.

Mixed potential resulting from an interfering redox reaction at membranes with finite electronic conductance

Kinetic domain

The combination of the Butler-Volmer equation

$$i = i_0 \left[\exp\left(\frac{\alpha \cdot z \cdot F \cdot \eta}{RT}\right) - \exp\left(\frac{-1(1-\alpha)\, z \cdot F \cdot \eta}{RT}\right)\right] \tag{4}$$

with: i_0 = standard exchange current density
α = transfer coefficient

with the condition for a stationary potential (2) and condition (3) leads to the following equations, describing the interference of the reaction of species A by a parallel redox reaction B:

with: $\eta_A \ll \frac{RT}{F}$ and $\eta_B \ll \frac{RT}{F}$ one obtains:

$$\Delta\Phi_{mix} \cong \Delta\Phi_{0,A} - \frac{i_{0,B} \cdot z_B}{i_{0,A} \cdot z_A}\eta_B + \frac{RT}{z_A F} \ln \frac{a_{A,s}}{a_{A,m}} \tag{5}$$

with: $\Delta\Phi_{0,A}$ = standard equilibrium potential of species A
$i_{0,A}$ = standard exchange current density for the electrochemical reaction of species A (e.g., for the distribution kinetics across the interphase)
$i_{0,B}$ = standard exchange current density for the interfering electron transfer (redox reaction B)
z_A = charge of species A
z_B = amount of charges transferred by redox-reaction B
η_B = overpotential for reaction B
$a_{A,s}$ = activity of species A in the solution
$a_{A,m}$ = activity of species A in the membrane

in case $\eta_A \ll \frac{RT}{F}$ and $\eta_B \gg \frac{RT}{F}$ the following expression is obtained:

$$\Delta\Phi_{mix} \cong \Delta\Phi_{0,A} - \frac{RT}{z_A F}\frac{i_{0,B}}{i_{0,A}} \exp\frac{\alpha_B z_B F}{RT}\eta_B + \frac{RT}{z_A F} \ln \frac{a_{A,s}}{a_{A,m}} \tag{6}$$

and in the case $\eta_A \gg \frac{RT}{F}$ and $\eta_B \gg \frac{RT}{F}$:

$$\Delta\Phi_{mix} \cong \Delta\Phi_{0,A} + \frac{RT}{(1-\alpha_A)\,z_A F} \ln\left(-\frac{i_{0,B}}{i_{0,A}}\right) + \frac{\alpha_B z_B}{(1-\alpha_A)\,z_A}\eta_B + \frac{RT}{z_A F}\ln\frac{a_{A,s}}{a_{A,m}} \tag{7}$$

with: $\alpha_{A,B}$ = the corresponding transfer coefficient

In Eqs. (5)–(7) the equilibrium potential $\Delta\Phi_{e,A}$ of the potential-determining ion A was expressed according to:

$$\Delta\Phi_{e,A} = \Delta\Phi_{0,A} + \frac{RT}{z_A F}\ln\frac{a_{A,s}}{a_{A,m}} \tag{8}$$

For an interfering redox reaction at an ion-selective membrane, the overpotential η_B can be easily determined experimentally. It is the potential difference between the ion-selective membrane and an inert redox electrode in the same solution containing the measured ion and an interfering redox system.

Diffusion overpotential for the potential-determining reaction

In this case the activity of ion A at the membrane surface $a^s_{A(aq)}$ is different from the activity in the bulk of the solution $a^0_{A(aq)}$. The mixed potential can then be described as:

$$\Delta\Phi_{mix} = \Delta\Phi_{0,A} + \frac{RT}{z_A F}\ln\frac{a^s_{A(aq)}}{a_{A,m}} = \Delta\Phi_{e,B} + \eta_B \tag{9}$$

Using the equation for the diffusion current i under the conditions of stationary diffusion:

$$i = z_i F D_i \frac{a^0_i - a^s_i}{\delta_i} \tag{10}$$

with: D_i = diffusion coefficient

δ = diffusion layer thickness, assumed to be the same for all species;

With Eqs. (3) and (4) the following expressions for the mixed potentials are obtained:

If $\eta_B \ll \frac{RT}{F}$:

$$\Delta\Phi_{mix} \cong \Delta\Phi_{0\,A} + \frac{RT}{z_A F}\ln\frac{a^0_{A(aq)} + \frac{z_B i_{0,B}\,\delta_A}{z_A D_A RT}\eta_B}{a_{A,m}} \tag{11}$$

and if $\eta_B \gg \frac{RT}{F}$:

$$\Delta\Phi_{mix} \cong \Delta\Phi_{0,A} + \frac{RT}{z_A F} \ln \frac{a^0_{A(aq)} + \dfrac{i_{0,B}\,\delta_A \exp \dfrac{\alpha_B z_B F}{RT}\eta_B}{z_A D_A F}}{a_{A,m}} \tag{12}$$

Because equations (11) and (12) contain the thickness of the diffusion layer δ_A which depends on the hydrodynamic conditions, the stationary potential can vary with the stirring speed. It follows further that the redox sensitivity of such membranes (e.g. Ag_2S, Ag halogenide, Ag_2Se, Ag_2Te and Ag) is given solely by the standard exchange current density of the interfering redox reaction. The latter is higher if the membrane shows electronic conductance. Equations (5)–(7) and (11)–(12) allow as to guess the influence of corrosion of solid state membrane electrodes upon their functioning as reversible Nernstian sensors. If the influence of the kinetic parameters in this equation is greater than the relative potential change given by an activity difference, the detection limit of the particular electrode is reached. If small overpotentials are established for a potential-determining reaction, deviations from an ideal Nernstian behavior occur. The EMF vs. log C function can still be linear but the slope may be sub- or super-Nernstian.

Mixed potential in case of a solid state membrane
Dissolving without complexing agents

The membrane consists of a binary compound $A_{\nu_A}B_{\nu_B}$ of constant composition. Ion activities in the solid state are 1. At the membrane surface the activities of A and B in the aqueous phase are given by the solubility product:

$$S.P. = (a^s_A)^{\nu_A} \cdot (a^s_B)^{\nu_B} \tag{13}$$

In case of a steady state with a constant diffusion layer thickness for all ions it follows [for simplicity the surfix (aq) is ommitted]:

$$z_A D_A(a^0_A - a^s_A) + z_B D_B(a^0_B - a^s_B) = 0 \tag{14}$$

The stationary potential according to Eq. (2) is then by:

$$\Delta\Phi_{mix} = \Delta\Phi_{0,A} + \frac{RT}{z_A F}\ln a^s_A = \Delta\Phi_{0,B} + \frac{RT}{z_B F}\ln a^s_B \tag{15}$$

Combination of Eqs. (13) and (14) yields:

$$z_A D_A (a^s_A)^{\frac{\nu_A}{\nu_B}+1} - (z_A D_A a^0_A + z_B D_A a^0_B)(a^s_A)^{\frac{\nu_A}{\nu_B}} + z_B D_B (S.P.)^{\frac{1}{\nu_B}} = 0 \tag{16}$$

Since Eq. (16) is not solvable, only the special case

$$z_A = |z_B| = 1 \quad \text{and} \quad \nu_A = \nu_B = 1$$

will be discussed.

Then Eq. (16) reads:

$$D_A(a_A^s)^2 - (D_A a_A^0 - D_B a_B^0)\, a_A^s - D_B(\text{S.P.}) = 0 \tag{17}$$

and

$$a_A^s = \frac{D_A a_A^0 - D_B a_B^0 + \sqrt{(D_A a_A^0 - D_B a_B^0)^2 + 4 D_A D_B (\text{S. P.})}}{2 D_A} \tag{18}$$

Insertion of Eqs. (18) in (15) results in an Equation for a dissolving solid state membrane electrode in the absence of complexing agents. In case $a_A^0 = a_B^0 = 0$ (e.g., pure H_2O) the stationary potential can be expressed as:

$$\Delta\Phi_{mix} = \Delta\Phi_{0,A} + \frac{RT}{2z_A F} \ln \frac{D_B}{D_A} (\text{S.P.}) \tag{19}$$

Dissolving with complexing agents

For the reaction at a silver halide membrane:

$$AgX + 2Y^- \rightleftarrows [AgY_2]^- + X^-$$

the same treatment as above leads to:

$$\Delta\Phi_{mix} = \Delta\Phi_0' - \frac{RT}{F} \ln \left(a_{y^-}^0 + \frac{2D_{x^-}}{D_{y^-}} a_{x^-}^0 \right) \tag{20}$$

with

$$\Delta\Phi_0' = \Delta\Phi_{0,Ag} + \frac{RT}{F} \ln \frac{2D_{X^-}}{D_Y} (\text{S.P.}_{AgX}) \tag{21}$$

Such a relationship has indeed been found for CN^- at AgI membranes.

Mixed potential with interfering ions

In the diffusion-controlled domain (preferable in situations with large overpotentials) a diffusion layer is formed. This layer is found on the solution side of solid-state membranes; it is located with in the membrane surface of liquid and glass membranes.

Solid-state membranes

Hulanicki and Levenstam [125] already suggested a model with a diffusion layer inside the solution. The stationary-state treatment of Ilushenko and Mirkin in case of a membrane made up of the active compound AX which is in a solution containing the ions A and B (also forming an insoluble compound BX) yields, under the conditions of Eqs. (3) and (12):

$$K = \frac{a_A^s}{a_B^s} = \frac{S.P._{AX}}{S.P._{BX}} \tag{22}$$

and:

$$a_A^s = \frac{KD_A a_A^0}{KD_A + D_B} + \frac{KD_B a_B^0}{KD_A + D_B} \tag{23}$$

this inserted in Eq. (15) results in:

$$\Delta\Phi_{mix} = \Delta\Phi'_{0,A} + \frac{RT}{z_A F} \ln\left(a_A^0 + \frac{D_B}{D_A} a_B^0\right) \tag{24}$$

with:

$$\Delta\Phi'_{0,A} = \Delta\Phi_{0,A} + \frac{RT}{z_A F} \ln \frac{KD_A}{KD_A + D_B} \tag{25}$$

Amalgam electrodes, and liquid and glass membrane electrodes

As has been shown [82,85–88], the behavior of amalgam electrodes under conditions of cementation is very similar to that of liquid and glass membrane electrodes under stationary state conditions. Here, Eq. (2) should be written in the following way:

$$\Delta\Phi_{mix} = \Delta\Phi_{0,A} + \frac{RT}{z_A F} \ln \frac{a_{A(aq)}}{a^s_{A(m)}} = \Delta\Phi_{0,B} + \frac{RT}{z_B F} \ln \frac{a_{B(aq)}}{a^s_{B(m)}} \tag{26}$$

with: $a^s_{A,B(m)}$ = activity of A or B in the surface layer of the amalgam or membrane

Since the surface layer is in equilibrium with the solution the exchange reaction $A_{(m)} + B_{(aq)} \rightleftarrows A_{(aq)} + B_{(m)}$ can take place.

For isovalent ions the law of mass action results in:

$$\frac{a_{A(aq)}}{a^s_{A(m)}} = K \frac{a_{B(aq)}}{a^s_{B(m)}} = \frac{a_{A(aq)} + Ka_{B(aq)}}{a^s_{A(m)} + a^s_{B(m)}} \tag{27}$$

Combined with Eq. (25) this expression yields:

$$\Delta\Phi_{mix} = \Delta\Phi_{0,A} + \frac{RT}{z_A F} \ln \frac{a_{A(aq)} + Ka_{B(aq)}}{a^s_{A(m)} + a^s_{B(m)}} \tag{28}$$

If the condition

$$a^s_{A(m)} + a^s_{B(m)} = \text{const.} \tag{29}$$

is valid, the potential of, for example, an amalgam electrode is independent of the ion activities in the electrode phase.

The conductivity of liquid and glass membranes is determined by ion-migration (absence of an excess of supporting electrolyte is assumed) in the diffusion layer. Equation (25) should then be written as:

$$\Delta\Phi_{mix} = \eta_{Ohm} + \Delta\Phi_{0,A} + \frac{RT}{z_A F} \ln \frac{a_{A(aq)}}{a^s_{A(m)}} \tag{30}$$

with: η_{Ohm} = overpotential due to ohmic polarization within the diffusion layer

If the stationary-state condition [Eq. (3)] is valid, the ohmic polarization in the diffusion layer can be expressed by the diffusion potential ε_D

$$\eta_{Ohm} = \varepsilon_D \tag{31}$$

Therefore, the Henderson approach [117] for monovalent ions of the same sign for calculating ε_D can be applied:

$$\eta_{Ohm} \cong \frac{RT}{F} \ln \frac{u_A a^s_{A(m)} + u_B a^s_{B(m)}}{u_A a^0_{A(m)}} \tag{32}$$

with: $a^0_{A(m)}$ = activity of ion A in the bulk of the membrane phase

$u_{A,B}$ = mobility of ions A, B within the diffusion layer

It is assumed that the activity of the interfering ion B in the bulk of the membrane phase $a^0_{B(m)} = 0$. Since $a^0_{A(m)}$ = const. Eqs. (28) and (32) yield:

$$\eta_{Ohm} \cong -\frac{RT}{F} \ln a^0_{A(m)} + \frac{RT}{F} \ln \frac{\left[a_{A(aq)} + K \frac{u_B}{u_A} a_{B(aq)}\right] \left[a^s_{A(m)} + a^s_{B(m)}\right]}{a_{A(aq)} + K a_{B(aq)}} \tag{33}$$

If Eqs. (28) and (32) are inserted into Eq. (30) the mixed potential of the ion-selective membrane for an ion exchange process of only monovalent ions of the same sign is obtained:

$$\Delta\Phi_{mix} = \Delta\Phi'_0 + \frac{RT}{F} \ln \left[a_{A(aq)} + \frac{u_B}{u_A} K a_{B(aq)}\right] \tag{34}$$

with

$$\Delta\Phi'_0 = E\Phi_{0,A} - \frac{RT}{F} \ln a^0_{A(m)} \tag{35}$$

Table A1. Survey of some commercially available ion-selective electrodes

Glass membrane electrodes

Electrode Type	Active Phase	Potential-determining Ions	Working range [M]	Selectivity coefficient[1] k^{Pot}_{M-I}	Recom-mended pH-range	Tempe-rature range [°C]	Electrical resistance at 25 °C [MΩ]	Recom-mended reference electrode	Manufac-turer
pNa	NAS_{11-18} $LAS_{26.2-12.4}$ etc. glasses	$Ag^+ > H^+ >$ $Na^+ > K^+$	1 to 10^{-8}	$Ag^+ \sim 500$; $H^+ \sim 10^3$; $K^+ \sim 10^{-3}$; $Li^+ \sim 10^{-3}$; $Cs^+ \sim 10^{-3}$; $Tl^+ \sim 2 \cdot 10^{-3}$; $Rb^+ \sim 3 \cdot 10^{-5}$; $NH_4^+ \sim 3 \cdot 10^{-5}$	7 to 10 ca. 4 pH units above pNa-value	0 to 100	~ >100	Ag/AgCl with double salt bridge (1 M NH_4NO_3)	Beckman, Corning, EIL, Ingold, Metrohm, Orion, Philips, Polymetron, Radelkis, Radiometer, Schott & Gen., Tacussel
pCation	NAS_{27-4} $KABS_{20-5-9}$ etc. glasses	$H^+ > Ag^+ > K^+$; $NH_4^+ > Na^+ >$ Li^+ Rb^+, Cs^+, Tl^+	1 to $5 \cdot 10^{-6}$	$Na^+ \sim 0.1$; $NH_4^+ \sim 0.3$; $Rb^+ \sim 0.5$; $Li^+ \sim 0.05$; $Cs^+ \sim 0.03$ (normalized to: $K^+ = 1$)	7 to 13 (for pK) 4 to 10 (for pAg) ca. 2 pH above pK-value	0 to 100	~ >100	Ag/AgCl with double salt bridge (1 M Li-tri-chloro-acetate)	Beckman, Corning, EIL, Ingold, Philips, Tacussel

Homogeneous solid-state membrane electrodes

Electrode type	Active Phase	Potential-determining Ions	Working range [M]	Selectivity coefficient[1] k^{Pot}_{M-1}	Recommended pH-range	Temperature range [°C]	Electrical resistance at 25 °C [MΩ]	Recommended reference electrode	Manufacturer
pAg	Ag_2S pellet	Ag^+, S^{2-},	1 to 10^{-7} $<10^{-17}$, buffered	Traces of Hg^{2+} interfere, prolonged contact with Hg^{2+}-containing solutions requires surface treatment $Cu^{2+} \sim 10^{-6}$ $Pb^{2+} \sim 10^{-10}$	2 to 9	−5 to 100	<1	Ag/AgCl with double salt bridge 1 M KNO_3	Beckman, Electrofact, Coleman, Corning, Foxboro, HNU, Metrohm, Orion, Philips, Polymetron, Radiometer, Schott & Gen., Tacussel
pAg	Ag_2S-single crystal	Ag^+, S^{2-}	s. above	$Cu^{2+} \sim 10^{-5}$; $Pb^{2+} \sim 10^{-6}$; $H^+ \sim 10^{-7}$; Hg^{2+} interferes	2 to 9			s. above	Crytur
pCu	CuS/Ag_2S	Ag^+, S^{2-}, Hg^{2+}, Cu^{2+}	1 to 10^{-8} $<10^{-17}$, buffered	Ag^+, Hg^{2+} must be absent $Fe^{3+} \sim 10$; $Cu^{1+} \sim 1$; Cl^-, Br^- at higher concentration	0 to 14	0 to 100	s. above	s. above	Beckman, Electrofact, Coleman, Corning, HNU, Polymetron, Radiometer, Metrohm, Orion, Tacussel, Philips
	Cu_2S	Cu^{2+}			3 to 14	0 to 70			
	CuSe-single crystal	Cu^+, Cu^{2+}	1 to 10^{-6} $<10^{-17}$ buffered	$Pb^{2+} \sim 10^{-4}$; $Cd^{2+} \sim$ $\sim 10^{-5}$; $Cu^+ \sim 10^{11}$; $Ag^+ \sim 10^6$; $Hg^{2+} \sim 10^4$	0 to 14	−5 to 60	s. above	s. above	
pCd	CdS/Ag_2S	Ag^+, S^{2-}, Cu^{2+}, Cd^{2+}	0.1 to 10^{-7} $<10^{-10}$, buffered	Ag^+, Hg^{2+}, Cu^{2+} must be absent $Fe^{2+} \sim 200$; $Tl^+ \sim 120$; $Pb^{2+} \sim 6$ $Mn^{2+} \sim 3$	1 to 14	0 to 100	s. above	s. above	Electrofact, Metrohm, Orion, Philips
pPb	PbS/Ag_2S	Ag^+, S^{2-}, Cu^{2+}, Pb^{2+}	0.1 to 10^{-7} $<10^{-10}$, buffered	Ag^+, Hg^{2+}, Cu^{2+} must be absent $Fe^{3+} \sim 1$; Cd ~ 1	2 to 14	0 to 100	s. above	s. above	Corning, Electrofact, Metrohm, Orion

Table A1. (continued)

Electrode type	Active Phase	Potential-determining Ions	Working range [M]	Selectivity coefficient[1] k^{Pot}_{M-I}	Recommended pH-range	Temperature range [°C]	Electrical resistance at 25 °C [MΩ]	Recommended reference electrode	Manufacturer
pS	Ag_2S-pellet	Ag^+, S^{2-}	1 to 10^{-6}, $<10^{-17}$, buffered	Traces of Hg^{2+} interfere prolonged contact with Hg^{2+}-containing solutions requires surface treatment	13 to 14	−5 to 100	<1	Ag/AgCl with double salt bridge 1 M KNO_3	Beckman, Orion, Coleman, Polymetron, Corning, Radiometer, Foxboro, Schott & Gen., HNU, Tacussel, Metrohm
	Ag_2S-single crystal	Ag^+, S^{2-}	s. above	s. above $Cu^{2+} \sim 10^{-5}$; $Pb^{2+} \sim 10^{-6}$	s. above			s. above	Crytur
pF	LaF_3-single crystal	F^-	1 to 10^{-6}	$OH^- \sim 0.1$; other halides, NO_3^-, HCO_3^-, $SO_4^{2-} \sim <10^{-3}$	4 to 8	−5 to 100	~0.2 to 5	usual reference electrode	Beckman, Foxboro, Coleman, Corning, Crytur, Metrohm, Polymetron, Orion, Radiometer, Philips
pCl	$AgCl/Ag_2S$-mixed pellet	Ag^+, Cl^-	1 to 10^{-5}	$Br^- \sim 10^2$; $I^- \sim 10^6$; $OH^- \sim 10^{-2}$; $CN^- \sim 10^4$; S^{2-} must be absent	2 to 11	~0 to 80	10 to 30	Ag/AgCl with double salt bridge 1 M KNO_3	Beckman, Orion, Corning, Polymetron, Electrofact, Foxboro, HNU, Metrohm
	AgCl-single crystal	Ag^+, Cl^-	1 to 10^{-5}	$Br^- \sim 2$; $I^- \sim 2$; $CN^- \sim 8$; $OH^- \sim 10^{-2}$; $NH_3 \sim 0.1$	0 to 14	−5 to 60	<25	s. above	Radiometer, Crytur

	AgCl	Ag^+, Cl^-	1 to 10^{-5}	$Br^- \sim 1$; $I^- \sim 10^2$; $CN^- \sim 400$; $S_2O_3^{2-} \sim 60$; $OH^- \sim 10^{-2}$; $CO_3^{2-} \sim 10^{-3}$	1 to 10	0 to 100	<1	s. above	Philips Schott & Gen. Tacussel
pBr	$AgBr/Ag_2S$-mixed pellet	Ag^+, Br^-	1 to $5 \cdot 10^{-6}$	$I^- \sim 5 \cdot 10^3$; $CN^- \sim 10^2$; $Cl^- \sim 5 \cdot 10^{-3}$; $OH^- \sim 10^{-5}$; S^{2-} must be absent	2 to 12	s. above	s. above	s. above	Beckman Corning Electrofact Foxboro Metrohm Orion
	AgBr-single crystal	Ag^+, Br^-	1 to 10^{-6}	$I^- \sim 2$; $CN^- \sim 1$; $Cl^- \sim 5 \cdot 10^{-3}$; $OH^- \sim 10^{-4}$; $NH_3 \sim 4 \cdot 10^{-3}$	0 to 14	—5 to 60	<25	s. above	Radiometer
	AgBr	Ag^+, Br^-	1 to 10^{-6}	$I^- \sim 20$; $CN^- \sim 25$; $Cl^- \sim 6 \cdot 10^{-3}$; $OH^- \sim 10^{-3}$; $S_2O_3^{2-} \sim 1.5$; $CO_3^{2-} \sim 2 \cdot 10^{-3}$	1 to 11	0 to 50	<1	s. above	Philips Schott & Gen. Tacussel
pI	AgI/Ag_2S-mixed pellet	Ag^+, S^{2-}, I^-	1 to $5 \cdot 10^{-8}$	$S^{2-} \sim 30$; $S_2O_3^{2-} \sim 3 \cdot 10^{-2}$; $CN^- \sim 10^{-2}$; $Br^- \sim 10^{-4}$; $Cl^- \sim 10^{-6}$; $SCN^- \sim 10^{-4}$; $OH^- \sim 10^{-7}$; $NH_3 \sim 3 \cdot 10^{-5}$	0 to 14	0 to 80	<0.5	s. above	Beckman Electrofact Corning Foxboro HNU Metrohm Orion Radiometer
	AgI	Ag^+, I^-		$CN^- \sim 0.34$; $CrO_4^{2-} \sim 4 \cdot 10^{-3}$; $S_2O_3^{2-} \sim 7 \cdot 10^{-4}$; $CO_3^{2-} \sim 10^{-4}$; $Br^- \sim 6 \cdot 10^{-5}$; $Cl^- \sim 6 \cdot 10^{-6}$; S^{2-} must be absent	1 to 12	0 to 50	<0.5	s. above	Crytur Philips Schott & Gen. Tacussel

Table A1. (continued)

Electrode type	Active Phase	Potential-determining Ions	Working range [M]	Selectivity coefficient[1] k^{pot}_{M-I}	Recommended pH-range	Temperature range [°C]	Electrical resistance at 25 °C [MΩ]	Recommended reference electrode	Manufacturer
pSCN	AgSCN/Ag_2S-mixed pellet	Ag^+, SCN^-	1 to $5 \cdot 10^{-6}$	$I^- \sim 10^3$; $Br^- \sim 10^2$; $CN^- \sim 10^2$; $S_2O_3^{2-} \sim 10^2$; $NH_3 \sim 10$; $Cl^- \sim 0.1$; $OH^- \sim 10^{-2}$; S^{2-} must be absent	2 to 12	0 to 95	<100	s. above	Metrohm Orion
pCN identical with pI	AgI/Ag_2S-mixed pellet	Ag^+, I^-, CN^-		$I^- \sim 10^2$; $Br^- \sim 10^{-4}$; $Cl^- \sim 10^{-6}$; $OH^- \sim 10^{-8}$; S^{2-} must be absent	11 to 13	0 to 80	<30	s. above	Beckman, Metrohm Electrofact, Orion Corning, Polymetron HNU, Radiometer
	AgI	Ag^+, I^-, CN^-		$I^- \sim 3$; $CrO_4^{2-} \sim 10^{-2}$; $S_2O_3^{2-} \sim 10^{-3}$; $CO_3^{2-} \sim 10^{-4}$; $Br^- \sim 10^{-4}$; $Cl^- \sim 10^{-5}$	10 to 12	0 to 50	0.5	s. above	Crytur Philips Schott & Gen.
Liquid and solid matrix membrane electrodes									
pCa, liquid	Ca-salt of dialkyl phosphoric acid in dioctylphenyl-phosphonate	Zn^{2+}, Ca^{2+}, Fe^{2+}, Pb^{2+}	1 to 10^{-5}	$Zn^{2+} \sim 3.2$; $Ca^{2+} = 1.0$; $Fe^{2+} \sim 0.8$; $Pb^{2+} \sim 0.63$; $Cu^{2+} \sim 0.27$; $Ni^{2+} \sim 0.080$; $Sr^{2+} \sim 0.27$; $Mg^{2+} \sim 0.014$; $Ba^{2+} \sim 0.010$; $Na^+ \sim 10^{-3}$; $K^+ \sim 10^{-3}$	5 to 9	0 to 50 10 to 60	<25 <500	normal Ag/AgCl	Corning

pCa PVC matrix	Ca-dioctyl-phenyl-phosphate	s. above	1 to 10^{-5}	Zn^{2+} ~ 1–5; Ca^{2+} = 1.0; Al^{3+} ~ 0.90; Mn^{2+} ~ 0.38; Cu^{2+} ~ 0.070; Fe^{2+} ~ 0.045; Co^{2+} ~ 0.042; Mg^{2+} ~ 0.032	5 to 11	0 to 60	~2		Radiometer
	unknown			Ba^{2+} ~ 0.020; Na^+ ~ 10^{-5}; K^+ ~ 10^{-6}; Li^+ ~ 10^{-4}					HNU
pCa PVC matrix or liquid	Ca^{2+}-carrier (Lit. 75)	Ca^{2+}	1 to 10^{-6}	Sr^{2+} ~ 10^{-2}; Li^+ ~ 10^{-3}; Cs^+ ~ 10^{-3}; Na^+, K^+, Ba^{2+}, Zn^{2+} ~ 10^{-4}; NH_4^+ ~ 10^{-5}	3 to 10	0 to 50	<10	s. above	Philips Orion
pMe^{2+} liquid (water hardness)	Ca-salt of dialkyl phosphoric acid in decanol	Zn^{2+}, Fe^{2+}, Cu^{2+}, Ni^{2+}, Ca^{2+}, Mg^{2+}, Ba^{2+}, Sr^{2+}	1 to 10^{-5}	Zn^{2+} ~ 3.5; Fe^{2+} ~ 3.5; Cu^{2+} ~ 3.1; Ni^{2+} ~ 1.35; Ca^{2+} ~ 1.0; Mg^{2+} ~ 1.0; Ba^{2+} ~ 0.94; Sr^{2+} ~ 0.54; Na^+, K^+ ~ 0.01	s. above	0 to 50 10 to 60	<25 <500	s. above	Orion Corning
pBa liquid	not given	Ba^{2+}	not given	Ca^{2+} ~ 0.03; K^+ ~ 10^{-2}; Na^+ ~ $4 \cdot 10^{-4}$	not given	10 to 60	<500	s. above	Corning
pBa PVC matrix	Ba^{2+}-carrier (Lit. 75)	Ba^{2+}	1 to 10^{-6}	H^+ ~ 0.06; Sr^{2+} ~ 0.03; K^+, Rb^+ ~ 10^{-2}; NH_4^+, Cs^+, Na^+ ~ $3 \cdot 10^{-3}$; Li^+, Ca^{2+} ~ 10^{-4}; Mg^{2+} ~ 10^{-5}	3 to 11	0 to 50	<25	s. above	Philips

Table A1. (continued)

Electrode type	Active Phase	Potential-determining Ions	Working range [M]	Selectivity coefficient[1] k^{Pot}_{M-I}	Recommended pH-range	Temperature range [°C]	Electrical resistance at 25 °C [MΩ]	Recommended reference electrode	Manufacturer
$pClO_4$ liquid	$Fe(o\text{-}phen)_3^{2+}$ in p-nitro-cymene	ClO_4^-, OH^-	0.1 to 10^{-5}	$OH^- \sim 1.0$; $I^- \sim 1.2 \cdot 10^{-2}$; $NO_3^- \sim 1.5 \cdot 10^{-3}$; $Br^- \sim 5.6 \cdot 10^{-4}$; $OAc^- \sim 5.1 \cdot 10^{-4}$; $HCO_3^- \sim 3.5 \cdot 10^{-4}$; $F^- \sim 2.5 \cdot 10^{-4}$; $Cl^- \sim 2.2 \cdot 10^{-4}$; $SO_4^{2-} \sim 1.6 \cdot 10^{-4}$	3 to 10	0 to 50	~25	s. above	Orion
pNO_3 liquid	$Ni(o\text{-}phen)_3^{2+}$ in p-nitro-cymene	ClO_4^-, I^-, ClO_3^-, NO_3^-	1 to 10^{-5}	$ClO_4^- \sim 10^3$; $I^- \sim 20$; $ClO_3^- \sim 2$; $Br^- \sim 0.9$; $S^{2-} \sim 0.57$; $NO_2^- \sim 6 \cdot 10^{-2}$; $CN^- \sim 2 \cdot 10^{-2}$; $HCO_3^- \sim 2 \cdot 10^{-2}$	3 to 10	0 to 50	~25	s. above	Orion
	tridodecyl-hexadecyl-ammonium nitrate in n-octyl-2-nitro-phenyl ether				1 to 9	10 to 60	<500		Corning
pNO_3 PVC matrix	tetradodecyl ammonium nitrate			$Cl^- \sim 6 \cdot 10^{-3}$; OAc^-, CO_3^{2-}, $S_2O_3^{2-}$; $SO_3^{2-} \sim 6 \cdot 10^{-3}$	3 to 11	5 to 40	~25		EIL
	unknown			$F^- \sim 9 \cdot 10^{-4}$; $SO_4^{2-} \sim 6 \cdot 10^{-4}$; $H_2PO_4^-$, $PO_4^{3-} \sim 3 \cdot 10^{-4}$; $HPO_4^- \sim 8 \cdot 10^{-5}$	1 to 9	0 to 50	~25		Philips HNU
pBF_4 liquid	$Ni(o\text{-}phen)_3^{2+}$ in p-nitro-cymene	I^-, BF_4^-	0.1 to 10^{-5}	$I^- \sim 20$; $NO_3^- \sim 0.1$; $Br^- \sim 4 \cdot 10^{-2}$; OAc^-; $HCO_3^- \sim 4 \cdot 10^{-3}$; F^-, Cl^-, $SO_4^{2-} \sim 10^{-3}$	2 to 12	0 to 50	~25	s. above	Orion

pLi PVC matrix	Li^+-carrier (s. Fig. 12)	Li^+, H^+	0.1 to 10^{-5}	$H^+ \sim 1$; Na^+, $NH_4^+ \sim 0.05$; $K^+ \sim 7 \cdot 10^{-3}$; $Rb^+ \sim 4 \cdot 10^{-3}$; Cs^+, Ca^{2+}, Sr^{2+}, Mg^{2+}, $Ba^{2+} \sim 10^{-4}$	3 to 11	0 to 50	~25	s. above	Philips
pK liquid	Valinomycin	Cs^+, Rb^+, K^+	1 to 10^{-5}	$Cs^+ \sim 1.0$; $NH_4^+ \sim 3 \cdot 10^{-2}$; $H^+ \sim 1 \cdot 10^{-2}$; $Ag^+ \sim 1 \cdot 10^{-3}$; $Na^+ \sim 1 \cdot 10^{-2}$; $Li^+ \sim 1 \cdot 10^{-4}$	2 to 11	0 to 50	<25	Ag/AgCl with double salt bridge (1 M Li-tri-chloro-acetate)	Orion
	tetrakis-(p-chloro-phenyl)borate	Rb^+, K^+, Cs^+	1 to 10^{-5}	$NR_4^+ \sim >1$; $Rb^+ \sim 2$; $Cs^+ \sim 0.4$; $Cs^+ \sim 0.4$; $NH_4^+ \sim 0.3$; Na^+, $Li^+ \sim 10^{-3}$; Ca^{2+}, $Mg^{2+} \sim 10^{-3}$	s. above	15 to 50	~100		Corning
PVC matrix	Valinomycin	Rb^+, K^+, Cs^+	1 to 10^{-6}	$Rb^+ \sim 1.9$; $Cs^+ \sim 0.38$; $NH_4^+ \sim 10^{-2}$; $Sr^{2+} \sim 10^{-3}$; Li^+, $Ba^{2+} \sim 10^{-4}$; H^+, Na^+, $Ca^{2+} \sim 10^{-5}$	s. above	0 to 50	<2		Philips Radelkis
pNH_4 PVC matrix	Nonactin- Monactin	NH_4^+	1 to 10^{-6}	$K^+ \sim 0.12$; $Rb^+ \sim 4.3 \cdot 10^{-3}$; $H^+ \sim 1.6 \cdot 10^{-2}$; $Cs^+ \sim 4.8 \cdot 10^{-3}$; $Li^+ \sim 4.2 \cdot 10^{-3}$; $Na^+ \sim 2 \cdot 10^{-3}$; $Sr^{2+} \sim 10^{-4}$; Ba^{2+}, $Ca^{2+} \sim 10^{-5}$; $Mg^{2+} \sim 10^{-5}$	4 to 10	0 to 50	<1	s. above	Philips

[1] according to the manufacturer; gas-sensors for NH_3, SO_2, CO_2, NO_x are available from most manufacturers of pH-glass electrodes

Table A2. Examples of bio-sensors [111]

Substrate (compound to be determined)	Enzyme (activity to be determined)	Products (sensed species)	Buffer and optimum pH	Indicating electrode	Range
Acetylcholine (indirectly: phosphororganic pesticides in the ng/ml-range)	Acetylcholinesterase	Choline, acetic acid	physiol, salt solution; pH 7.2	Acetylcholine-Liquid membrane	$10^{-2} \cdot 10^{-4}$ M
5′-Adenosine Monophosphate (5′-AMP)	AMP-Deaminase	Inosine 5′-monophosphate, NH_3	0.05 M Tris; pH 7.5	NH_3-Gas-Sensor	$10^{-2} \cdot 10^{-4}$ M
Albumin (Human)	Anti-Human-Serum-Albumin (Immuno-Reaction)	not reacted antibody	0.1 M NaOH	Ag_2S	0.5–30 μg/ml
Amygdalin	Glucosidase	Glucose, Benzaldehyde, CN^-	NaH_2PO_4; pH 6.4	Cyanide	$10^{-2} \cdot 10^{-5}$ M
L-Arginine	Arginase + Urease (double reaction)	NH_3, CO_2	Tris; pH 7	NH_3-airgap	
Asparagine	Asparaginase	Asparaginic acid, NH_4^+	Tris; pH 7	NH_4^+ glass	
Catechol	Polyphenoloxidase	Quinone	Acetate; pH 5.9	Copper	
Cholesterol	Cholesterolester-hydrolase + Cholesteroloxidase (double reaction)	Cholest-4-en-3-one, H_2O_2	Na_2HPO_4; NaH_2PO_4; pH 6.8	Iodide	50–400 mg%
Creatinine	Creatine deaminase	N-Methylhydantoin, NH_4^-	Tris/NaH_2PO_4; pH 8.5	NH_3-Gas-Sensor	1–100 mg%
Cyanide/Thiosulfate	Rhodanase	SCN^-	NaH_2PO_4; pH 7.9	SCN od. CN	
L-Cysteine/Cyanide	Cyanoalaninesynthase	Cyanoalanine, HS^-	Tris-Acetate; pH 8.5	Ag_2S	
Diphenylcarbamylfluoride	Chymotrypsin	DPC-chymotrypsin, F^-	pH 7.5	Fluoride	0.03 μM
Glucose	Glucoseoxidase	Gluconic acid, H_2O_2	KH_2PO_4/K_2HPO_4; pH 5.7	Iodide Fluoride (add. of Fluorphenol)	1–300 mg% 1–100 mM
Glutamine	Glutaminase	Glutamic acid, NH_4^-	Tris; pH 7	NH_4^+-glass	
L-Leucine	L-Amino-acid oxidase	$RCOCOO^-$, NH_4^+, H_2O_2	Tris; pH 7	NH_4^+-glass	

Mannan	Concanvalin A	—	Phthalate; pH 3.5	PVC-wire	
D-Methionine	D-Amino-acid oxidase	$RCOCOO^-$, NH_4^-, H_2O_2	Tris; pH 7	NH_4^+-glass	
Nitrite	Nitritereductase + Methylviologen	NH_4^+	Phosphate; pH 7.2	NH_4^+-liquid membrane	10^{-3} M
Penicillin	Penicillinase	correspond. acid	pH 6.9	pH-glass	3–1100 µg/ml
L-Phenylalanine	L-Amino-acid oxidase	$RCOCOO^-$, NH_4^+, H_2O_2		Iodide	
Sulfate	Sulfate-reductase	HS^-		Ag_2S	
D,L-Tyrosine	D,L-Amino-acid oxidase	$RCOCOO^-$, NH_4^+, H_2O_2	Tris; pH 7	NH_4^+-glass	
Tyrosine	Tyrosine decarboxylase	Tyramine, CO_2	Na-citrate; pH 5.5	CO_2-Sensor	
Urea	Urease	NH_3, CO_2	Phosphate; pH 7–8.5	NH_4^+-glass NH_3, CO_2-Gas-Sensor	

F References

1. Lark-Horovitz, K.: Naturwiss. *19*, 397 (1931)
2. Nicolsky, B. P., Tolmacheva, T. A.: Zh. Fiz. Khim. *10*, 495 (1937)
3. Schindler, J. G., Schindler, M. M.: Bioelektrochemische Membran-Elektroden, Berlin, de Gruyter 1983
4. Guilbaut, G. G.: Enzyme electrodes in analytical chemistry, in: Comprehensive Analytical Chemistry (ed.) Svehla, G. S., Amsterdam, Pergamon Press 1977
5. Rechnitz, G. A.: Science *214*, 287 (1981)

5a. Keating, M. Y., Rechnitz, G. A.: Anal. Chem. *56*, 801 (1984)

6. Eisenman, G. (ed.): Glass Electrodes for Hydrogen and other Cations, New York, M. Dekker 1967
7. Durst, R. A. (ed.): Ion-Selective Electrodes, Washington, Natl. Bur. of Standards Spec. Publ. *314*, 1969
8. Buck, R. P.: Potentiometry, pH measurements and ion-selective electrodes, in: Physical Methods of Chemistry, part IIa (eds.) Weissberger, A., Rossiter, B. W., New York, Interscience 1971
9. Moody, G. J., Thomas, J. D. R.: Selective Ion Sensitive Electrodes, Watford, Merrow 1971
10. Eisenman, G. (ed.): Membranes, Vol. 1+2, New York, M. Dekker 1972/73
11. Cammann, K.: Das Arbeiten mit ionenselektiven Elektroden, Berlin—Heidelberg—New York, Springer 1973 (2nd ed. 1977, English ed. 1979)
12. Koryta, J.: Ion-Selective Electrodes, Cambridge, Cambridge Press 1975
13. Lakshminarayanaiah, N.: Membrane Electrodes, New York, Academic Press 1976
14. Bailey, P. L.: Analysis with Ion-Selective Electrodes, London, Heyden 1976
15. Freiser, H. (ed.): Ion-Selective Electrodes in Analytical Chemistry, New York, Plenum Press 1978
16. Covington, A. K. (ed.): Ion-Selective Electrode Methodology, Boca Raton, CRC Press
17. Cremer, M.: Z. Biol. *47*, 562 (1906)
18. Haber, F., Klemensiewicz, Z.: Z. Phys. Chem. (Leipzig) *67*, 385 (1909)
19. Haber, F., Beutner, R.: ibid. *26*, 327 (1908)
20. Nernst, W., Riesenfeld, E. H.: ibid. *8*, 600 (1902)
21. Michaelis, L.: Die Wasserstoffionen-Konzentration, Berlin, Springer 1922
22. Horovitz, K.: Z. Physik *15*, 369 (1923)
23. Schiller, H.: Ann. Physik *75*, 105 (1924)
24. MacInnes, D. A., Dole, M.: J.A.C.S. *52*, 29 (1930)
25. Lengyel, B., Blum, E.: Trans. Faraday Soc. *30*, 461 (1934)
26. Sokolof, S. I., Passynsky, A. H.: Z. Physik Chem., *A160*, 366 (1932)
27. Beutner, R.: Die Entstehung elektrischer Ströme in lebenden Geweben, Stuttgart, F. Enke 1920
28. Osterhout, W. J. V.: Proc. Nat. Acad. Sci. U.S., *21*, 125 (1935)
29. Michaelis, L.: Bull. Natl. Research Council, U.S., *69*, 119 (1929)
30. Teorell, T.: Proc. Natl. Acad. Sci. U.S. *21*, 152 (1935)
31. Meyer, K. H., Sievers, J. F.: Helv. Chim. Acta *19*, 649, 665, 987 (1936)
32. Kolthoff, I. M., Sanders, H. L.: J.A.C.S. *59*, 416 (1937)
33. Marshall, C. E.: J. Phys. Chem. *43*, 1155 (1939), ibid. *48*, 67 (1944)
34. Marshall, C. E.: Soil Sci. Soc. Am. Proc. *7*, 182 (1942)
35. Marshall, C. E., Eime, L. O.: J.A.C.S. *70*, 1302 (1948)
36. Sollner, K.: J.A.C.S. *65*, 2260 (1943)
37. Gregor, H. P., Sollner, K.: J. Phys. Chem. *58*, 409 (1954)
38. Sollner, K.: Ann. N.Y. Acad. of Sci. *148*, 154 (1968)
39. Eisenman, G., Rudin, D. O., Casby, J. U.: Science *126*, 831 (1957)
40. Pungor, E., Havas, J., Tóth, K.: Acta Chim. Hung., *41*, 239 (1964)
41. Pungor, E., Havas, J., Tóth, K.: Z. Chem. *5*, 9 (1965)
42. Frant, M. S., Ross, J. W.: U.S. Patent No. 3.431.182 (1972)
43. Frant, M. S., Ross, J. W.: U.S. Patent No. 3.672.962 (1972)
44. Ross, J. W.: Science *156*, 1378 (1967)
45. Shatkay, A.: Anal. Chem. *39*, 1056 (1967)

46. Coetzee, C. F., Freiser, H.: Anal. Chem. *40*, 2071 (1968); ibid. *41*, 1128 (1969)
47. Štefanac, Z., Simon, W.: Chimia *20*, 436 (1966); Microchem. J. *12*, 125 (1967)
48. Moody, G. J., Oke, R. B., Thomas, J. D. R.: Analyst, *95*, 910 (1970)
49. Simon, W. et al.: Pure Appl. Chem. *44*, 613 (1975)
50. Ammann, D. et al. in: Ion and Enzyme Electrodes in Biology and Medicine (eds.) Kessler, M. et al., München—Berlin—Wien, Urban & Schwarzenberg 1976
51. Güggi, M. et al.: Helv. Chim. Acta *58*, 2417 (1976)
52. Güggi, M., Pretsch, E., Simon, W.: Anal. Chim. Acta *91*, 107 (1977)
53. Sollner, K.: Z. Elektrochemie *36*, 36 (1930)
54. Teorell, T.: Trans. Faraday Soc. *33*, 1053 (1937)
55. Meyer, K. H., Sievers, J. F.: ibid. *33*, 1073 (1937)
56. Donnan, F. G.: Z. Elektrochemie *17*, 572 (1911)
57. Donnan, F. G.: Chem. Rev. *1*, 73 (1924)
58. Morf, W. E.: The Principles of Ion-Selective Electrodes and of Membrane Transport, Amsterdam—Oxford—New York, Elsevier 1981
59. Donnan, F. G., Guggenheim, E. A.: Z. Phys. Chem. (Leipzig) *162*, 346 (1932)
60. Schlögl, R.: Z. Phys. Chem. (Frankfurt a. Main) *1*, 305 (1954)
61. Koryta, J.: Anal. Chim. Acta *61*, 329 (1972)
62. Nernst, W.: Z. Phys. Chem. *4*, 129 (1889)
63. Planck, M.: Ann. Phys. *39*, 161 (1890); ibid *40*, 561 (1890)
64. Johnson, F. H., Eyring, H., Polissar, M. J.: The Kinetic Basis of Molecular Biology, New York, Wiley 1954
65. Prigogine, I.: Introduction to Thermodynamic of Irreversible Processes, 2nd ed, New York, Wiley 1961
66. van der Linden, W. E.: Ion selective electrodes, in Comprehensive Analytical Chemistry, Vol. XI (ed.) Svehla, G., Amsterdam—Oxford—New York, Elsevier 1981
67. Pure Appl. Chem. *48*, 127 (1976)
68. Buck, R. P.: Anal. Chem. *50*, 17R (1978)
69. James, H., Carmack, G., Freiser, H.: ibid. *44*, 856 (1972)
70. Růžička, J., Lamm, C. L., Thell, J. C.: Analyt. Chim. Acta *62*, 15 (1972)
71. Bergveld, P.: IEEE Trans. Biomed. Eng. *17*, 70 (1970); ibid. *19*, 342 (1972)
72. Moss, S. D., Janata, J., Johnson, C. C.: Anal. Chem. *47*, 2238 (1975)
73. Janata, J., Huber, R. J.: Ion-Selective Electrode, Rev., vol. 1, 31 (1979)
74. Nicolsky, B. P., Shults, M. M., Peschechonowa, N. V., Belijustin, A. A.: Ber. Akad. Wiss. USSR *14*, 641 (1961)
75. Weber, E.: Kontakte (Darmstadt) 1984 (1), 26 (1984)
76. Baucke, F. G. K.: Non-Crystalline Solids *14*, 13 (1974)
77. Butler, J. A. V.: Trans. Faraday Soc. *19*, 729 (1924)
78. Erdey-Gruz, T., Volmer, M.: Z. Phys. Chem. (Leipzig) *150*, 203 (1930)
79. Bockris, J. O'M, Reddy, A. K. N.: Modern Electrochemistry, Vol. 2, New York, Plenum Press 1970
80. Cammann, K.: Dissertation, Ludwig-Maximilians-Universität, München 1975
81. Cammann, K.: Drug. Res. *28* (I), 6 (1978)
82. Cammann, K.: A mixed-potential ISE-theory in: Conference on Ion-Selective Electrodes, Budapest 1977 (ed.) Pungor, E., Budapest, Akadémiai Kiado 1978
83. Cammann, K., Rechnitz, G. A.: Anal. Chem. *48*, 856 (1976)
84. Cammann, K.: ibid. *50*, 936 (1978)
85. Akimov, G. V.: Osnovy Ucheniya o Korrozi i Zashchite Metallov, Academy of Sciences of the USSR, Moscow (1946)
86. Kozlovskii, M. T.: Vestn. Akad. Nauk Kaz SSR *4*, 103 (1954)
87. Kozlovskii, M. T.: Tr. Soveshch. Elektrokhim. Academy of Sciences of the USSR, Moscow, October 4th 1956 (1959)
88. Kozlovskii, M. T., Zebreva, A. I., Gladyshev, V. P.: Amal'gamy i ikh Primenenie, Alma-Ata, Science 1971
89. Evans, U. R.: The Corrision and Oxidation of Metals — Scientific Principles and Practical Application:, St. Martin's Press, Inc., New York 1960
90. Gerischer, H.: Z. Phys. Chem. (Frankfurt) *26*, 223 (1960; ibid. *27*, 48 (1961)

91. Guastalla, J.: Nature *227*, 485 (1970)
92. Gavach, C., Mlodnicka, T., Guastalla, J.: C.R. Acad. Sci. Ser. C., *266*, 1196 (1968)
93. Gavach, C., Henry, F.: J. Electroanal. Chem., *54*, 361 (1975)
94. Armstrong, R. D., Covington, A. K., Evans, G. P.: ibid. *159*, 33 (1983)
95. Armstrong, R. D., Covington, A. K., Evans, G. P., Handyside, T.: Electrochimica Acta, Vol. 29, No. *8*, 1127 (1984)
96. Nowak, P., Pomianowski, A.: paper presented at the 4th. scientific session on ion-selective electrodes, Matrafüred, Oct. 8–12, Hungary 1984; proceedings: Akadèmiài Kiado, Budapest, in press
97. Turaeva, M. S., Lyalin, O. O.: ibid.
98. Lyalin, O. O., Turaeva, M. S.: ibid.
99. Samec, Z., Mareček, V., Koryta, J., Khalil, W.: J. Electroanal. Chem. Interfacial Electrochem. *83*, 393 (1977)
100. Koryta, J., Vanýsek, P.: in Advances in Electrochemistry and Electrochemical Engineering, ed. Gerischer, H., Tobias, C. W., Wiley-Interscience, New York, 1981, vol. *12*, pp. 113–176
101. Koryta, J.: Electrochim. Acta *29*, 445 (1984)
102. Koryta, J.: Electrolysis at the interface of two Immiscible Electrolyte Solutions and its Analytical Aspects, in: Ion-Selective Electrodes. 3rd. Symposium held at Matrafüred, Hungary 1980, ed. Pundor, E., Elsevier, Amsterdam—Oxford—New York 1981
103. Oesch, U., Simon, W.: Anal. Chem. *52*, 692 (1980)
104. Cammann, K., Rechnitz, G. A.: paper presented at the 4th. scientific session on ion-selective electrodes, Matrafüred, Oct. 8–12, Hungary 1984; proceedings: Akadèmiài Kiado, Budapest, in press
105. Stark, G.: Biochem. Biophys. Acta *600*, 233 (1980)
106. Iljuschenko, M. A., Mirkin, W. A.: Chabarschysy Vestnik, Kasakstan academy of Science *4*, 41 (1981)
107. Zubay, G.: Biochemistry, Addison-Wesley Plublishing Company, Reading, Massachusetts, U.S.A. (1983)
108. Kostyuk, P. G.: Electrical events during active transport of ions through biological membranes, in: Topic in Bioelectrochemistry and Bioenergetics, Vol. 2, (ed.) Milàzzo, G., New York, Wiley 1978
109. Neumann, E.: Principles of electric field effects in chemical and biological systems, in: Topics in Bioelectrochemistry and Bioenergetics, Vol. 4, (ed.) Milazzo, G., New York, Wiley 1981
110. Jardetzky, O.: Nature, *211*, 969 (1966)
111. Hodgkin, A. L., Huxley, A. F.: J. Physiol. (London) *116*, 449; ibid. *117*, 500 (1952)
112. Chizmadzhev, Yu. A., Pastushenko, V. F.: Mechanisms of Membrane Excitability, in: Topics in Bioelectrochemistry and Bioenergetics Vol. 2, (ed.) Milazzo, G., New York, Wiley 1978
113. Stewart, F. C.: Trans. Faraday Soc. *33*, 1006 (1937)
114. Booth, I. R., Mitchell, W. J., Hamilton, W. A.: Biochem. J. *182*, 687 (1979)
115. Skou, J. C.: Mitochondria: biogenesis and bioenergetics, Biomembranes: molecular arrangements and transport mechanisms, Vol. 28, S. 39, North-Holland, American Elsevier 1972
116. Eisenman, G., Sandblom, J. P., Walker, J. L.: Science *155*, 965 (1967)
117. Miller, I. R.: Structural and energetic aspects of charge transport in lipid layers and in biological membranes, in: Topics in Bioelectrochemistry and Bioenergetics, Vol. 4 (ed.) Milazzo, G., New York, Wiley 1981
118. Schiebe, M.: Dissertation Universität Ulm 1982
119. Leslie, R. B., Chapman, D., Hart, C. J.: Biochim. Biophys. Acta *135*, 797 (1967)
120. Rosenberg, B., Jendrasiak, G. L.: Chem. Phys. Lipids *2*, 47 (1968)
121. Rosenberg, B.: Disc. Faraday Soc. *51*, 1 (1971)
122. Carmack, G. D., Freiser, H.: Anal. Chem. *45*, 1977 (1973)
123. Gavach, C. et al. in: Electrical Phenomenes at the Biological Membrane Level, p. 419 (ed.) Roux, E., Amsterdam, Elsevier 1976
124. Cope, F. W.: Physiol. Chem. Phys. *2*, 545 (1970)
125. Hulanicki, A., Levenstam, A.: Talanta *24*, 171 (1977)

Author Index Volumes 101–128

Contents of Vols. 50–100 see Vol. 100
Author and Subject Index Vols. 26–50 see Vol. 50

The volume numbers are printed in italics

Bei Fragen zur Produktsicherheit wenden Sie sich bitte an:
If you have any questions regarding product safety,
please contact:

Walter de Gruyter GmbH
Genthiner Straße 13
10785 Berlin
productsafety@degruyterbrill.com